Welding Science and Technology

Welding Science and Technology

Md. Ibrahim Khan
BSc. Engineering (Mech.), ME (Prod.)
PhD, FIE
Former Professor and Head, Mechanical Engineering
Department Faculty of Engineering
Integral University, Lucknow

Publishing Globally

NEW AGE INTERNATIONAL (P) LIMITED, PUBLISHERS
LONDON • NEW DELHI • NAIROBI
Bangalore • Chennai • Cochin • Guwahati • Hyderabad • Kolkata • Lucknow • Mumbai
Visit us at **www.newagepublishers.com**

Published by New Age International (P) Ltd., Publishers
First Edition: 2007
Reprint: 2018

GLOBAL OFFICES

- **New Delhi** **NEW AGE INTERNATIONAL (P) LIMITED, PUBLISHERS**
7/30 A, Daryaganj, New Delhi-110002, (INDIA)
Tel.: (011) 23253771, 23253472, **Telefax:** 23267437, 43551305
E-mail: contactus@newagepublishers.com • Visit us at www.newagepublishers.com
- **London** **NEW AGE INTERNATIONAL (UK) LTD.**
27 Old Gloucester Street, London, WC1N 3AX, UK
E-mail: info@newacademicscience.co.uk • Visit us at www.newacademicscience.co.uk
- **Nairobi** **NEW AGE GOLDEN (EAST AFRICA) LTD.**
Ground Floor, Westlands Arcade, Chiromo Road (Next to Naivas Supermarket)
Westlands, Nairobi, KENYA, **Tel.:** 00-254-713848772, 00-254-725700286
E-mail: kenya@newagepublishers.com

BRANCHES

- **Bangalore** 37/10, 8th Cross (Near Hanuman Temple), Azad Nagar, Chamarajpet, Bangalore- 560 018
Tel.: (080) 26756823, **Telefax:** 26756820, **E-mail: bangalore@newagepublishers.com**
- **Chennai** 26, Damodaran Street, T. Nagar, Chennai-600 017, **Tel.:** (044) 24353401, **Telefax:** 24351463
E-mail: chennai@newagepublishers.com
- **Cochin** CC-39/1016, Carrier Station Road, Ernakulam South, Cochin-682 016, **Tel.:** (0484) 2377303, **Telefax:** 4051304
E-mail: cochin@newagepublishers.com
- **Guwahati** Hemsen Complex, Mohd. Shah Road, Paltan Bazar, Near Starline Hotel, Guwahati-781 008,
Tel.: (0361) 2513881, **Telefax:** 2543669, **E-mail: guwahati@newagepublishers.com**
- **Hyderabad** 105, 1st Floor, Madhiray Kaveri Tower, 3-2-19, Azam Jahi Road, Near Kumar Theater, Nimboliadda Kachiguda, Hyderabad-500 027, **Tel.:** (040) 24652456, **Telefax:** 24652457
E-mail: hyderabad@newagepublishers.com
- **Kolkata** RDB Chambers (Formerly Lotus Cinema) 106A, 1st Floor, S N Banerjee Road, Kolkata-700 014
Tel.: (033) 22273773, **Telefax:** 22275247, **E-mail: kolkata@newagepublishers.com**
- **Lucknow** 16-A, Jopling Road, Lucknow-226 001, **Tel.:** (0522) 2209578, 4045297, **Telefax:** 2204098
E-mail: lucknow@newagepublishers.com
- **Mumbai** 142C, Victor House, Ground Floor, N.M. Joshi Marg, Lower Parel, Mumbai-400 013
Tel.: (022) 24927869, **Telefax:** 24915415, **E-mail: mumbai@newagepublishers.com**
- **New Delhi** 22, Golden House, Daryaganj, New Delhi-110 002, **Tel.:** (011) 23262368, 23262370, **Telefax:** 43551305
E-mail: sales@newagepublishers.com

ISBN: 978-81-224-2073-9

C-17-04-10425

Printed in India at Glorious Printers, Delhi.
Typeset at Goswami Associates, Delhi.

NEW AGE INTERNATIONAL (P) LIMITED, PUBLISHERS
7/30 A, Daryaganj, New Delhi-110002
Visit us at **www.newagepublishers.com**
(CIN: U74899DL1966PTC004618)

Preface

The last four decades have seen tremendous developments in the art, science and technology of welding. During the second war the use of welding was limited to the repair and maintenance jobs. Now it is used to weld structures of serious structural integrity like space-crafts and fission chambers of atomic power plants. The developments in welding are taking place at a fantastic rate. It has now become a group activity requiring skills from different disciplines. Some major contributors are: metallurgists, designers, engineers, architects, physicists, chemists, safety engineers etc. A lot of descriptive and quantitative material is available in the welding textbooks. The major goal of the present book is to provide the welding engineers and managers responsible for activities related to welding with the latest developments in the science and technology of welding and to prepare them to tackle the day-to-day problems at welding sites in a systematic, scientific and logical manner. This need the author has felt during his past 30 years of teaching this subject both at undergraduate and graduate level and giving refresher and short-term courses to the practicing engineers. The book completely covers the syllabus of "Advanced Welding Technology"—an elective course of UPTU, Lucknow in addition to covering a wide spectrum of other important topics of general interest to the practicing engineers and students of mechanical, production and industrial and industrial metallurgy engineering branches.

Special topics like welding pipelines and piping, underwater welding, welding of plastics, welding of dissimilar metals, hardfacing and cladding have also been covered. Standard codes and practices have also been described. Materials and experimental results have been considered from a number of sources and in each case the author tried to acknowledge them throughout the book. Numerical problems have been solved at appropriate places in the text to demonstrate the applications of the material explained.

In order to achieve the goals set forth and still limit the physical size of the book, all supporting materials not directly falling in the welding area have not been covered. It has also been kept in mind that the present work is not an encyclopaedia or handbook and is not intended to be so, therefore, a list of selected references for further reading have been provided at the end of the text. It is hoped that the book will serve the intended purpose of benefiting the students of the subject and the practicing engineers. I earnestly look forward to suggestions from readers for the improvements to make it more useful.

—M.I.K.

Acknowledgements

The author would like to express his deepest gratitude to his wife and children for their patience and sacrificing their family time during the preparation of this book. The author acknowledges the books and references given at the end of the text which were consulted during its preparation. The author is really grateful to Prof. S.W. Akhtar, V.C. and Prof. S.M. Iqbal, P.V.C. of Integral University for their kind support and encouragements. The author expresses his deep sense of gratitude to his old colleagues and friends, especially to Prof. Emeritus (Dr.) P.C. Pandey and Dr. S.M. Yahya for their excellent suggestions and comments and Prof. (Dr.) B.K. Gupta and Prof. (Dr.) R.C. Gupta for their encouragements.

The author is thankful to M/s New Age International for their marvelous efforts to print this book in record time with an excellent get-up.

Contents

Chapter 1

Introduction to Welding Technology

1.1 DEFINITION AND CLASSIFICATION

Welding is a process of permanent joining two materials (usually metals) through localised coalescence resulting from a suitable combination of temperature, pressure and metallurgical conditions. Depending upon the combination of temperature and pressure from a high temperature with no pressure to a high pressure with low temperature, a wide range of welding processes has been developed.

Classification of Welding Process

American Welding Society has classified the welding processes as shown in Fig. 1.1. Various welding processes differ in the manner in which temperature and pressure are combined and achieved.

Welding Processes can also be classified as follows (based on the source of energy):

1. Gas Welding
   - — Oxyacetylene
   - — Oxy hydrogen
2. Arc Welding
   - — Carbon Arc
   - — Metal Arc
   - — Submerged Arc
   - — Inert-gas-Welding

   TIG and MIG
   - — Plasma Arc
   - — Electro-slag
3. Resistance Welding
   - — Spot
   - — Seam
   - — Projection

— Butt Welding
— Induction Welding
4. Solid State Welding
— Friction Welding
— Ultrasonic Welding
— Explosive Welding
— Forge and Diffusion Welding
5. Thermo-chemical Welding
— Thermit Welding
— Atomic H_2 Welding
(also arc welding)
6. Radiant Energy Welding
— Electron Beam Welding
— Laser Beam Welding

In order to obtain coalescence between two metals there must be a combination of proximity and activity between the molecules of the pieces being joined, sufficient to cause the formation of common metallic crystals.

Proximity and activity can be increased by plastic deformation (solid-state-welding) or by melting the two surfaces so that fusion occurs (fusion welding). In solid-state-welding the surfaces to be joined are mechanically or chemically cleaned prior to welding while in fusion welding the contaminants are removed from the molten pool by the use of fluxes. In vacuum or in outer space the removal of contaminant layer is quite easy and welds are formed under light pressure.

1.2 CONDITIONS FOR OBTAINING SATISFACTORY WELDS

To obtain satisfactory welds it is desirable to have:

- a source of energy to create union by FUSION or PRESSURE
- a method for removing surface CONTAMINANTS
- a method for protecting metal from atmospheric CONTAMINATION
- control of weld METALLURGY

1.2.1 Source of Energy

Energy supplied is usually in the form of heat generated by a flame, an arc, the resistance to an electric current, radiant energy or by mechanical means (friction, ultrasonic vibrations or by explosion). In a limited number of processes, pressure is used to force weld region to plastic condition. In fusion welding the metal parts to be joined melt and fuse together in the weld region. The word fusion is synonymous with melting but in welding fusion implies union. The parts to be joined may melt but not fuse together and thus the fusion welding may not take place.

1.2.2 Surface Contaminants

Surface contaminants may be organic films, absorbed gases and chemical compounds of the base metal (usually oxides). Heat, when used as a source of energy, effectively removes organic films and adsorbed gases and only oxide film remains to be cleaned. Fluxes are used to clean the oxide film and other contaminants to form slag which floats and solidifies above the weld bead protecting the weld from further oxidation.

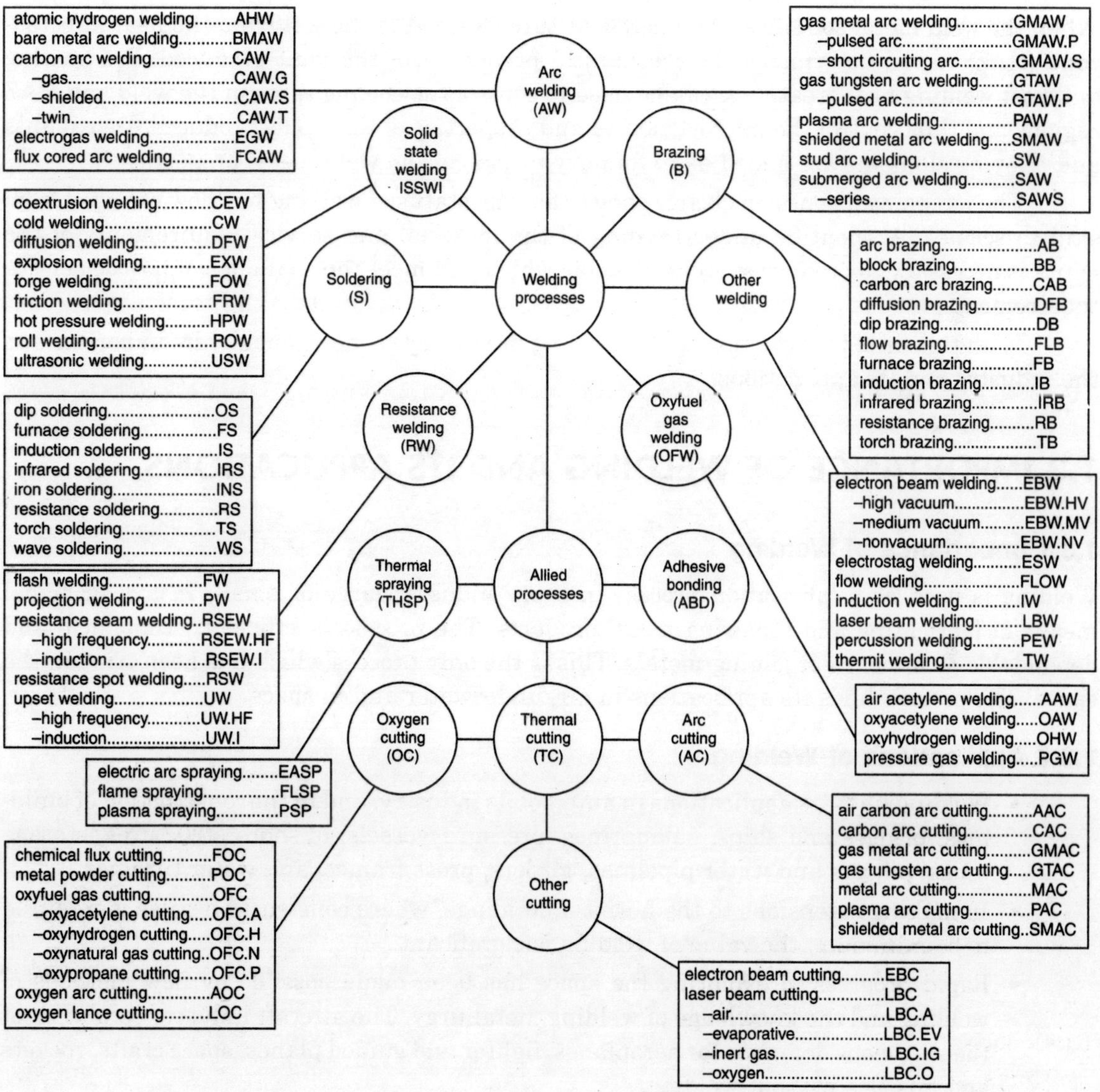

Fig. 1.1 Master Chart of Welding and Allied Processes

1.2.3 Protecting Metal From Atmospheric Contamination

To protect the molten weld pool and filler metal from atmospheric contaminants, specially the oxygen and nitrogen present in the air, some shielding gases are used. These gases could be argon, helium or carbon-dioxide supplied externally. Carbon dioxide could also be produced by the burning of the flux coating on the consumable electrode which supplies the molten filler metal to the weld pool.

1.2.4 Control of Weld Metallurgy

When the weld metal solidifies, the microstructures formed in the weld and the heat-affected-zone (HAZ) region determines the mechanical properties of the joint produced. Pre-heating and post welding heat-treatment can be used to control the cooling rates in the weld and HAZ regions and thus control the microstructure and properties of the welds produced. Deoxidants and alloying elements are added as in foundry to control the weld-metal properties.

The foregoing discussion clearly shows that the status of welding has now changed from skill to science. A scientific understanding of the material and service requirements of the joints is necessary to produce successful welds which will meet the challenge of hostile service requirements.

With this brief introduction to the welding process let us now consider its importance to the industry and its applications.

1.3 IMPORTANCE OF WELDING AND ITS APPLICATIONS

1.3.1 Importance of Welding

Welding is used as a fabrication process in every industry large or small. It is a principal means of fabricating and repairing metal products. The process is efficient, economical and dependable as a means of joining metals. This is the only process which has been tried in the space. The process finds its applications in air, underwater and in space.

1.3.2 Applications of Welding

- Welding finds its applications in automobile industry, and in the construction of buildings, bridges and ships, submarines, pressure vessels, offshore structures, storage tanks, oil, gas and water pipelines, girders, press frames, and water turbines.
- In making extensions to the hospital buildings, where construction noise is required to be minimum, the value of welding is significant.
- Rapid progress in exploring the space has been made possible by new methods of welding and the knowledge of welding metallurgy. The aircraft industry cannot meet the enormous demands for aeroplanes, fighter and guided planes, space crafts, rockets and missiles without welding.
- The process is used in critical applications like the fabrication of fission chambers of nuclear power plants.
- A large contribution, the welding has made to the society, is the manufacture of

household products like refrigerators, kitchen cabinets, dishwashers and other similar items.

It finds applications in the fabrication and repair of farm, mining and oil machinery, machine tools, jigs and fixtures, boilers, furnaces, railway coaches and wagons, anchor chains, earth moving machinery, ships, submarines, underwater construction and repair.

1.4 SELECTION OF A WELDING PROCESS

Welding is basically a joining process. Ideally a weld should achieve a complete continuity between the parts being joined such that the joint is indistinguishable from the metal in which the joint is made. Such an ideal situation is unachievable but welds giving satisfactory service can be made in several ways. The choice of a particular welding process will depend on the following factors.

1. Type of metal and its metallurgical characteristics
2. Types of joint, its location and welding position
3. End use of the joint
4. Cost of production
5. Structural (mass) size
6. Desired performance
7. Experience and abilities of manpower
8. Joint accessibility
9. Joint design
10. Accuracy of assembling required
11. Welding equipment available
12. Work sequence
13. Welder skill

Frequently several processes can be used for any particular job. The process should be such that it is most, suitable in terms of technical requirements and cost. These two factors may not be compatible, thus forcing a compromise. Table 2.1 of chapter 2 shows by "*x*" marks the welding process, materials and material thickness combinations that are usually compatible. The first column in the table shows a variety of engineering materials with four thickness ranges. The major process currently in use in industry are listed across the top of the table. The information given is a general guide and may not necessarily be valid for specific situations.

1.5 WELDING QUALITY AND PERFORMANCE

Welding is one of the principle activities in modern fabrication, ship building and offshore industry. The performance of these industries regarding product quality, delivery schedule and productivity depends upon structural design, production planning, welding technology adopted and distortion control measures implemented during fabrication. The quality of welding depends on the following parameters:

1. Skill of Welder
2. Welding parameters
3. Shielding medium and
4. Working environment
5. Work layout
6. Plate edge preparation
7. Fit-up and alignment
8. Protection from wild winds during-on-site welding
9. Dimensional accuracy
10. Correct processes and procedures
11. Suitable distortion control procedures in place

Selection of Welding Process and Filler Metal:

The welding process and filler metal should be so selected that the weld deposit will be compatible with the base metal and will have mechanical properties similar to or better than the base metal.

Comparison of high energy density welding processes and TIG welding for plate thickness 6 mm.

Parameter	*TIG*	*Plasma*	*Laser*	*EB*
Power input to workpiece	2 kW	4 kW	4 kW	5 kW
Total power used	3 kW	6 kW	50 kW	6 kW
Traverse Speed	2 mm/s	5.7 mm/s	16 mm/s	40 mm/s
Positional Welding	Good penetration	Good penetration	Yes Requires optics to move the beam	Requires mechanism to move the beam
Distortion Shrinkage	Nominal Significant in V-shaped weld	Nominal significant in V-shaped weld	Small Minimum	Minimum Minimum
Special Process Requirements	Normal Light Screening	Normal Light Screening	Safety interlock against misplaced beam reflection	Vacuum chambers, X-ray Screening
Surface Geometry	Underside Protrusion	Underside protrusion	Very fine ripples	Ruffled swarf on back face

■ QUESTIONS

1.1 Define 'Welding'. Explain the meaning and signification of coalescence and fusion in regard to welding. Why is it easier to obtain quality welds in space than in air?

1.2 Explain the conditions for obtaining satisfactory welds. Discuss the importance of welding and state its applications.

1.3 Discuss the factors which are considered in choosing a welding process for a specific application.

CHAPTER 2

Review of Conventional Welding Processes

In the following paragraphs distinguishing features, attributes, limitations and comparisons where applicable will be discussed for the commonly used welding processes. This introduction to the welding processes will help the modern welding engineers to consider alternative processes available for the situation. This aspect may otherwise be overlooked. A major problem, frequently arises when several processes can be used for a particular application. Selection could be based upon fitness for service and cost. These two factors, sometimes, may not be compatible. Process selection is also affected by such factors as:

(*a*) production quantity, (*b*) acceptability of installation costs, (*c*) joint location, (*d*) joint service requirements, (*e*) adaptability of the process to the location of the operation, (*f*) availability of skill/experience of operators.

In this review of conventional welding processes we shall be discussing Gas Welding, Arc Welding, Shielded Metal Arc, Submerged Arc, Tungsten Inert Gas, Metal Inert Gas, Metal Active Gas Welding, Resistance Welding, Electroslag Welding, Spot, Seam and Projection Welding, Flash Butt and Upset Butt Welding, and high Frequency Welding.

Advanced welding processes such as Electron Beam welding, Laser Beam Welding, Plasma Arc Welding, Explosive Welding, Friction Welding, Ultrasonic Welding and Underwater Welding are discussed in chapter 4. Now let us start to review the conventional welding processes, starting with gas welding.

2.1 GAS WELDING

Gas welding includes all the processes in which fuel gases are used in combination with oxygen to obtain a gas flame. The commonly used gases are acetylene, natural gas, and hydrogen in combination with oxygen. Oxyhydrogen welding was the first commercially used gas process which gave a maximum temperature of 1980°C at the tip of the flame. The most commonly used gas combination is oxyacetylene process which produces a flame temperature of 3500°C. This process will be discussed in detail in the following paragraphs.

1. Oxyacetylene welding flame uses oxygen and acetylene. Oxygen is commercially made by liquefying air, and separating the oxygen from nitrogen. It is stored in cylinders as

shown in Fig. 2.1 at a pressure of 14 MPa. Acetylene is obtained by dropping lumps of calcium carbide in water contained in an acetylene generator according to the following reaction.

$$CaC_2 + 2H_2O = Ca(OH)_2 + C_2H_2$$

Calcium carbide + Water = Slaked lime + Acetylene gas

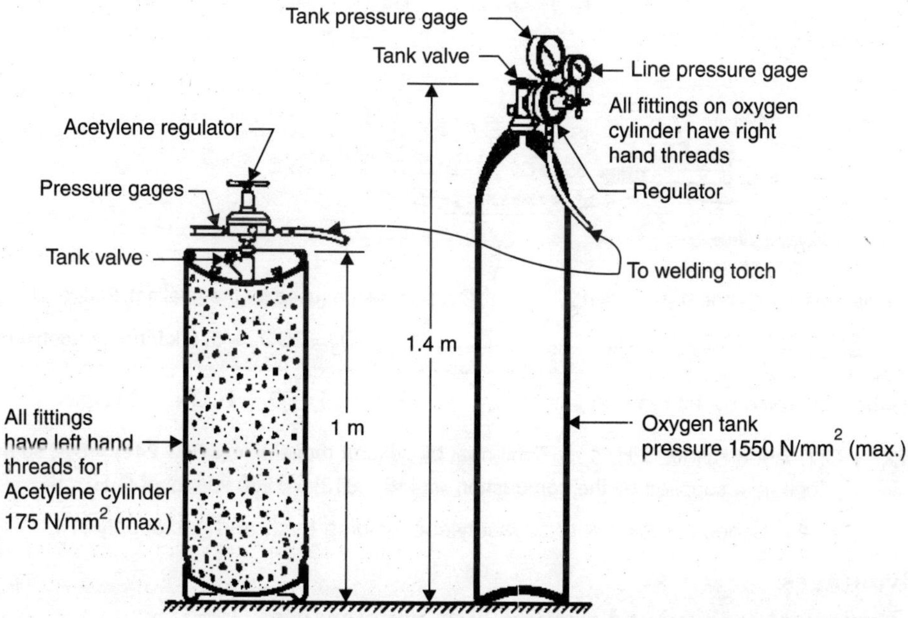

Fig. 2.1 Cylinders and regulators for oxyacetylene welding [1]

2. Concentrated heat liberated at the inner cone is 35.6% of total heat. Remaining heat develops at the outer envelope and is used for preheating thus reducing thermal gradient and cooling rate improving weld properties.
3. 1 Volume O_2 is used to burn 1 Volume of acetylene, in the first reaction. This oxygen is supplied through the torch, in pure form $1\frac{1}{2}$ Volume of additional oxygen required in the second reaction is supplied from the atmosphere.
4. When oxygen is **just enough** for the first reaction, the resulting flame is **neutral**. If less than enough, → the flame is said to be reducing flame. If more than enough oxygen is supplied in the first reaction, the flame is called an oxidizing flame.
5. Neutral flame has the widest application.
   - Reducing flame is used for the welding of **monel metal**, nickel and certain alloy steels and many of the non-ferrous, hardsurfacing materials.
   - Oxidising flame is used for the welding of brass and bronze.

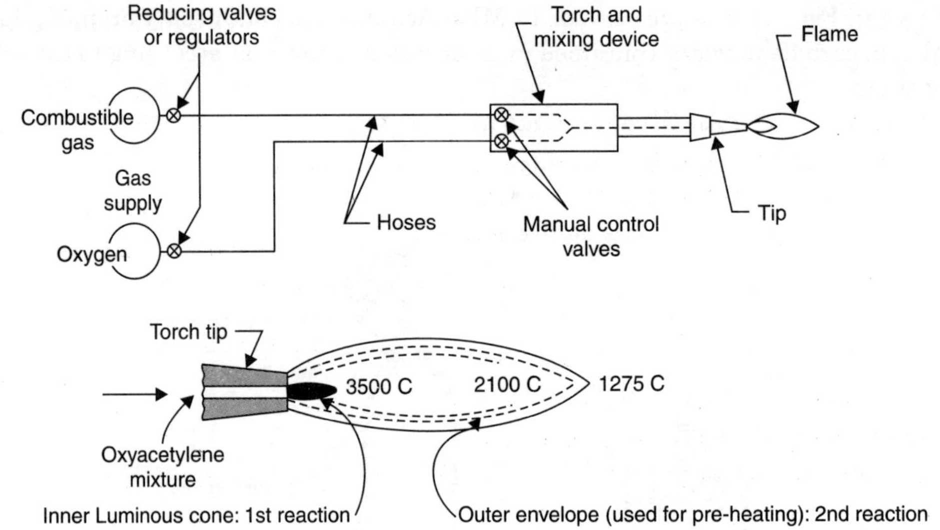

$C_2H_2 + O_2 \longrightarrow 2\,CO + H_2$

$2CO + O_2 = 2CO_2 + 570$ kJ/mol of acetylene

Total heat liberated by 1st reaction

$H_2 + \frac{1}{2}O_2 = H_2O + 242$ kJ/mol

$(227 + 221) = 448$ kJ/mol C_2H_2

Total heat by second reaction = $(570 + 242) = 812$ kJ/mol of C_2H_2

Total heat supplied by the combustion = $(448 + 812) = 1260$ kJ/mol of C_2H_2

Fig. 2.2 Schematic sketch of oxyacetylene welding torch and gas supply [1].

Advantages:

1. Equipment is cheap and requires little maintenance.
2. Equipment is portable and can be used in field/or in factory.
3. Equipment can be used for cutting as well as welding.

Acetylene is used as a fuel which on reaction with oxygen liberates concentrated heat sufficient to melt steel to produce a fusion weld. Acetylene gas, if kept enclosed, decomposes into carbon and hydrogen. This reaction results into increase in pressure. **At 0.2 N/mm² pressure,** the mixture of carbon and hydrogen may cause **violent explosion** even in the absence of oxygen, when exposed to spark or shock. To counter this problem, acetylene is dissolved in acetone. **At 0.1 N/mm²** one volume of acetone dissolves **twenty volumes of acetylene.** This solubility **linearly increases to 300** volumes of acetylene per one volume of acetone, **at 1.2 N/mm².**

An excess of oxygen or acetylene is used depending on whether oxidising or reducing (carburizing) flame is needed.

Oxidizing (decarburizing) flame is used for the welding of brass, bronze and copper-zinc and tin alloys, while reducing (carburising) flame is used for the welding of low carbon and alloy steels monel metal and for hard surfacing. Neutral flame is obtained when the ratio of oxygen to acetylene is about 1 : 1 to 1.15 : 1. Most welding is done with neutral flame. The process has the advantage of control over workpiece temperature, good welds can therefore be obtained. Weld and HAZ, being wider in gas welding resulting in considerable distortion. Ineffective shielding of weld-metal may result in contamination. Stabilised methyl acetylene propadiene (MAPP) is replacing acetylene where portability is important. It also gives higher energy in a given volume.

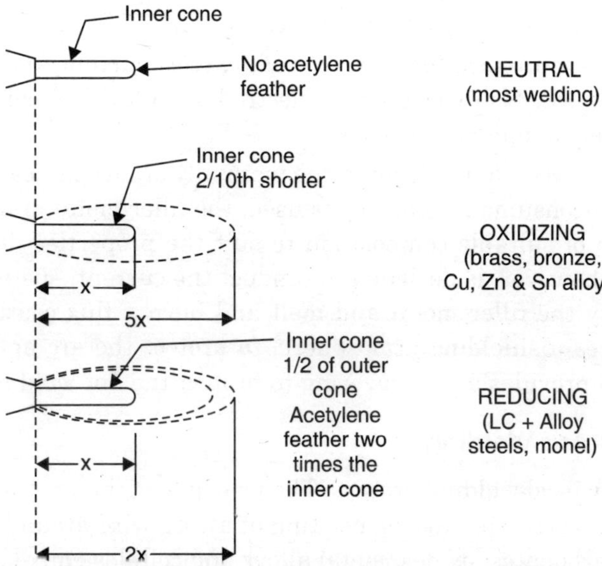

Fig. 2.3 Neutral, oxidizing and reducing flames

2.2 ARC WELDING

An arc is a sustained electric discharge in a conducting medium. Arc temperature depends upon the energy density of the arc column. Arc could be used as a source of heat for welding.

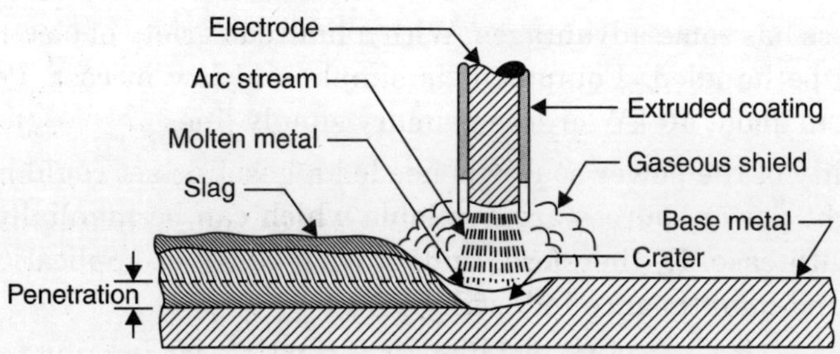

Fig. 2.4 Diagrammatic sketch of arc flame

Arc welding is a group of welding processes that use an electric arc as a source of heat to melt and join metals, pressure or filler metal may or may not be required. These processes include

- Shielded metal arc welding (SMAW)
- Submerged arc Welding (SAW)
- Gas metal arc (GMA, MIG, MAG)
- Gas tungsten arc (GTA, TIG)

- Plasma arc welding (PAW)
- Electroslag/Electrogas Welding

Arc is struck between the workpiece and the electrode and moves relative to the workpiece, manually or mechanically along the joint.

Electrode, may be **consumable** wire or rod, carries current and sustains the arc between its tip and the work. Non consumable electrodes could be of carbon or tungsten rod. Filler metal is separately supplied, if needed.

The electrode is moved along the joint line manually or mechanically with respect to the workpiece. When a non-consumable elecrode is used, the filler metal, if needed, is supplied by a separate rod or wire of suitable composition to suit the properties desired in the joint. A consumable electrode, however, is designed to conduct the current, sustain the arc discharge, melt by itself to supply the filler metal and melt and burn a flux coating on it (if it is flux coated). It also produces a shielding atmosphere, to protect the arc and weld pool from the atmospheric gases and provides a slag covering to protect the hot weld metal from oxidation.

2.2.1 Shielded Metal Arc Welding

It is the most commonly used welding process. The principle of the process is shown in Fig. 2.4. It uses a consumable covered electrode consisting of a core wire around which a flux coating containing fluorides, carbonates, oxides, metal alloys and cellulose mixed with silicate binders is extruded.

- This covering provides arc stabilizers, gases to displace air, metal and slag to support, protect and insulate the hot weld metal.
- Electrodes and types of coating used are discussed in more detail in chapter 4. The electrodes are available in diameters ranging from 2 mm (for thin sheets) to 8 mm (for use at higher currents to provide high deposition rates). Alloy filler metal compositions could be formulated easily by using metal powders in the flux coating.
- This process has some advantages. With a limited variety of electrodes many welding jobs could be handled. Equipment is simple and low in cost. Power source can be connected to about 10 kW or less primary supply line.
- If portability of the power source is needed a gasoline set could be used. Solid-state, light weight power sources are available which can be manually carried to desired location with ease. It, therefore, finds a wide range of applications in construction, pipe line and maintenance industries.
- The process is best suited for welding plate thicknesses ranging from 3 mm to 19 mm. Greater skill is needed to weld sections less than 3 mm thickness.
- Hard surfacing is another good application of this process.

SMAW is used in current ranges between 50-300 A, allowing weld metal deposition rates between 1-8 kg/h in flat position.

- Normally a welder is able to deposit only 4.5 kg of weld metal per day. This is because usually in all position welding small diameter electrodes are used and a considerable electrode manipulation and cleaning of slag covering after each pass is necessary. This makes the labour cost quite high. Material cost is also more because only 60% of the electrode material is deposited and the rest goes mainly as stub end loss.

- Inspite of these deficiencies, the process is dominant because of its simplicity and versatility. In many situations, however, other more productive welding processes such as submerged arc and $C0_2$ processes are replacing SMAW technique.

Brief details regarding electrode flux covering, its purpose and constituents are given below:

SMA Welding uses a covered electrode core wire around which a mixture of silicate binders and powdered materials (*e.g.* carbonates, fluorides, oxides, cellulose and metal alloys) is extruded and baked producing a dry, hard concentric covering.

Purpose of covering: 1. stabilizes arc 2. produces gases to shield weld from air, 3. adds alloying elements to the weld and 4. produces slag to protect and support the weld 5. Facilitate overhead/position welding 6. Metallurgical refining of weld deposit, 7. Reduce spatter, 8. Increase deposition efficiency, 9. Influence weld shape and penetration, 10. Reduce cooling rate, 11. Increase weld deposition by adding powdered metal in coating.

Coating constituents:

1. Slag formers: SiO_2, MnO_2, and $\overbrace{FeO.Al_2O_3}$ (sometimes).
2. Improving Arc characteristics: Na_2O, CaO, MgO and TiO_2.
3. Deoxidizers: Graphite, Al and woodflour.
4. Binders: Sodium silicate, K-silicate and asbestus.
5. Alloying elements: to enhance strength: V, Ce, Co, Mo, Al, Zr, Cr, Ni, Mn, W.

Contact electrodes have thick coating with high metal powder content, permit DRAG or CONTACT welding and high deposition rates.

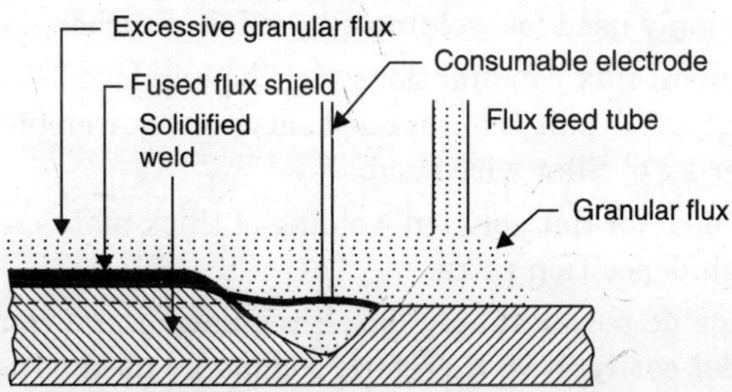

Fig. 2.5 Submerged arc welding-working principle

2.2.2 Submerged Arc Welding

Submerged arc welding (SAW) is next to SMAW in importance and in use. The working of the process is shown in Fig. 2.5. In this process the arc and the weld pool are shielded from atmospheric contamination by an envelope of molten flux to protect liquid metal and a layer of unfused granular flux which shields the arc. The flux containing CaO, CaF_2 and SiO_2 is sintered to form a coarse powder. This flux is then spread over the joint to be made.

- Arc is covered. Radiation heat loss is eliminated and welding fumes are little.
- Process is mechanized or semi-automatic. High currents (200–2000 A) and high deposition rates (27-45 kg/h) result in high savings in cost.

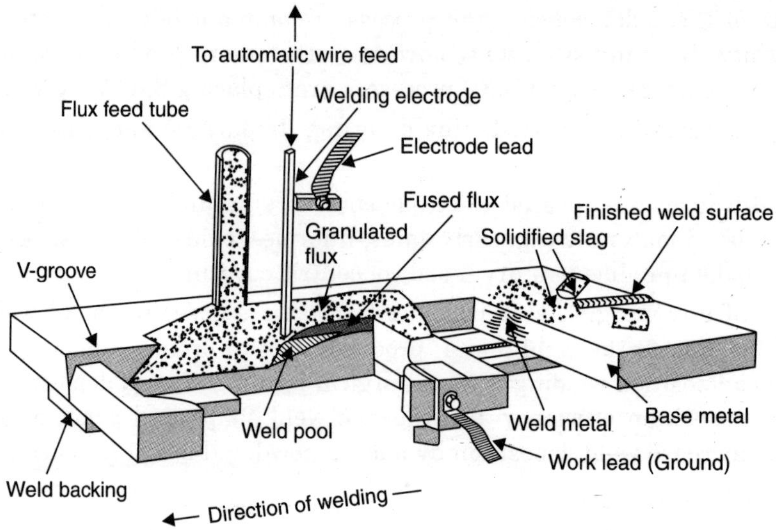

Fig. 2.5 Submerged arc welding process

- Power sources of 600-2000 A output, automatic wire feed and tracking systems on mechanized equipment permit high quality welds with minimum of manual skill. Welding speeds up to 80 mm/s on thin gauges and deposition rates up to 45 kg/h on thick sections are major advantages of this process.
- Plate thicknesses up to 25 mm could be welded in a single pass without edge preparation using dcep.
- Process is commonly used for welding all grades of carbon, low alloy and alloy steels.
- Various filler metal-flux combinations may be employed to obtain desired weld deposit characteristics to suit the intended service requirements. Nearly one kg of flux is consumed per kg of filler wire used.
- The process is ideal for flat position welding of thick plates requiring consistent weld quality and high deposition rates.
- Constant voltage dc power supply is self regulating and could be used on constant-speed wire feeder easily. It is, therefore, commonly used power source and is the best choice for high speed welding of thin gauge steels.

2.2.3 Tungsten inert gas (Tig) Welding

- In TIG welding an arc is maintained between a non-consumable tungsten electrode and the work-piece, in inert gas medium, and is used as a heat source. Filler metal is fed from outside. The principle of operation of the process is shown in Fig. 2.6.
- Direct current is normally used with electrode negative polarity for welding most metals except aluminium, magnesium and their alloys, because of the refractory oxide film on the surface which persists even when the metal beneath melts. With electrode positive, cathode spots form on aluminium surface and remove oxide film due to ionic bombardment, but excessive heat generates at the electrode.

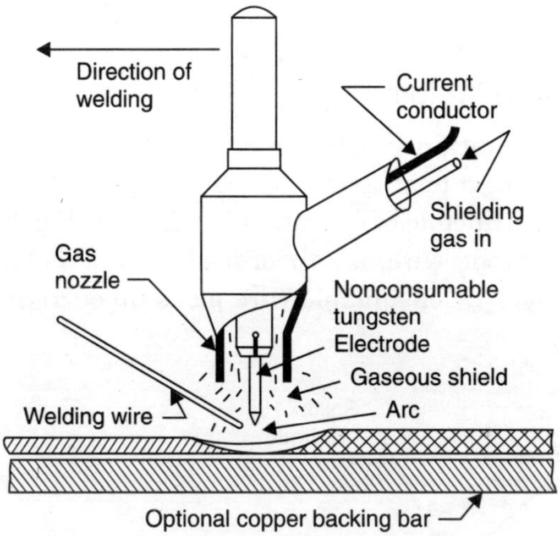

Fig. 2.6 Tungsten Inert Gas (TIG) Welding

- Welding aluminium is best achieved by using alternating current. Large heat input to the workpiece is supplied during the electrode negative half of the cycle. During electrode positive half cycle the oxide film is removed. Since a high reignition voltage is required when the work is negative various means are used to compensate for this effect. Oxide fails to disperse if such means are not used.
- Electrode material could be pure tungsten for d. c. s. p. Thoriated tungsten or zirconated tungsten can work with a.c. as well as with d.c. welding. In a. c. welding, heat input to the electrode is higher, the tip invariably melts. Electrodes containing thoria or zirconia give steadier arc due to their higher thermionic emissivity compared to the pure tungsten electrode.
- Shielding gases used are: argon, helium, and argon helium mixtrure. For very reactive metals welding should be done in an argon filled chamber to obtain ductile welds. In open-air welding with normal equipment some contamination with argon always occurs. Deoxidants are added to the filler metal as a consequence when welding rimming or semi-skilled carbon steel, monel metal, copper, cupro-nickel and nickel.
- Copper can be welded with nitrogen as a shielding gas. Nitrogen reacts with liquid tungsten and not with copper. Thoriated tungsten electrode with straight polarity should be employed. With nitrogen atmosphere anode heat input per ampere is higher compared to argon atmosphere. It is good for high conductivity metal as copper.
- The process is costly and is used only where there is a definite technical advantage *e.g.* welding copper, aluminium, magnesium and their alloys up to 6 mm thick; alloy steels, nickel and its alloys up to 2.5 mm thick, and for the reactive metals.
- Argon spot welds could be made with a torch having the nozzle projecting beyond the electrode tip; it is held against the work, arc is struck and maintained for a preset time and argon is cut-off after a delay. A molten pool forms on the top sheet and fuses into the sheet underneath, producing a plug/spot weld. This welding is ideal for situations having access to one side of the joint only. The equipment required is light

and portable. Process is slow and not adaptable to fully mechanised control as spot welding.

2.2.4 Metal Inert Gas (MIG) Welding

In MIG welding the arc is maintained between a consumable electrode and the workpiece in inert gas medium. It is used as a heat source which melts the electrode and thus supplies the filler metal to the joint. The principle of operation is shown in Fig. 2.7. The apparatus consists of a coil of consumable electrode wire, a pair of feed rolls, a welding torch having a control switch and an inert gas supply. Consumable wire picks up current while it passes through a copper guide tube.

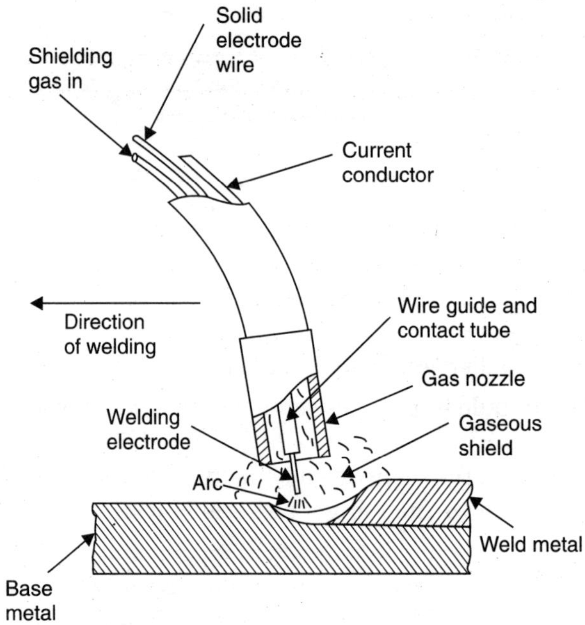

Fig. 2.7 Metal Inert Gas (MIG) Welding

- Electrode wire diameter is between 1 .5 mm to 3.0 mm and current used is between 100 to 300 A for welding aluminium, copper, nickel and alloy steels (current density is of the order of 100A per mm square: thus projected transfer occurs). The arc projects in line with the wire axis and metal also transfers in the same line.
- Projected transfer occurs within a range of current. Below the lower limit the transfer is gravitational and above the upper limit, for aluminium, the metal flow is unstable resulting in the formation of dross, porosity and irregular weld profile.
- Welding may be done below the threshold current and conditions could be adjusted to get short-circuit transfer. Wires of 0.75 mm diameter or less with wire reel directly mounted on the gun itself could be used with short circuit or dip transfer. Such a welding is called fine-wire welding and is suitable for joining sheet metals.
- Dcrp is commonly used and a power source with flat characteristics is preferred for both projected and short circuiting transfer, as it gives more consistent arc-length.

Welding of aluminium is only possible with dcsp. Drooping characteristic power sources may also be used with a choke incorporated in the circuit to limit the short circuit current and prevent spatter.

- Shielding gas is normally argon, but argon-oxygen mixtures (oxygen: 20%) are sometimes used for welding austenitic stainless steels in order to impove weld profile. Similarly 80% Ar + 20% CO_2 improves weld profile of carbon steel and sheet metal and is cheaper and better than pure argon. CO_2 shielding can also be used.
- The process is suitable for welding high alloy steels, aluminium, copper, nickel and their alloys. it is complementary to TIG, being particularly suited to thicker sections and fillet welds.
- MIG spot welding gives deeper penetration and is specially suitable for thick materials and for the welding of carbon, low alloy and high alloy steels.

2.2.5 Metal Active Gas (MAG) Welding

This process differs from MIG in that it uses CO_2 instead of inert gases (argon or helium) both the normal and fine-wire machines could be used. The differences are: metal transfer mode, power source, cost and field of application. The process is schematically shown in Fig. 2.8.

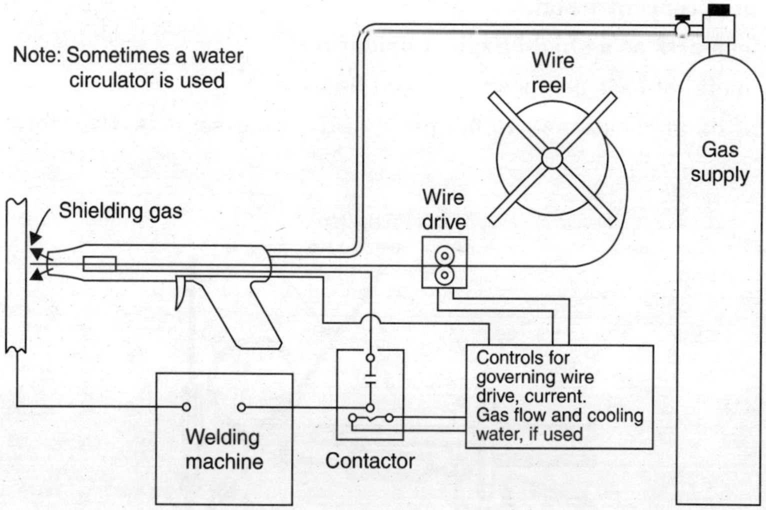

Fig. 2.8 Schematic diagram of MIG/MAG (CO_2) welding

- In CO_2 welding there is no threshold current to change transfer mode from gravitational to projected type. At low currents the free flight transfer is of repelled type and there is excessive scatter loss. This situation is quite common in fine wire welding but can be overcome by adjusting welding parameters to obtain short-circuiting mode of transfer (the drop comes in contact with the weld pool and is detached from the wire by surface tension and electromagnetic forces before it can be projected laterally). If the current is excessive during short-circuiting, detachement will be violent and will cause spatter.
- To get rid of this problem the power source is modified either by adjusting the slope of a drooping characteristic machine or by inserting a reactance in the circuit of a flat

characteristic machine. Thus the short circuit current is limited to a suitable level. At currents in excess of 200 A using 1.5 mm or thicker wires the process is sufficiently regular permitting free flight transfer but welding is to be done in flat position only.

- At arc temperature carbon di-oxide dissociates to carbon monoxide and oxygen. To save metal from oxidation, deoxidized wire for welding carbon steel is essential, otherwise 40% of the silicon and manganese content may be lost.
- This process finds its main application in the welding of carbon and low alloy steels.

2.2.6 Atomic Hydrogen Welding

In atomic hydrogen welding a single phase AC arc is maintained between two tungsten electrodes and hydrogen gas is introduced into the arc. Hydrogen molecules absorb heat from the arc and change into atomic hydrogen. This atomic hydrogen when comes in contact with the plates to be welded recombines into molecular hydrogen, liberating a large amount of intense heat giving rise to a temperature of 6100°C. Weld filler, metal may be added using welding rod as in oxy-acetylene welding. It differs from SMAW in that the arc is indendent of base metal (work) making electrode holder a mobile without arc getting extinguished. Thus heat input to the weld could be controlled by manually to control weld metal properties. The process has the following special features:

1. High heat concentration.
2. Hydrogen acts as a shield against oxidation.
3. Filler metal of base composition could be used.
4. Most of its applications can be met by MIG process, it is, therefore, not commonly used.

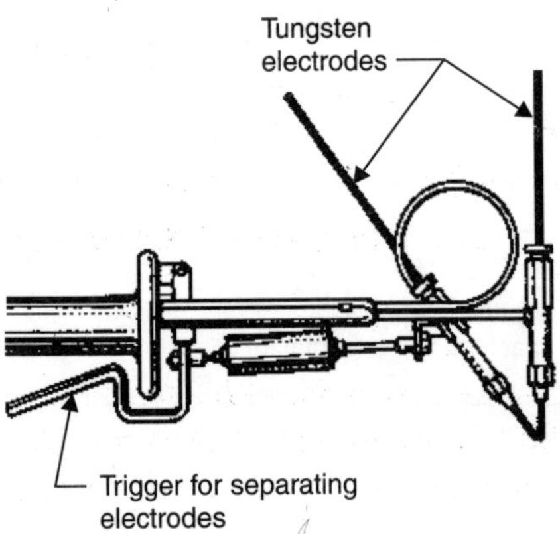

Fig. 2.8 Atomic hydrogen welding torch

2.3 RESISTANCE WELDING

In the following proceses, ohmic resistance is used as a heat source.

2.3.1 Electroslag Welding

The electroslag welding is used for welding thick plates. The plates have square edge preparation and are set vertically up with about 25 mm gap in between as shown in Fig. 2.9. A starting piece is provided at the bottom. Some flux and welding wire electrodes are fed into the gap between the edges. Arc starts and the slag melts. Molten slag is conductive, the arc is short circuited and heat is generated due to the passage of heavy currents through the slag. The slag agitates vigorously and the parent metal and the filler metal melt, forming a liquid metal pool covered by a layer of liquid slag. This pool is retained by water cooled copper dams. A little flux is added from time to time to maintain a slag pool of constant depth. A number of electrodes could be used depending upon the plate thikness.

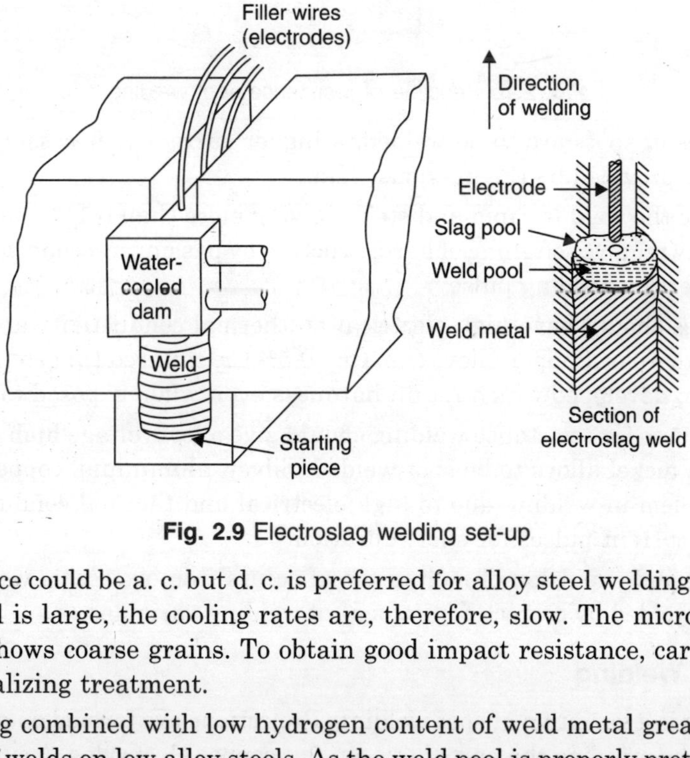

Fig. 2.9 Electroslag welding set-up

Power source could be a. c. but d. c. is preferred for alloy steel welding. Welding speed is low and weld pool is large, the cooling rates are, therefore, slow. The microstructure of weld metal and HAZ shows coarse grains. To obtain good impact resistance, carbon and low alloy steels need normalizing treatment.

Slow cooling combined with low hydrogen content of weld metal greatly minimizes the risk of cracking of welds on low alloy steels. As the weld pool is properly protected from atmospheric contamination, the use of deoxidized wire is not essential.

Electroslag welding is used for the vertical welding of plate and sections over 12 mm thick in carbon and low alloy steels and has been used for the welding of high alloy steels and titanium.

2.3.2 Spot Welding

- In this process, the parts to be joined are normally overlapped and the metal at the interface fuses due to resistance heating. The principle of operation of the process is shown in Fig. 2.10. The workpieces are clamped between two water cooled copper electrodes. On the passage of a high transient current the interface melts over a spot and forms a weld. The cooling of the electrode limits the size of the spot. A very high current (10,000 amp or more) is used for a short duration (fraction of a second) to complete the weld. The interfaces to be joined are initially cleaned by various methods: grinding, scratch brushing or vapour degreasing. A spot weld normally contains small porosity (due to shinkage) in the weld center which is usually harmless.

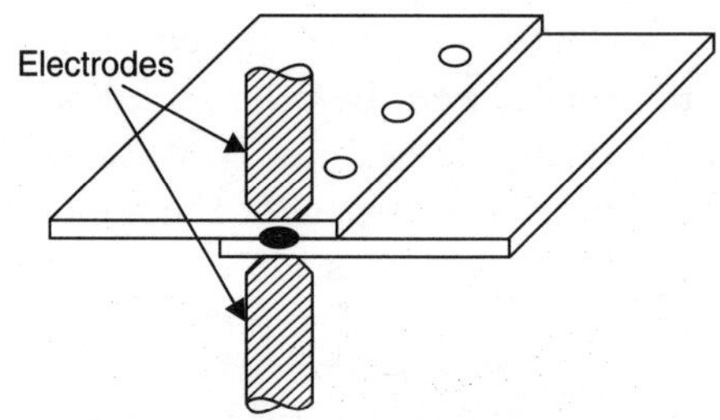

Fig. 2.10 Principle of resistance spot welding

- If a series of spots are to be welded, a higher current is necessary in view of short circuiting provided by the previous weld.
- Cooling of the weld is rapid and steels having more than 0.15% carbon and low alloy steels may require softening of hard structure by passing a second, less intense currect pulse after the welding pulse.
- Electrodes should have high electrical an thermal conductivity and should have resistance to wear. Copper alloys (*e.g.* Cu– 0.5% Cr, sintered tungsten copper compacts) have been developed which retain hardness even when exposed to welding heat.
- Power source for resistance welding should give a low voltage high current output for steel and nickel alloys to be spot welded. Silver, aluminium, copper and their alloys pose problem in welding due to high electrical and thermal conductivity necessitating high current pulses for short duration.
- Cracking and expulsion of molten metal occurs from excessive welding current and may be avoided by correct adjustment of welding variables.

2.3.3 Projection Welding

Projection welding is a variation of spot welding. Projections are formed on one of the pieces to be joined, usually by pressing the parts between flat copper electrodes. A current pulse makes the weld at the tip of the projection leaving clean surfaces without indentations. Schematic of the set-up is shown in Fig. 2.11.

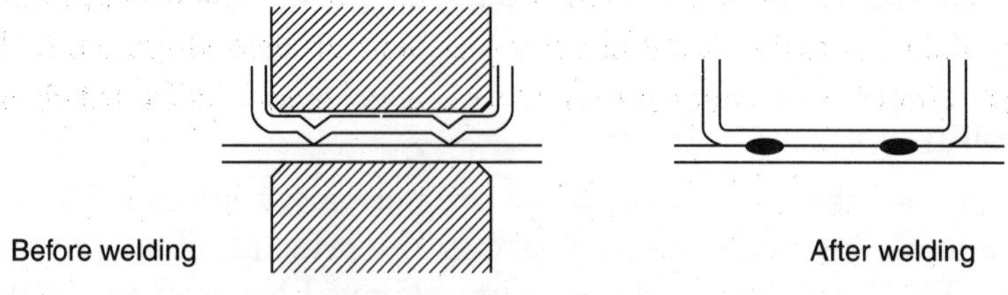

Fig. 2.11 Projection welding

2.3.4 Seam Welding

Seam welding is a continuous spot welding process where overlapped parts to be welded are fed between a pair of copper alloy (roller disc shaped) electrodes (Fig. 2.12).

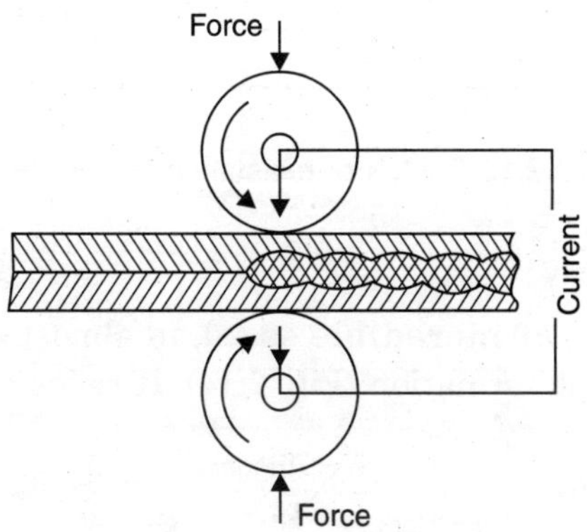

Fig. 2.12 Sketch of seam welding

2.3.5 Flash Welding

It is classified as a resistance welding process as the heat is generated at the faying surfaces of the joint by resistance to the flow of electric current, and by arcs across the interface. A thin layer of liquid metal forms at the faying surfaces. When the parts are forced together to form a joint, the layer of liquid metal on the faces alongwith the impurities is expelled, the hot metal upsets and forms a flash. No external filler metal is added during welding. Welds can be made in sheet and bar thicknesses ranging from 0.2 to 25 mm (sheets) and 1 to 76 mm (bars). Machines are available in capacities ranging from 10 kVA to 1500 kVA. The distance by which the pieces get shortened due to upsetting is called flashing allowance. The process is used for joining rails, steel strips, window frames, etc.

2.3.6 Butt (Upset) Welding

The principle of the process is shown in Fig. 2.13. Here the workpiece temperature at the joint is raised by resistance to the passage of an electric current across the interface of the joint. The parts to be joined (wires or rods usually) are held in clamps, one stationary and the other movable which act as conductors for the low voltage electric supply and also apply force to form the joint. Force is applied only after the abutting surfaces reach near to the melting temperature. This causes up-setting. Uniform and accurately mating surfaces are desirable to exclude air and give uniform heating.

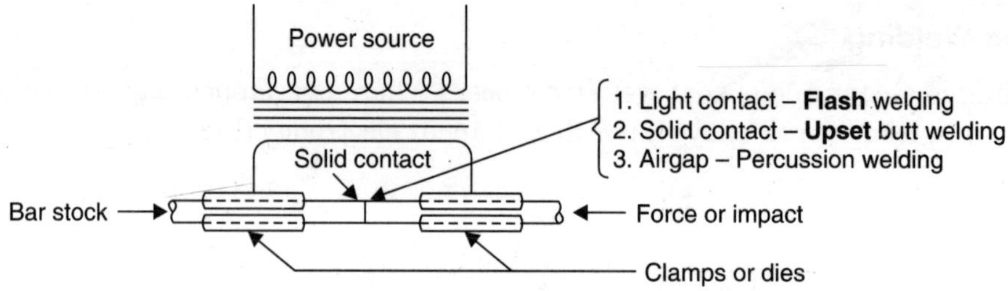

Fig. 2.13 Sketch of resistance butt welding

2.3.7 Percussion Welding

This process makes butt welds at **incredible** speed, in almost any combination of dissimilar materials and without the flash formation (Fig. 2.14). It relies on arc effect for heating.

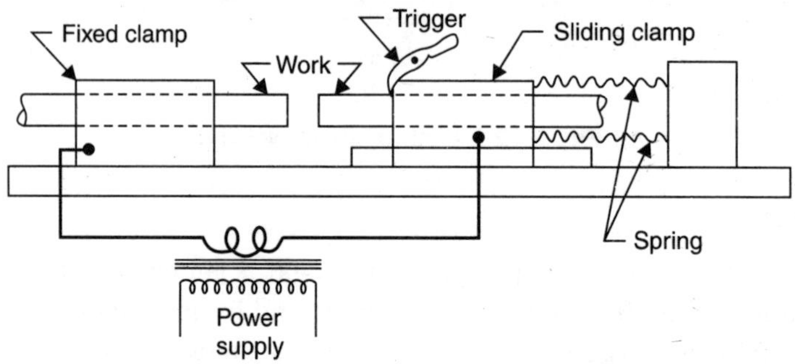

Fig. 2.14 Principle of percussion welding

The pieces to be joined are kept apart, one in a stationery holder and the other in a moveable clamp held against a heavy spring pressure. When the movable clamp is released the part to be welded moves towards the other part. Arcing occurs when the gap between the pieces to be welded is 1.6 mm. The ends to be welded are prepared for accurate mating. An extremely heavy current impulse flows for a short duration (0.001 to 0.1 second) across the gap between the pieces **forming an arc.** The intense heat developed for a very short duration causes superficial melting over the entire end surfaces of the bars. Immediately after this current pulse, the pieces are brought together with an impact blow (hence the name percussion) to complete the weld.

The electric energy for the discharge is built-up in one of two ways. In the electrostatic method, energy is stored in a capacitor, and the parts to be welded are heated by the sudden discharge of a heavy current from the capacitor. The electromagnetic welder uses the energy discharge caused by the collapsing of the magnetic field linking the primary and secondary windings of a transformer or other inductive device. In either case intense arcing is created which is followed by a quick blow to make the weld.

Special Applications:

- Heat treated parts can be joined without affecting the heat treatment.
- Parts having different thermal conductivities and mass can be joined successfully. For example stellite tips to tool shanks, copper to alluminium or stainless steel. Silver

contact tips to copper, cast iron to steet, zinc to steel. These welds are produced without flash or upset at the joint.

Limitation:

The limitation of the process is that only small areas upto 650 mm^2 of nearly regular sections can be welded.

2.3.8 High Frequency Resistance Welding

In high frequency resistance welding shown in Fig. 2.15, welding current of 200–450,000 Hz frequency passes between the electrodes in contact with the edges of a strip forming a tube when it passes through forming rolls. The rolls also apply welding pressure. The amount of upset is regulated by the relative position of the welding electrodes and the rolls applying the upset force. The required welding heat is governed by the current passing through the work and the speed of tube movement.

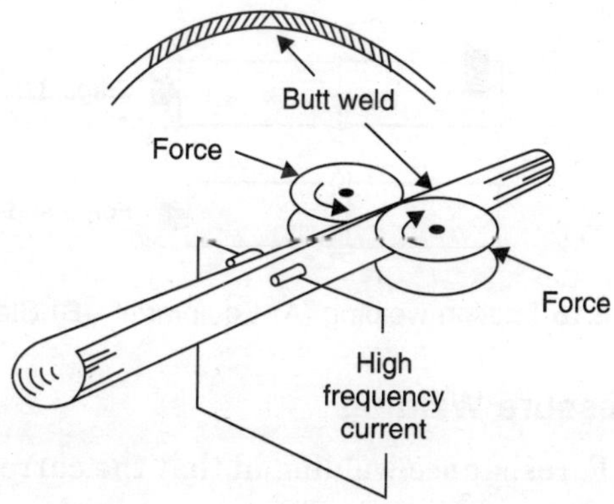

Fig. 2.15 Sketch of high frequency resistance welding

2.4 SOLID PHASE WELDING

This group of welding processes uses pressure and heat (below the melting temperature) to produce coalescence between the pieces to be joined without the use of filler metal. The processes under this category include: Diffusion Bonding, Cold Welding, Explosive Welding, Friction Welding, High Frequency Pressure Welding, Forge Welding, Hammer Welding, Ultrasonic Welding, etc. The important ones will now be discussed.

2.4.1 Friction Welding

Friction heat between two sliding/rotating surfaces is employed in this process to form a joint. The principle of working of the process is shown in Fig. 2.16. The pieces to be joined are clamped in chucks. One chuck rotates against a stationary one. Pressure is used to generate enough heat to reach a bonding temperature within a few seconds. At this stage the rotation is stopped and pressure is retained or increased to complete the weld. To accomodate awkward or very long parts, an intermediate slug or disc is rotated in between the sections to be joined.

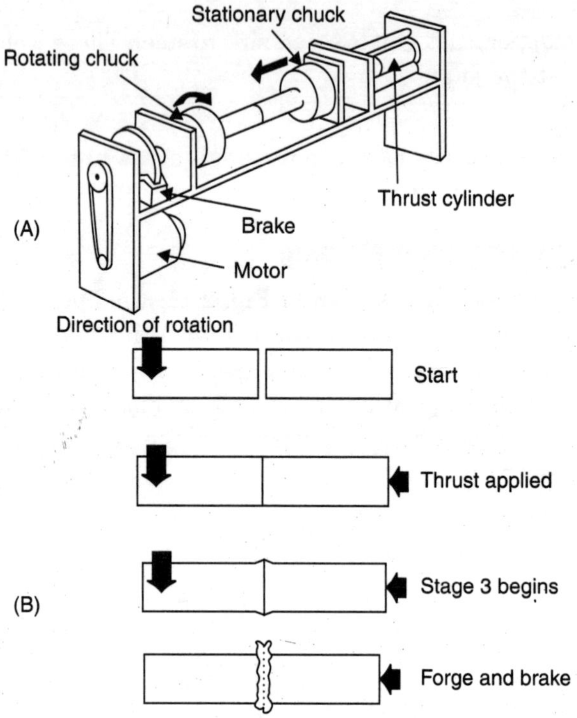

Fig. 2.16 Friction welding (A) Equipment (B) Stages

2.4.2 High Frequency Pressure Welding

This process differs from H.F. resistance welding in that the current is induced in the surface layer by a coil wound around the workpiece. This causes surface layer to be heated. Weld is formed by a forging action of the joint (Fig. 2.17). It is used in the manufacture of tubes. The process is also termed as H.F. Induction Welding.

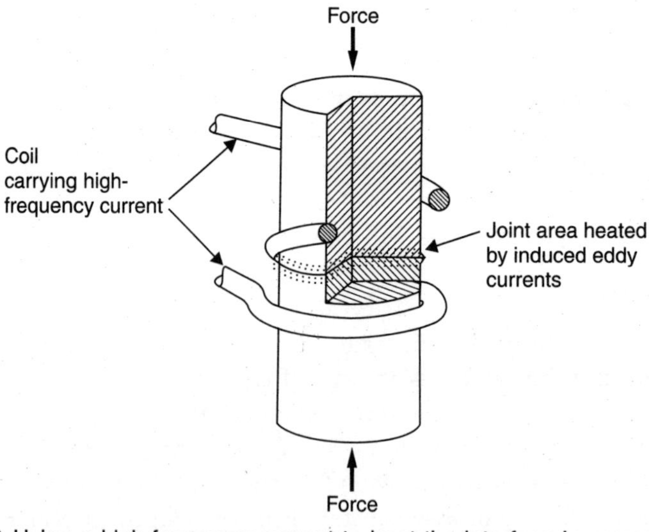

Fig. 2.17(a) Using a high-frequency current to heat the interface in pressure welding

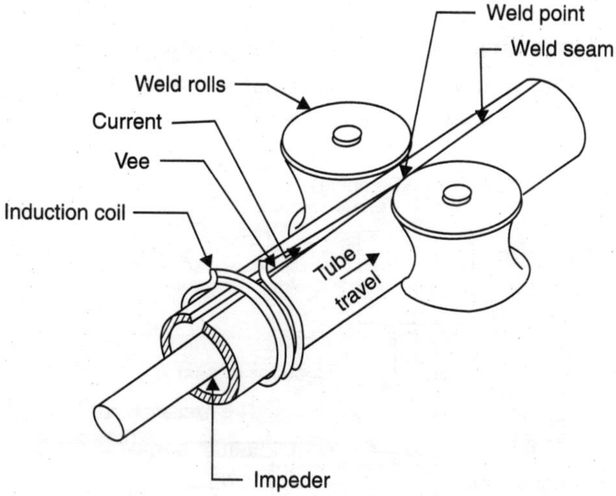

Fig. 2.17(b) Sketch of high-frequency pressure welding

2.4.3 Ultrasonic Welding

- Ultrasonic process of welding is shown in Fig. 2.18. The core of magnetostrictive ultrasonic vibrations generator (15-60 kHz) is connected to the work through a horn having a suitable shaped welding tip to which pressure is applied. The combination of ultrasonic vibrations with moderate pressure causes the formation of a spot weld or seam weld (with modified apparatus). The deformation caused is less than 5 percent.

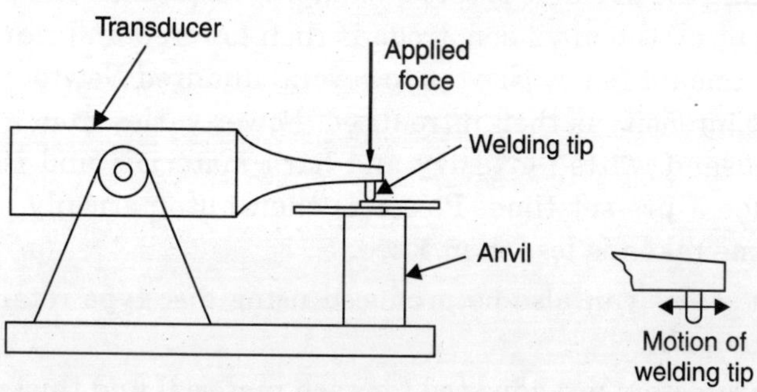

Fig. 2.18(a) Ultrasonic welding

- Friction between the interface surfaces, along the axis of the welding tip, causes the removal of surface contaminants and oxide film exposing the clean metallic surface in contact with each other which weld together due to applied pressure. Weld produced is as strong as parent metal.
- Some local heating may occur and some grains may cross the interface but not melting or bulk heating occurs.

The process is briefly discussed in the following paragraphs:

1. It is solid state joining process for similar or dissimilar metals in the form of thin strips or foils to produce, generally lap joints.

2. H.F. (15000 – 75000 Hz) vibratory energy gets into the weld area in a plane parallel to the weldment surface producing oscillating shear stresses at the weld interface, breaking and expelling surface oxides and contaminants.
3. This interfacial movement results into metal-to-metal contact permitting coalescence and the formation of a sound welded joint.

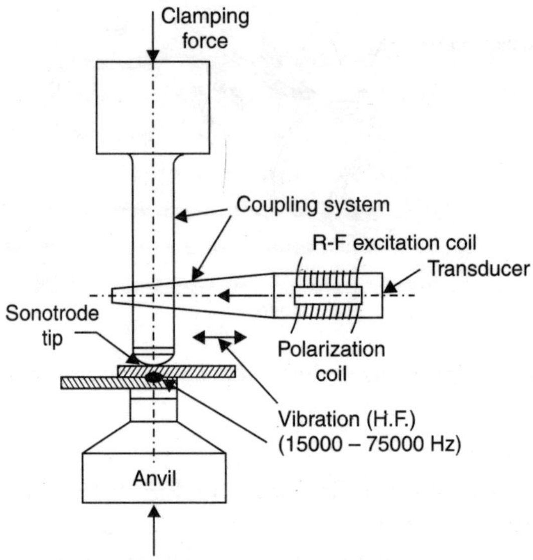

Fig. 2.18(b) Ultrasonic welding (detailed sketch)

4. Before welding the machine is set for clamping force, time and power and overlapping plates are put on the anvil sonotrode is then lowered and clamping force is built to the desired amount (a few Newton to several hundred Newton) and ultrasonic power of sufficient intensity is then introduced. Power varies from a few watts for foils to several thousand watts for heavy and hard materials and is applied through the sonotrode for a pre-set time. Power is then automatically, cutoff and weldment released, time taken is less than 1 sec.
5. Continuous seams can also be produced using disc type rotary sonotrode and disc type or plain anvil.
6. Machine parameters are adjusted for each material and thickness combination.
7. Materials from very thin foils and plates upto 3 mm thickness can be welded.
8. Advantages and applications include.
   (*a*) The process is excellent for joining **thin sheets** to **thicker** sheets.
   (*b*) Local plastic deformation and mechanical mixing result into sound welds.
   (*c*) Ring-type continuous welds can be used for hermetic sealing.
   (*d*) Many applications in electrical/electronic industries, sealing and packaging, air craft, missiles, and in fabrication of nuclear reactor components.

(*e*) Typical applications of the process include: welding of ferrous metals, aluminium, copper, nickel, titanium, zirconium and their alloys, and a variety of dissimilar metal combinations. It is applicable to foils and thin sheets only.

(*f*) Other applications include: almost all commonly used armatures, slotted commuters, starter motor armatures, joining of braded brush wires, to brush plates, and a wide variety of wire terminals.

(*g*) With newly developed solid-state frequency converters, more than 90% of the line power is delivered electrically as high frequency power to the transducer.

(*h*) In the case of ceramic transducers as much as 65 – 70% of the input electrical line power may be delivered to the weldmetal as acoustical power.

Energy required to weld

Energy required to weld a given meterial increases with material hardness and thickness. This relationship for spot welding is given by

$$E_a = 63\, H^{3/2}\, t^{1.5}$$

where E_a = acoustical energy in joules

H = Vicker's microhardness number

t = material thickness adjascent to active in inches.

This equation is valid for Aluminium, Steel, Nickel and Copper for thicknesses upto 0.81 mm.

2.4.4 Explosive Welding

Explosive welding is a welding process that uses a controlled application of enormous pressure generated by the detonation of an explosive. This is utilized to accelerate one of the components called the flyer to a high velocity before it collides with the stationary component. At the moment of impact the kinetic energy of the flyer plate is released as a compressive stress wave on the interface of the two plates. The pressure generated is on the order of thousands of megapascals. The surfaces to be joined must be clean. The surface films, if any, are liquefied, scarfed off the colliding surfaces leaving clean oxide free surfaces. This impact permits the normal inter-atomic and intermolecular forces to affect a bond. The result of this process is a cold weld **without** a HAZ. Combination of dissimilar metals, copper to stainless steel, aluminium to steel or titanium to steel can be easily obtained by this process. EW is well suited to cladding application. The principle of operation is shown in Fig. 2.19.

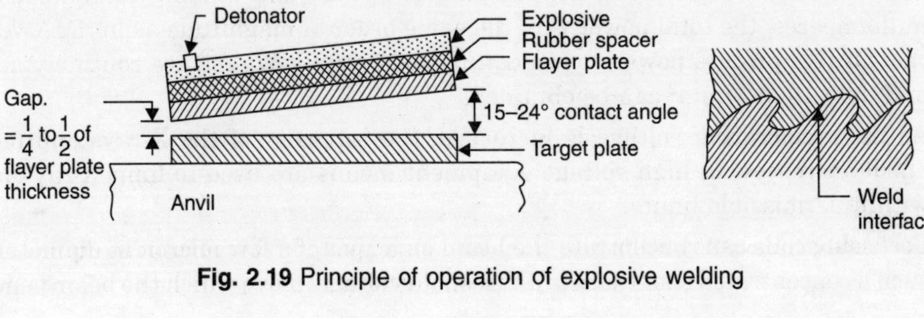

Fig. 2.19 Principle of operation of explosive welding

The main features of the process are listed below :

1. It joins plates face-to-face.
2. One of the plates called the target plate is kept fixed on anvil. The other plate called the flayer plate is kept at an angle of 15 – 24° to the target plate. The minimum gap is $\frac{1}{4}$ to $\frac{1}{2}$ the flayer plate thickness.
3. A layer of explosive charge is kept on the flayer plate with intervening layer of rubber spacers.
4. When explosive charge is detonated the flayer plate comes down and hits the target plate with a high velocity (2400 – 3600 m/s) and the plates get welded face-to-face.
5. The process can be used to join dissimilar materials and the weld interface is seen to be wavy as shown in figure.
6. The various oxides/films present on metal surfaces are broken up or dispersed by the high pressure.
7. Areas from 0.7 to 2 m^2 have been bonded by this process.
8. Process is simple, rapid and gives close thickness tolerance.
9. Low melting point and low impact resistance materials cannot be welded by this process effectively.
10. Explosive detonation velocity should be approx 2400 – 3600 m/s. The velocity depends on the thickness of explosive layer and its packing density.
11. Low melting point and low impact resistance materials cannot be welded effectively by this process.

2.5 HIGH ENERGY DENSITY WELDING PROCESSES

2.5.1 Electron Beam welding

- Electron beam welding uses the kinetic energy of a dense focussed beam of high velocity electrons as a heat source for fusion. In the equipment for this process, electrons are emitted by a cathode, accelerated by a ring-shaped anode, focussed by means of an electromagnetic field and finally impinge on the workpiece as shown schematically in Fig. 2.20. The operation takes place in a vacuum of about 10^{-3} mm of mercury. Accelerating voltages are in the range of 20-200 kV and welding currents are a few milliamperes, the total power is of the same order of magnitude as in SMAW, except that in this process power concentrations of 1–100 kW/mm^2 are routinely achieved and upto 10 MW/mm^2 can be obtained.
- As the accelerating voltage is increased, the intensity of the X-rays emitted from anode increases. In high voltage equipment means are used to limit X-ray emission within permissible limits.
- Focussing coils can concentrate the beam on a spot of a few micron in diameter. With such a concentrated spot there is a threshold voltage above which the beam penetrates

the metal and when the work is traversed relative to the beam a weld bead of exceedingly narrow width relative to the plate thickness is formed.

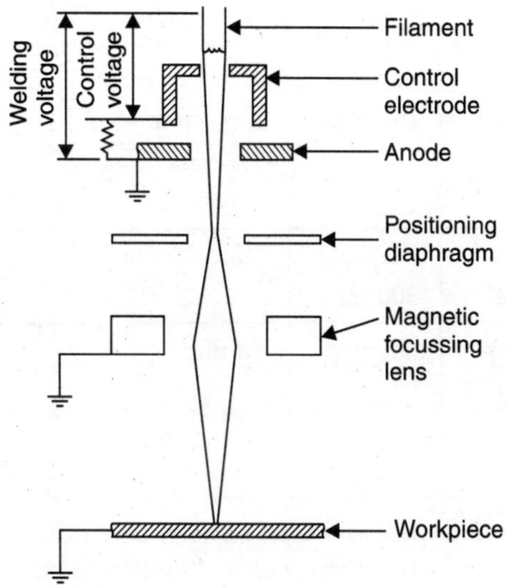

Fig. 2.20 Principle of electron beam welding

- This type of weld could be used for welding dissimilar materials and it is used when the effect of welding heat is to be minimized (distortion is minimum).
- The beam may be defocussed and could be used for pre-heating or post-welding heat treatment. Periodic defocussing could be useful for metals having high vapour pressure at the melting point. The process is applicable to metals that do not excessively vaporize or emit gas when melted. Can weld metals sensitive to interstitial embrittlement.
- The process is specially suitable for welding dissimiiar metals and reactive metals (super alloys (previously impossible to weld)) and for joints requiring accurate control of weld profile and penetration and for joining turbine and aircraft engine parts where distortion is unacceptable. Its major limitation is the need for a vacuum chamber. It can join plate thicknesses from thin foils to 50 mm thick plates. The gun is placed in a vacuum chamber, it may be raised lowered or moved horizontally. It can be positioned while the chamber is evacuated prior to welding. The circuit is energised and directed to the desired spot. Usually the beam is stationary and the job moves at a desired speed.
- Temperatures attained can vaporise any known metal (even tungsten). There are three commercial versions of the EBW process, depending upon the degree of vacuum used as given in the following table:

Table 2.1 Commercial versions of EBW process

S. No.	*EBM Type*	*Vacuum pressure*	*Working distance limit*	*Thickness range for single pass weld*	*Systems power level*	*Special Applications*
1.	Hard vacuum process	10^{-4} torr (0.013 Pa)	Upto 750 mm.	A few thousand Angstrom to 225 mm	1 – 25 kW	Gives best properties when welding interstitially sensitive materials
2.	Soft vacuum process	10^{-1} torr (13 Pa)	Upto 300 mm	Upto 50 mm	15 kW	–do–
3.	Non-vacuum	100 kPa (1 atm.)	25 mm	13 mm	—	Cannot successfully weld interstitially sensitive materials

- Deep penetration, with depth-to-width ratio of 20 : 1, is a unique characteristic of this process. It is mainly due to high power densities achievable with electron beams, which cause instantaneous volatilization of metal. A needle like metal vapour filled cavity or keyhole is produced through the metal plate thickness. As the welding proceeds this key-hole moves forward alongwith the beam and gravity and surface tension act to cause molten metal to flow into the cavities just behind. The limited ability of the beam to traverse the metal thickness is a unique property that ensures full penetration through the metal thickness.
- The process can be adapted to numerical control and can be performed in air or under a blanket of CO_2 but the welds suffer from contamination.

2.5.2 Laser Beam Welding

Laser is the abbreviation of light amplification by stimulated emission of radiation. It is very strong coherent monochromatic beam of light, highly concentrated with a very small beam divergence. The beam exiting from the laser source may be 1–10 mm in diameter, when focussed on a spot has energy density of more than 10 KW/mm^2. Laser beam welding is a thermoelectric process accomplished by material evaporation and melting. Focussing is achieved by various lens arrangements while focusing of electron beam is achieved by electrostatic and magnetic means. Because of this focusing, high power densities are achieved by both the 'electron' and the 'Laser' beams.

- The process does not require a vacuum chamber, size of HAZ is smaller and the thermal damage to the adjascent part is negligible. Laser can be used to join dissimilar metals, difficult-to-weld metals *e.g.* copper, nickel, chromium, stainless steel, titanium and columbium. Currently the process is largely in use in aerospace and electronic industries.
- The principle of working of a Laser Welder is shown in Fig. 2.21(*a*). An intense green light is thrown on a speciai man-made ruby, 10 mm in diameter, containing about

0.05% by weight of chromium oxide. The green light pumps the chromium atoms to a higher state of energy. Each of these excited atoms emits red light that is in phase with the colliding red light wave.

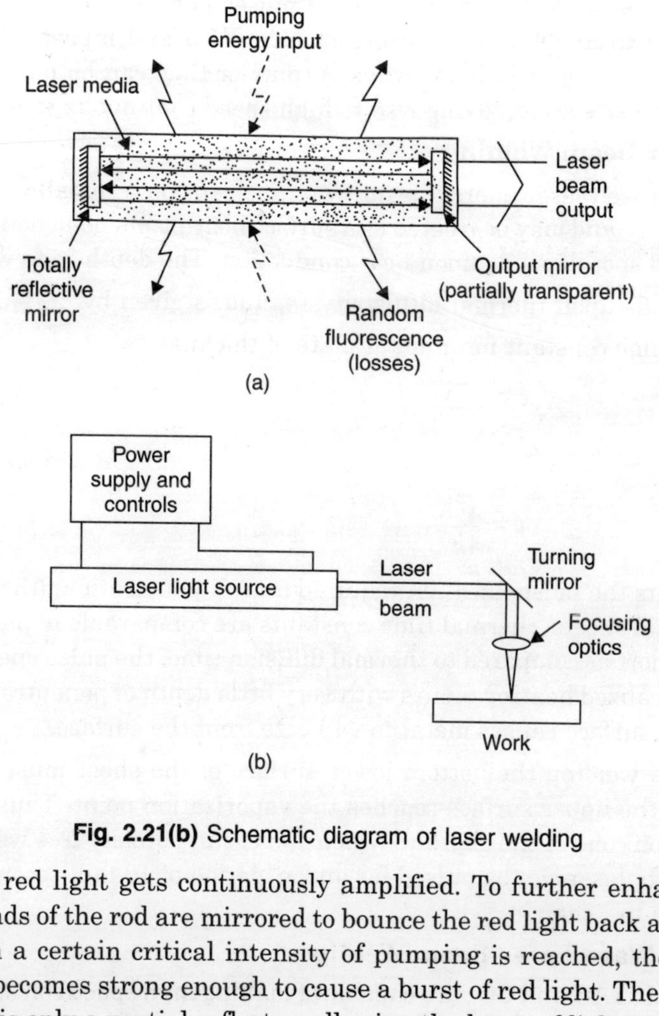

Fig. 2.21(b) Schematic diagram of laser welding

- Thus, the red light gets continuously amplified. To further enhance this effect the parallel ends of the rod are mirrored to bounce the red light back and forth within the rod. When a certain critical intensity of pumping is reached, the chain reaction of collisions becomes strong enough to cause a burst of red light. The mirror in the front of the rod is only a partial reflector, allowing the burst of light to escape through it.
- Lasers used for welding could be of two types:

  1. Solid-state lasers
  2. Gas Lasers (The chief gas Laser is CO_2 laser)

Solid-state lasers are ruby, Nd : Glass and Nd : YAG. The last two are the Lasers in which (Nd : Glass) or single crystals of Yttrium-Aluminium-Garnet (Nd : YAG) are doped with Nd (neodymium) ions as the active medium. The chief gas laser is CO_2 laser.

- Ruby and Nd: Glass are capable of high **energy pulses** but are limited in maximum repetition rate, Nd YAG and CO_2 Lasers can be continuous wave or pulsed at very high repetition rate.

- Incident laser radiations do reflect back from metallic surfaces in appreciable amounts, sufficient energy is still absorbed to maintain a continuous molten puddle. Ruby and Nd: Glass lasers, because of their high energy outputs per pulses, overcome this reflectivity problem.
- Due to inherently low pulse rates 1–50 pulses per second, welding speeds for thin sheets are extremely slow. In contrast Nd : YAG and inparticular CO_2 lasers are capable of very high continous wave outputs or they can be pulsed at several thousand pulses per second, giving rise to high speed continuous welding.

Pulsed Laser Beam Welding

A pulse of focussed laser energy beam when incident on a metallic surface is absorbed within a very small area and may be treated as a surface heating phenomenon. Thermal response beneath the focussed spot depends upon heat conduction. The depth 'x' to which the energy is felt in time 't' depends upon thermal diffusivity, k, and is given by $\sqrt{4kt}$. This leads to the concept of thermal time constant for a metal plate of thickness 'x'.

$$x = \sqrt{4kt}$$

$$x^2 = 4kt$$

$$t = \frac{x^2}{4k}$$

This represents the pulse duration required for full panetration. (through melting). For 0.13 to 0.25 mm metal sheets, thermal time constants are comparable to pulse duration. If the laser pulse is very short as compared to thermal diffision time, the pulse energy remains at the surface and rapid localized heating occurs with very little depth of penentration. This accumulation of heat at the surface causes metal to vaporize from the surface.

In laser beam welding the bottom lower surface of the sheet must reach the melting temperature before the upper surface reaches the vaporization point. Thus, thermal diffusivity and pulse duration control the depth to which successful porosity free welds could be made. Typically a solid-state laser can be pulsed for an 'on' period of 10 milliseconds. This limits the depth of penetration to 1 mm.

Continuous Wave Laser Beam Welding

Lasers like Nd : YAG and CO_2 are capable of making high speed continuous metal welds. Laser's, more than 500 watts capacity are capable of welding steel sheets 0.25 mm thick at several mm/second. CO_2 lasers of 10 kW continuous wave output power can produce deep penetration welds in 13 mm thick steel plates at 25 mm/s.

When heating or melting a metal with a Laser beam, the concept of energy absorbed per unit volume of metal becomes a controlling parameter. The energy absorbed can be written in dimensions of J/mm^3. This parameter becomes a measure of power dersity/welding speed. For example

$$W/mm^2 \times S/mm = J/mm^3$$

The focused spot size '*d*' of a laser beam is given by

$$d = f\theta$$

where f is the focal length of the lens and θ is the full angle beam divergence. The power density, *PD*, at the focal plane of the lens is given by

$$PD = \frac{4P_1}{\pi d^2}$$

where P_1 is the input power, hence

$$PD = \frac{4P_1}{\pi (f\theta)^2}$$

Therefore power density depends upon the laser power and beam divergence. For a laser beam operating in the basic mode, the energy distribution across the beam is gaussian, the beam divergence is

$$\theta \propto \frac{\lambda}{a} \qquad \text{Thus } PD \propto \frac{4P_1\lambda^2}{\pi f^2 a^2}$$

where a is a characteristic dimension of the laser beam and λ is the wavelength of laser radiation. It can, therefore, be noted that the power density is inversely proportional to the square of the wavelength of the laser radiation.

This continuous power provided by continuous wave laser beam makes high power carbon dioxide laser with **deep** penetration capability. There is precise controt of energy delivery to highly localized regions. This is good for "narrow gap", geometries and permits welding without the need for filler metal. This results in savings in filler metal. Deep penetration welds made by this process are similar to the electron beam welds. The process offers the following advantages.

Advantages:

1. Vacuum environment is not required, reative metals can be protected from the atomosphere by inert gas shields.
2. X-rays are not generated by the beam.
3. Laser beam can be manipulated using the principles of optics. This permits easy automation.
4. Can successfully join a variety of metals and alloys.
5. Because of low energy inputs per unit weld length, the cooling rates are high. Cooling rates and associated problems could be modified by pre- or post heating.

Typical CO_2 Laser Beam Welding Performance

S. No.	*Laser Power Level*	*Plate material*	*Material thickness/penetration*	*Welding speed*
1.	5 kW	Carbon steel	2.5 mm	85 mm/s
		Stainless steel	5.0 mm	42 mm/s
2.	10 kW	Aluminium	5.0 mm	38 mm/s
		Titanium	5.00 mm	57 mm/s
3	15 kW	304 stainless steel	18 mm penetration	8 mm/s
			15 mm penetration	25 mm/s
4	6 kW	Steel	Thin gauge	1270 mm/s

6. Ruby lasers are used for spot welding of thin gauge metals, microelectronic components, tasks requiring precise control of energy input to work.
7. 100 kW pulses of one millisecond duration give a series of overlapping spot welds which could be used for special applications.
8. The electrical efficiency of the process is 10 – 20% only.
9. With slight modifications, the process could be used for gas assisted cutting and for surface heat treating and alloying applications.
10. Typically a solid state laser can be pulsed for an on period of 10 milliseconds. This limits the depth of penetration to 1.0 mm.

Table. Thermal time constants for laser beam welding, seconds

Material	*Time in seconds*		
	Thickness 0.18 mm	*Thickness 0.64 mm*	*Thickness 2.5 mm*
Copper	0.035	0.884	14.1
Aluminium	0.047	1.170	18.8
1% C-steel	0.333	8.330	133.3
Stainless steel	1.004	25.10	401.7
Titanium	0.593	14.8	237.3
Tungsten	0.060	1.509	34.1

2.5.3 Plasma Arc Welding

Plasma is the fourth state of matter (other three being: solid, liquid and gas). It is hot ionized arc vapour. In arc welding this arc plasma is blown away by moving gas streams, but in a plasma torch it is contained and used effectively giving rise to the following processes:

- Plasma arc welding
- Micro-plasma arc welding
- Plasma spraying

Plasma Welding

- Plasma welding is an extension of TIG welding. The main difference is the water cooled nozzle in between the electrode and the work. This causes constriction of the arc column, resulting in very high arc temperature between 16,600–3300°C. Fig. 2.22 shows two main types of torhes in common use: Transferred Arc and Non-transferred Arc. In the first type the tip of the tungsten electrode (d.c. negative) is located within the torch nozzle. The torch consists of an electrode, a watercooled nozzle, for arc constriction and a passage each for supply of water and gas. A power supply unit provides d.c. The welding area is blanketed by shielding, gas supplied through an outer gas cup. Transferred arc transfers heat directly from electrode in the torch to the workpiece.
- When the gas (argon) is fed through the arc it becomes heated to the plasma temperature range (16,600 – 33.000°C). The arrangement is such that the arc first strikes to the nozzle. The plasma so formed is swept out through the nozzle and the main current path is then formed between the electrode and the work piece. The transferred (constricted) arc may be used for cutting metals that are not so readily cut by oxy-acetylene torch (non-ferrous metals and stainless steel). For best cutting action argon/hydrogen or nitrogen hydrogen mixtures are used. This requires high output voltage welding machines.A non-transferred arc is established between the electrode and torch nozzle indpendent of the workpiece. The heat is carried by the hot gases (plasma) coming out from the torch. The transferred arc delivers heat more effectively to the workpiece as the heat is generated by the anode spot on the workpiece as well as the plasma jet heat. Thus it is most commonly used.

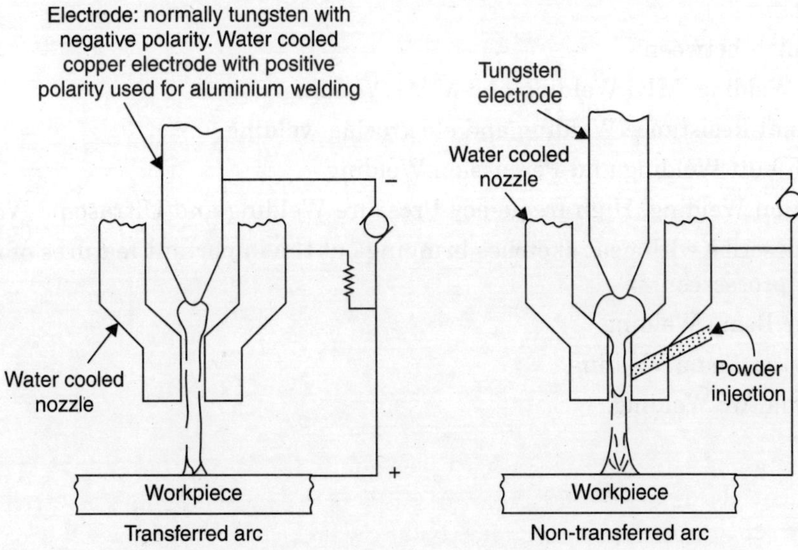

Fig. 2.22 Plasma arc welding

- Plasma welding makes use of the key-hole technique. When the plasma jet strikes metal it cuts or keyholes entirely through the workpiece making a small hole and molten metal in front of the arc flows around the arc column, and is drawn behind the hole by surface tension. Thus butt welds on 12.5 mm or larger thicknesses could be made in a single pass with full penetration. It is good for welding plates accessible from one side only.

- Plasma arc welding can weld carbon steels, stainless steels, copper, brass, aluminium, titanium, monel and inconel including hastalloys, molybdenum and tantalum etc.

Micro-Plasma Arc Welding is a modified process using currents between 0.1–10 A.

It is capable of welding extremely thin sheets and foils between 0.05–1.6 mm thickness. The precise control of heat is achieved through "Pulsed mode" operation.

Plasma Spraying: In non-transferred arc torch the arc is struck between electrode and nozzle. The rate of gas flow through this torch is moderately high and a jet of plasma issues from the nozzle. For spraying, powder or wire is injected inta the plasma stream which is hot enough to melt any solid that does not decompose or sublime. Thus ceramics may be sprayed on to a metal surface. When metal is sprayed, high density caating is obtained. Shielding gases could be either argon or nitrogen or 5-25% hydrogen mixed with nitrogen or argon. The non-transferred torch is also known as a plasma device. Plasma heat could also be used to melt metal for certain applications.

■ QUESTIONS

2.1 Why shielded metal arc welding process is most commonly used. Briefly describe the process. What are the advantages and limitations of this process?

2.2 With neat sketches, compare the processes of shielded metal arc and submerged arc welding.

2.3 Distingnish between:

(*a*) TIG Welding, MIG Welding and MAG. Welding

(*b*) Normal Resistance Welding and electroslag welding

(*c*) Flash butt Welding and Percussion Welding

(*d*) Friction Welding, High frequency Pressure Welding and Ultrasonic Welding.

2.4 Briefly describe with neat sketches bringing out the important features of the following welding processes:

(*a*) Laser Beam Welding

(*b*) Electron Beam Welding

(*c*) Plasma Arc Welding.

Chapter 3

Welding Science

3.1 INTRODUCTION

After a brief review of welding processes let us go into the science of welding. This will help us in the understanding of the further discussions regarding the welding applications and technologies that will follow. Most welding processes require the application of heat or pressure or both to produce a suitable bond between the pieces to be joined sufficient in strength to meet the demands of the task (the intended use).

Almost all the available and concievable high intensity heat sources have been used in welding. Externally used heat sources of technical importance include: arcs, electron beams, light beams. exothermic reactions and electrical resistance. A heat source must transfer sufficient energy at high intensity to produce local melting and fusion.

It has been the endeavour of welding engineers to evolve a welding heat source which provides high heat intensity (energy density per unit cross-sectional area of source—plasma arc, electron beam, laser beam, etc.) to cause melting. During welding, heat may be considered to be transferred from the source to the surface of the work and then by conduction, from the contact area to colder regions of the metal. These two processes are somewhat competitive. With high **intensity** heat sources, say electron beam, energy is delivered through the contact area so rapidly that local melting occurs before there is significant loss of heat by conduction. In Bunsen burner on the other extreme a large quantity of heat is lost by conduction to the workpiece without melting. Thus Bunsen burner is not suitable for welding.

3.2 CHARACTERISTICS OF WELDING POWER SOURCES

3.2.1 Arc Welding Power Sources

The various welding processes described in Chapter 2 require special power sources (having low voltage and high current for arc welding) to produce energy sufficient to make a good weld. Power sources could be a.c. (transformers), d.c. (generator/rectifiers) with constant current or constant voltage characteristics having current rating 70-400 amperes at 60% or 80% duty cycle. Heat input to the weld is a function of arc voltage, arc current and travel speed. Arc length is related to arc voltage.

The voltage supplied by the electrical generating stations for industrial use is 240 or 480 volt and the open circuit voltage for arc welding is between 50-80 V. Once the arc is struck the working voltage falls down to 10 to 30 V. As arc is the source of welding energy its study is, therefore, important.

3.2.2 Arc Characteristics

When the arc operates in a stable manner, the voltage and current are related. The relationship is shown in Fig. 3.1. It can be seen from this graph that the arc does not follow Ohm's law.

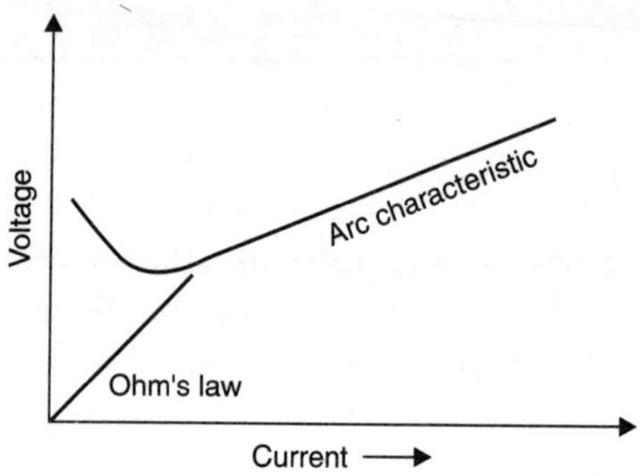

Fig. 3.1 Typical arc characteristic compared with Ohm's law

The arc voltage varies only slightly over a wide range of currents.

- The curve does not pass through the origin.
- The slope of the curve depends upon:
  (*i*) metals involved
  (*ii*) arc atmosphere
  (*iii*) arc length

3.2.3 Arc-length Control

For this discussion consider arc characteristics for four arc-lengths between tungsten and copper electrodes in argon atmosphere (Fig. 3.2). From this data we can plot a relation between arc-length and arc-voltage (Fig. 3.2). Suppose a welder uses GTA Welding process to weld copper sheets and makes a current setting of 150 A. The arc-characteristics (Fig. 3.2) show that for a 2 mm arc to be operating stable, the voltage should be 15 V. This value of arc voltage will be maintained as long as the power source delivers 150 A and the welder maintains an arc length of 2 mm. This is practically not possible during manual welding operation as the arc length may change, and consequently the voltage will rise or fall accordingly and the operating point will, therefore, shift from one characteristics to another. For arc to remain stable, the power-supply unit must allow the voltage to vary while keeping the current substantially constant (Fig. 3.3). Thus, the power-supply unit must meet the practical requirements for a specific process. A typical characteristics curve for manual GTA Welding operation is shown in Fig. 3.4.

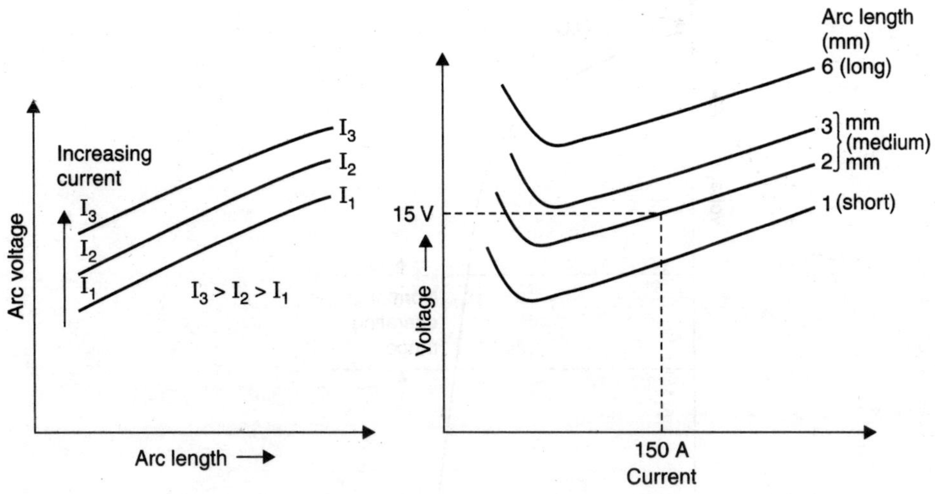

Fig. 3.2 Arc characteristics for welding copper (G.T.A. welding)

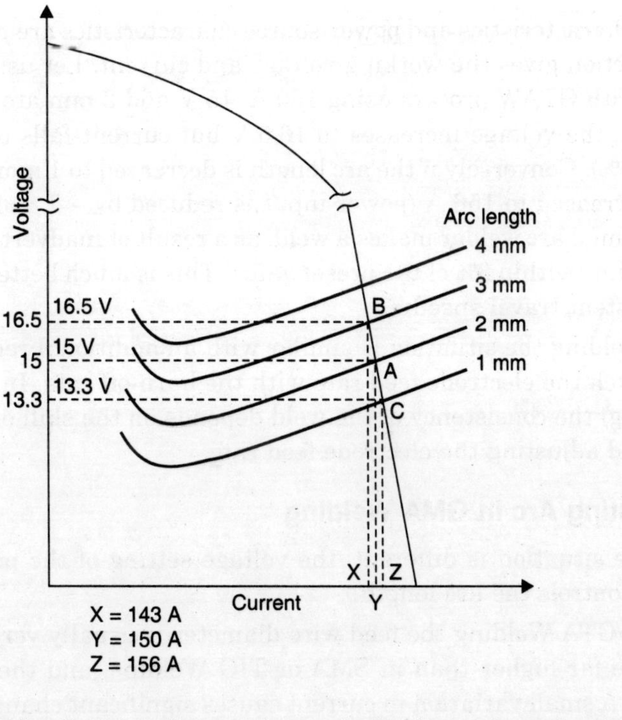

Fig. 3.3 Variations in voltage and current with change in arc-length

When welding is not taking place, no output current is drawn from the circuit. The voltage at the output is called open circuit voltage (O.C.V.) and it is of the order of 50–80 V. As the welding arc is struck and welding operation is carried out the voltage falls and over an operating range of 10-30 V the current varies only a little. Power-sources of this type of volt-ampere output are known as "drooping characteristics" units or 'constant-current' machines.

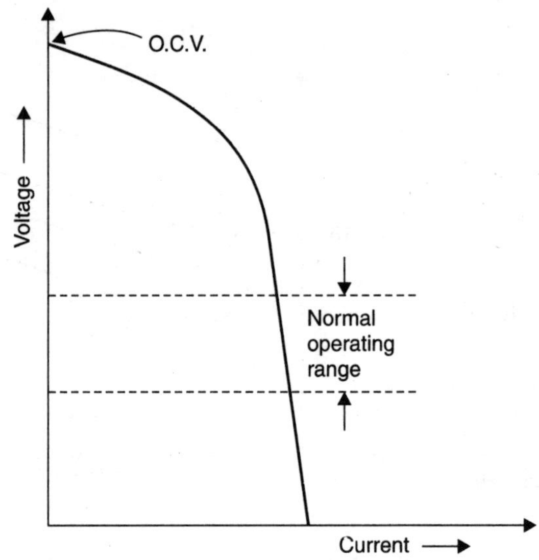

Fig. 3.4 Typical power supply characteristics used in manual GTA welding operation

If the arc-characteristics and power-source characteristics are plotted on one graph (Fig. 3.3) their intersection gives the working voltage and current. Let us, consider the example of welding copper with GTAW process using 150 A, 15 V and 2 mm arc length. If the arc length changes to 3 mm, the voltage increases to 16.5 V but current falls to 143 A. (power input is increased to + 4.8%). Conversely if the arc length is decreased to 1 mm, the voltage falls to 13.3 V and current increased to 156 A (power input is reduced by – 7.8%). It is important here to note that as a manual arc welder makes a weld, as a result of inadvertent hand movements the power input remains within 8% of the preset value. This is much better than requiring them to maintain a consistent travel speed.

In SMA Welding the situation is similar with an additional requirement on the part of the welder to match the electrode feed rate with the burn-off rate. In manual metal arc welding (SMA Welding) the consistency of the weld depends on the skill of the operator in judging the arc length and adjusting the electrode feed rate.

3.2.4 Self Adjusting Arc in GMA Welding

- Here the situation is different, the voltage setting of the power-source and not the welder controls the arc length.
- In GMA/GTA Welding the feed wire diameter is usually very small and the burn-off rates are far higher than in SMA or TIG Welding, and they vary much more with current. A small variation in current causes significant change in burn-off rate. Some typical burn-off curves for low-carbon-steel wires with carbon-di-oxide shielding are shown in Fig. 3.5. Change in burn-off rates with change in current are also shown. We find that the electrode burn off rate changes rapidly with change in current. Thus we should have a power source which can accomodate these large changes in the

burn-off rates. For a small change in voltage, there should be a large change in current. Special power-sources have been designed for this purpose.

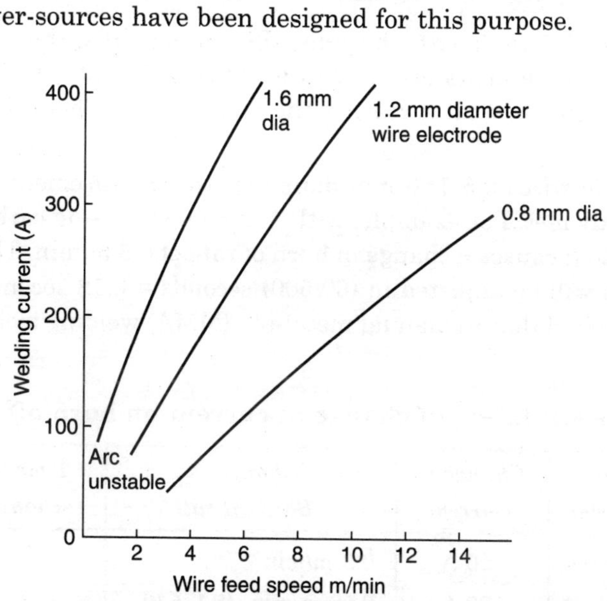

Fig. 3.5 Wire feed rate *Vs* current for three electrodes in CO_2 welding

- Some welding power sources are designed to give a flat volt-ampere characteristics as shown in Fig. 3.6 with a voltage falling by 2 V for each 100 A fall in current. This type of characteristics is also known as constant potential characteristics.

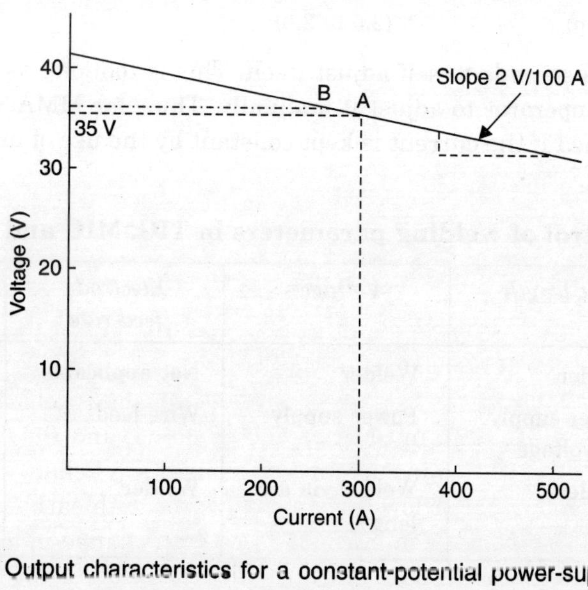

Fig. 3.6 Output characteristics for a constant-potential power-supply unit

- Consider an arc operating at 300 A, 35 V (point A in Fig. 3.6). If the arc length increases, voltage rises to point B (say). This causes significant decrease in current, giving lower burn-off rate. Arc length is immediately adjusted as the electrode tip in this situation will approach weld pool, and the arc length shortens. When this happens the current increases and the burnoff matches with wire feed rate. The system returns to equilibrium.

- Conversely, if the arc-length shortens, the voltage falls, the current rises, burn-off rate increases, wire melts faster than it is being fed into the area, arc length thus increases continuously till it reaches the preset value. This is called self-adjustment of the arc.
- With electrode wires 0.8-1 .6 mm diameter, this requirement for rapid self-adjustment is readily met. For example, with 1 .2 mm wire using carbon dioxide shielding, a change in 20 A causes a change in burn of rate of 0.5 m/min. Thus a change of 1 mm in arc length will be adjusted in (60/500) seconds = 0.12 seconds. Proceeding in the same way we find that in manual metal-arc (MMA) welding a change in arc length of 1 mm

Table 3.1. Effect of change in current on burn-off rate

Welding Process	*Wire diameter*	*Change in Current*	*Change in Burn-off rate*	*Time taken to adjust 1 mm change in arc length (sec)*
CO_2 Welding	1.6 mm	20 A*	0.3 m/min**	0.20 sec.
CO_2 Welding	1.2 mm	20 A	0.50 m/min. (5.1–5.6)	0.12 sec
CO_2 Welding	0.8 mm	20 A	1.1 m/min (10.4–11.5)	0.054 sec
SMA Welding (200 Amperes oper. current)	4 mm	20 A	0.02 m/min.	3.00 sec.

*(200–to–220 Amp) **(2.5 to 2.8)

will require 3 seconds to self-adjust itself. This is too long as compared to the time taken by the operator to adjust it manually. Thus, for MMA Welding better results will be obtained if the current is kept constant by the use of drooping characteristics power supply.

Table 3.2. Control of welding parameters in TIG, MIG and MMA Welding

Welding Process	*Arc length*	*Voltage*	*Electrode feed rate*	*Current*
TIG	Welder	Welder	Not applicable	Power supply
MIG	Power supply via voltage	Power supply	Wire feed	Electrode speed via wire feed motor
MMA	Welder	Welder via arc length	Welder	Power supply

3.3 ARC WELDING POWER SUPPLY EQUIPMENTS

An arc welding power supply equipment should have the following characteristics:

- must isolate the welding circuit from the mains supply.
- provide the required voltages and desired welding currents for the operation.
- provide the output volt-ampere characteristics which matches the arc system.
- incorporate a low-voltage supply for the operation of auxiliary units.
- if the work is to be carried out on site the unit should be self contained with a petrol or diesel engine driving a generator or alternator.

3.3.1 Alternating-current Welding Power Sources

Alternating current power sources are commonly used in manual metal arc welding of steels and GTA Welding of aluminium and its alloys. For a.c. welding the power supply is invariably a transformer with a control for current adjustment either by varying the inductance or by altering the magnetic coupling between primary and secondary windings of the transformer.

The flow of alternating current in welding circuit is regulated by placing an inductor in line between the transformer and the electrodes. By changing the inductance the current can be changed. For current control during welding a means of changing this inductance is necessary.

Three different types of reactors are available for changing this inductance for current control:

— tapped reactors
— moving core reactors
— saturable reactors

Tapped reactors. These consist of a copper cable wound on a laminated core. The windings are provided with tapping circuit as shown in Fig. 3.7. Coarse and fine controls are provided, but only a limited number of settings can be accomodated.

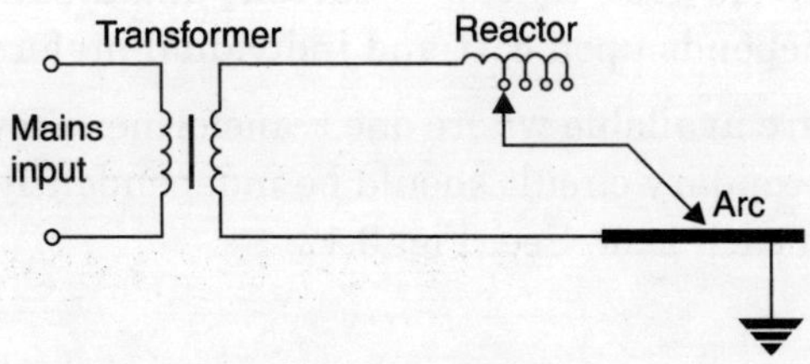

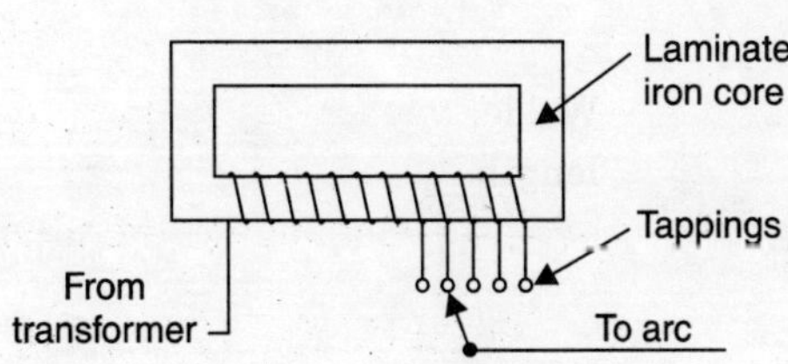

Fig. 3.7 Tapped reactors

Moving-core reactor. A laminated core is moved in or out of reactor coil, thus increasing or reducing the inductance of the winding. See Fig. 3.8. This system has the advantages of continuously variable adjustment.

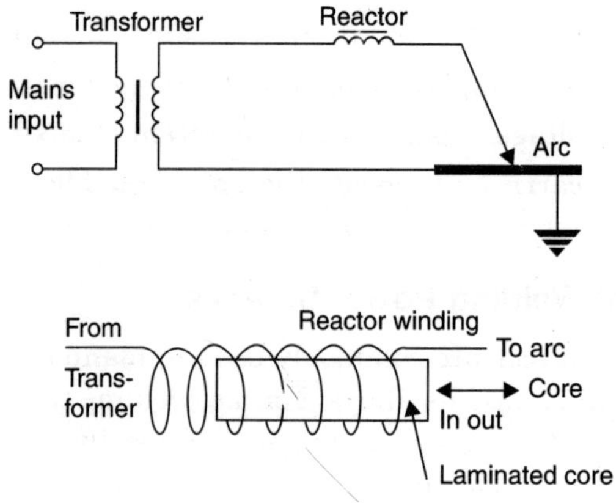

Fig. 3.8 Moving core reactor

Saturable reactors. Here welding current control is achieved by putting saturable reactor unit in the secondary circuit. See Fig. 3.9. Direct current supplied to this winding affects the impedance offered to alternating current flowing in the main coil. Thus welding current can be continuously regulated by changing direct current in the control winding. These reactors are costly but can be remotely controlled.

Moving coils. Changing the position of one coil along the core changes the magnetic coupling between primary and secondary. See Fig. 3.10.

Moving shunt-core. Movement of a shunt core in or out (instead of moving coils) changes the magnetic coupling between primary and secondary, and thus the welding current is controlled. See Fig. 3.11.

All these designs provide good control of current and a suitable output for MMA and GTA Welding. The choice depends upon cost and individual preferences.

Multi-operator sets are available where one transformer provides 3 or 6 outlets. In this case, the current in each secondary circuit should be independently controlled and a separate reactor must be included in each lead. See. Fig. 3.12.

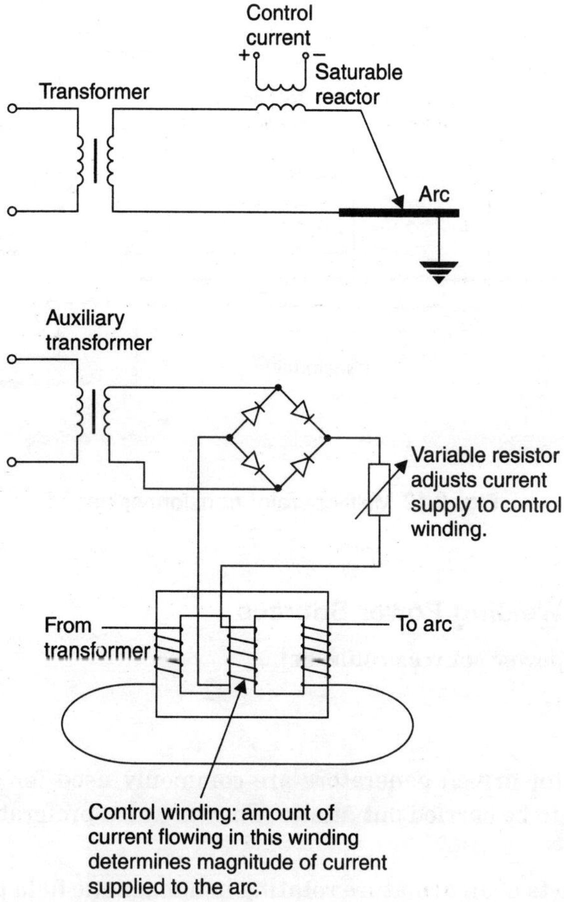

Fig. 3.9 Saturable reactor used to regulate welding current

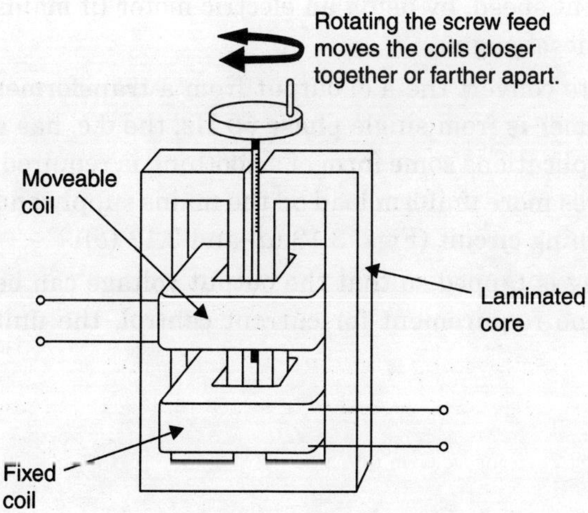

Fig. 3.10 Moving-coil transformer

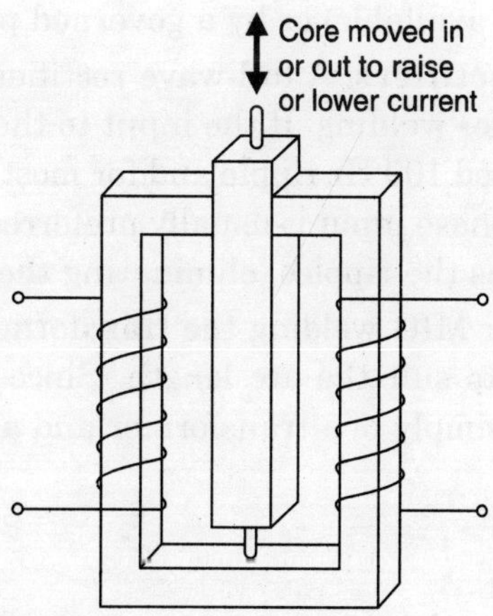

Fig. 3.11 Moveable-core transformer

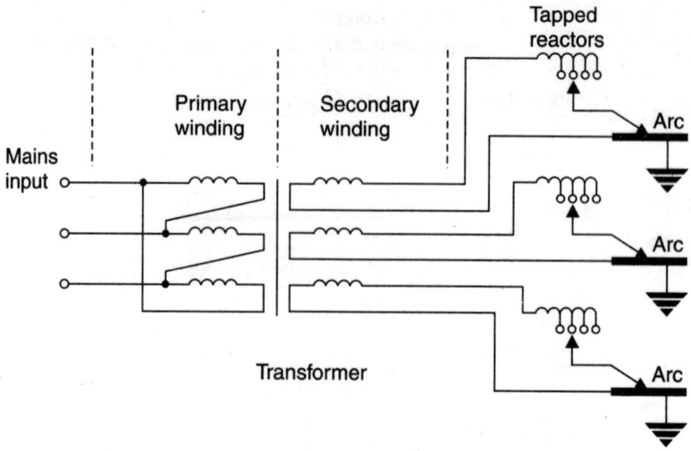

Fig. 3.12 Multi-operator transformer unit

3.3. 2 Direct-Current Welding Power Sources

Direct current welding power sources could be:

— generators

— rectifiers

Generators. Motor driven generators are commonly used for welding with d.c., specially when the work is to be carried out at site. They are also preferable if the line voltage is quite fluctuating.

A generator consists of an armature rotating in a magnetic field produced by coils which are connected in series and in parallel with the armature winding.

Generator output is regulated by regulating the current flowing in the series and shunt windings. The armature must rotate at a constant speed, by using an electric motor (if mains supply is available) or by a governed petrol or diesel engine.

Rectifiers. A full-wave rectifier is used to convert the a.c. output from a transformer into d.c. for welding. If the input to the transformer is from single phase 50 Hz, the d.c. has a pronounced 100 Hz ripple and for most of the applications some form of smoothing is required. A three-phase input is usually preferred as it gives more uniform load on the mains supply and smoothens the ripples, eliminating the smoothening circuit (Figs. 3.13 (*a*) and 3.13 (*b*)).

For MIG welding the transformer winding is tapped so that the output voltage can be selected to suit the arc length. Since there is no requirement for current control, the unit consists simply of a transformer and a rectifier.

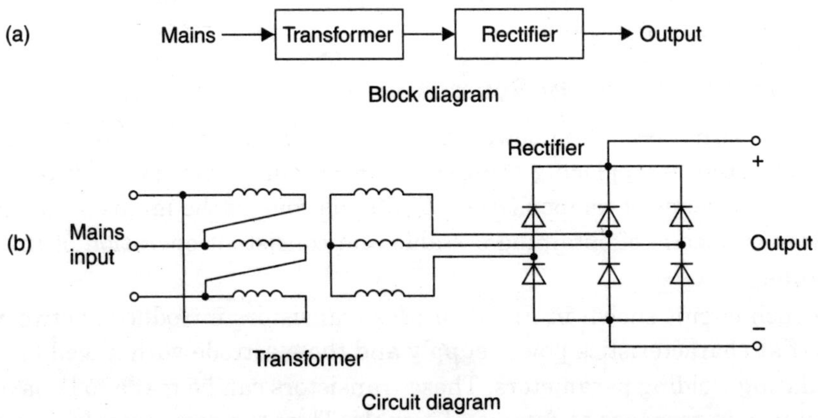

Fig. 3.13 Simple three-phase full-wave rectifier unit for welding

In case of manual metal arc and GTA welding a reactor is introduced into the a.c. line between the transformer and the rectifier to obtain drooping volt-ampere characteristics (Fig. 3.14). The reactor behaves in a similar way as in a.c. welding supply units. Saturable reactors are commonly used in most of the units because they are better suited to three-phase operation and can be remotely controlled. It is important to note that a reactor controls (opposes) a.c. only. In d.c. circuit it has no effect on steady flow of current: but it opposes any changes in current level, which is a good feature for low current GMA Welding.

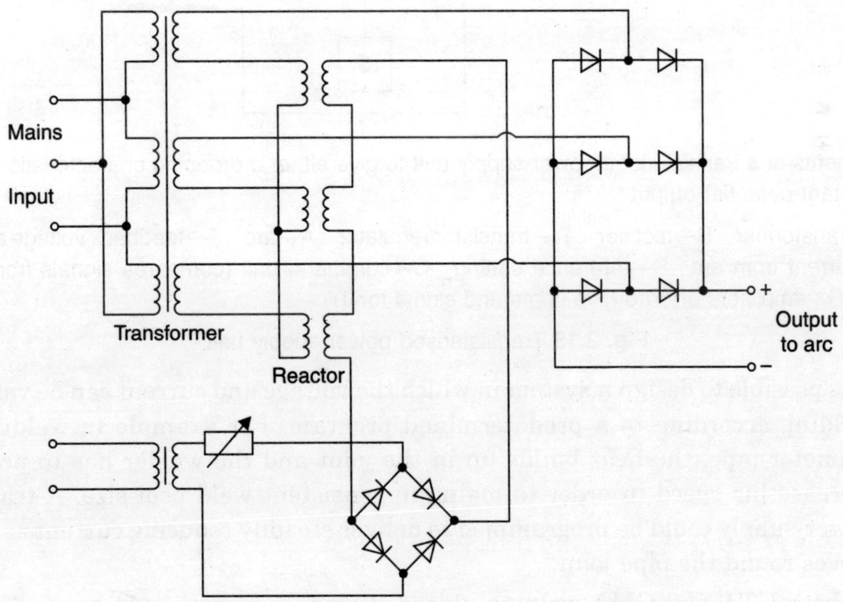

Fig. 3.14 Drooping characteristic output from rectifier unit

By providing extra taps to the output from the reactor in a transformer reactor set, it is possible to produce a combined a.c./d.c. unit suitable for MMA and GTA welding. This type of power unit is more useful when there is a mixed type of requirement in a job-shop, but it costs more than individual a.c. or d.c. unit.

3.3.3 Solid-state Welding Power Sources

- Many modern arc-welding power supply units contain solid-state circuits for regulating the output or replacing the reactors found in conventional systems, or in some cases, as a means of compensating for fluctuations in the mains output voltage. This provides a means of obtaining a stable and consistent operation of the arc in GMA Welding.
- One such circuit shown in Fig. 3.15 uses transistors introduced between the output from flat characteristics power-supply and the electrode with a feed back system for regulating welding parameters. These transistors can be made to behave as variable resistance in response to command signals. Thus the same supply unit can be made to work as a constant voltage source for GMA welding and then, simply by changing the command signals, it can be made to give a drooping characteristics output to suit GTA Welding.

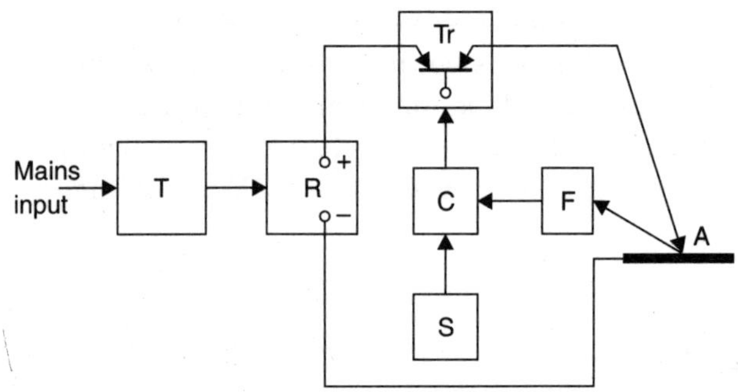

Elements of a transistorised power-supply unit to give either a drooping characteristic or a constant-potential output

T—transformer R—rectifier Tr—transistor regulator A—arc F—feedback voltage and/ or current from arc S—reference setting C—command unit (compares signals from F and S ; amplifiers error to give command signal for Tr)

Fig. 3.15 Transisterised power supply unit

- It is possible to design a system in which the voltage and current can be varied during welding according to a predetermined program. For example in welding a small diameter pipe, the heat builds up in the joint and the welder has to progressively increase his speed in order to maintain consistent weld pool size. A transistorised power-supply could be programmed to deliver steadily reducing current as the welder moves round the pipe joint.
- In both GTA and GMA welding pulsed current supplies could be used (as will be discussed later in this chapter). A transistorised power-unit provides accurately controlled current pulses. These power units offer the prospect of providing easily controlled universal power-supply units.

3.4 WELDING POWER-SOURCE SELECTION CRITERIA

The following factors must be considered when selecting a power source for welding.

1. Initial cost.
2. Cost of periodic maintenance and repair.
3. Mains supply available: 220 V, 440 V or not available.
4. Steady output current even with input voltage fluctuation.
5. Whether machine causes imbalance in the power load.
6. Machine's inherent power factor or needs capacitor to raise it.
7. Whether portability is needed.
8. Type of current needed a.c or d.c. or both.
9. Current rating required to accomodate all sizes of electrodes needed for the jobs
10. Machine's ability to strike and maintain stable arc for the type of electrodes to be used.
11. Type of volt-ampere characteristics (constant current or constant voltage) needed for the process employed.
12. Whether machine is required to give radiographic quality welds and impact strength with the type of electrodes used.
13. Whether the machine needs to serve several welding processes expected to be used in the shop.
14. Need for remote current control.
15. Machine's ability to stand shop environment (corrosive gases, moisture, dust, etc.).

3.5 WELDING ENERGY INPUT

3.5.1 Arc Energy Input

The energy input, "H", is computed as the ratio of total input power, 'P', of the heat source in watts to its travel velocity, 'V', in mm/second.

$$H = \frac{P}{V} \qquad ...(3.1)$$

If the source of heat is an electric arc

$$H = \frac{EI}{V} \qquad ...(3.2)$$

where E = voltage in volts and I = current in amperes.

Precisely speaking, net energy input would be

$$H_{\text{not}} = \frac{f_1 EI}{V} \qquad ...(3.3)$$

where, f_1 = the heat transfer efficiency which is from 80% to 90% for most consumable electrode arcs.

The primary function of the heat sources is to melt metal. In this regard it is useful to introduce the concept of melting efficiency, f_2, which is the ratio of energy used for melting metal to the total energy supplied.

$$f_2 = \frac{QA_w}{H_{net}} = \frac{QA_w \,.\, V}{f_1\, EI} \quad \text{...(3.4)}$$

where, Q = theoretical quantity of heat required to melt a given volume of metal. This is required to elevate the temperature of the solid metal to its melting point plus the heat of fusion to convert solid to liquid at the melting point.

A reasonable approximation of Q is

$$Q = (T_m + 273)^2/300{,}000 \text{ J/mm}^3$$

where, T_m = melting temperature, °C ...(3.5)

$$A_w = A_m + A_r \quad \text{...(3.6)}$$

A_m = plate cross-section melted

A_r = filler metal cross-section melted

A_w = total weld metal cross-section melted.

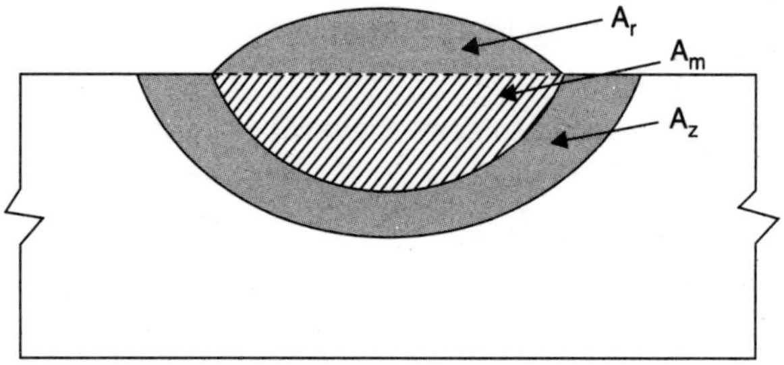

Fig. 3.16 Bead-on-plate cross section

$$A_w = \frac{f_1 f_2 EI}{QV} \quad \text{...(3.7)}$$

Let us take the example of submerged arc welding, when an arc weld is made on steel plate under the following conditions:

$E = 20$ V	$f_1 = 0.9$
$I = 200$ A	$f_2 = 0.3$
$V = 5$ mm/s	$Q = 10$ J/mm^3

The weld cross-sectional area-can be estimated on the basis of equation (3.7)

$$A_w = \frac{0.9 \times 0.3 \times 20 \times 200}{5 \times 10}$$

$$= 21.6 \text{ mm}^2$$

3.6 ENERGY SOURCES FOR WELDING

Welding energy sources can be grouped into the following five categories:

— Electrical sources

— Chemical sources

— Optical sources

— Mechanical sources

— Solid state sources.

Of the above sources, electrical sources of energy are more commonly used. Arc and resistance welding will now be highlighted in the following paragraphs.

3.6.1 Arc Welding

- A large number of welding processes use the electric arc as source of heat for fusion. The electric arc consists of a relatively high current discharge sustained through a theramally ionized gaseous column called plasma.
- Power dissipation of the arc is *EI* (*EI* cos ϕ for A.C. welding). Not all of the heat generated in the arc is effectively utilized in the arc welding process. Values of heat utilization may vary from 20 to 85 percent. Efficiency of heat utilization is usually low for GTAW, intermediate for SMAW and high for SAW.
- With higher travel speeds the efficiency of heat transfer in the fusion zone is increased. Thus for the same arc energy input, the volume of fused metal increases as travel speed is increased.

3.6.2 Resistance Welding

- The resistance welding process employ a combination of force and heat to produce a weld between the workpieces. The heat generated by the current flow may be expressed by:

$$H = I^2 Rt \qquad ...(3.8)$$

where H = heat generated, in Joules (watt. seconds)

I = current, in amperes

R = resistance, in ohms

t = time of current flow in seconds.

- The welding current and time can be easily measured. The resistance is a complex factor and difficult to measure. It consists of:

  — the contact resistance between the electrodes and the work

  — the contact resistance between the workpieces

  — the body resistance of the workpieces

  — the resistance of the electrodes

- In general the resistances involved are of the order of 100 $\mu\,\Omega$. As a result, the currents are large running into thousands and tens of thousands of amperes. In the case of capacitor-discharge power supplies the currents may be as high as 200,000 A.

Example. Two sheets of steel 1.0 mm thick are to be spot welded. In ordinary spot welding machine a current of 10,000 A is required for 0.1 second, while with a capacitor discharge power source making a projection weld between the same sheets, the current pulse of 30,000 A was required for 0.005 seconds. Compare the two processes. Assume effective resistance of 100 μ Ω (micro-ohm).

(*a*) $H = (10{,}000)^2\ (0.0001)\ (0.1) = 1000$ J (for ordinary spot welding machine)

(*b*) $H = (30{,}000)^2\ (0.0001)\ (0.005) = 450$ J (for capacitor discharge power source)

Approx. 1381 J are required to melt 1 g of steel.

Assume that the fusion zone of the above weld is a cylinder of 5 mm diameter and 1 .5 mm height. Weight of metal melted will be $(\pi/4)(5)^2 \times (1.5) \times \rho = 0.246$ g. To heat and melt this mass would require 339 J assuming $\rho = 8.356 \times 10^{-3}$ g/mm^3.

- Thus the capacitor discharge power source utilises energy more effectively.

3.7 ARC CHARACTERISTICS

3.7.1 Introduction

- For all practical purposes a welding arc may be regarded as a gaseous conductor which **converts electrical energy into heat**.
- Arc is a heat source for many welding processes because it produces heat at HIGH INTENSITY. The heat can be easily controlled by controlling the electrical parameters.
- In welding, the arc removes surface oxides and also controls the **transfer of metals**.
- The welding arcs may be of the following types:
  (*a*) Steady Arc—electrical discharge between two electrodes.
  (*b*) Unsteady Arc—arc interrupted due to electrical short circuiting during metal transfer.
  (*c*) Continuously Non-steady Arc: This is due to alternating directional flow of current.
  (*d*) Pulsed Arc: Intermittent current pulses are superimposed on a regular arc to obtain spray type of metal transfer during the pulse intervals.

3.7.2 The Plasma

- The current is carried by the PLASMA, the ionized state of gas composed of nearly equal number of electrons and ions.
- The electrons flow from negative to positive terminal.
- Other states of matter including molten metal, vapour slags, neutral and excited gaseous atoms and molecules.

- The formation of plasma is governed by the concept of the Ideal Gas Law and Law of Mass Action. A basic equation is given below:

$$\frac{n_e n_i}{n_0} = \frac{2Z_i(2\pi m_e Kt)^{3/2}\, e}{Z_0 h^3} - \frac{V_i}{Kt} \qquad ...(3.9)$$

where n_e, n_i, n_0 = particle densities (number per unit volume for electrons, ions and neutral atoms resp.)

V_i = the ionisation potential

t = temperature in degrees absolute

Z_i and Z_0 = partition functions for ions and neutral particles.

h = Plank's constant

m_e = electron mass

K = Boltzmann's constant

- The heated gas of the arc attains a temperature of between 5000 and 50,000 K depending upon the kind of gas and intensity of the current carried by it.
- In the region very near to the arc terminals the current-conducting electrons are accelerated so suddenly that the required number of collisions does not occur. Current conduction based wholly on thermal ionization does not hold in this region.

3.7.3 Arc Temperature

- Arc temperature can be determined by measuring the spectral radiation emitted. The measured values of arc temperatures normally fall between 5000 and 30,000 K, depending upon the nature of plasma and current conducted by it.
- In covered electrodes, due to the presence of easily ionized materials such as sodium and potassium in coatings the maximum temperatures reached are about 6000 K. In pure inert gas arcs the axial temperature may rise to 30,000 K.
- An isothermal map of a 200 A, 12.1 V Argon Arc between tungsten cathode and a watercooled copper anode is shown below.

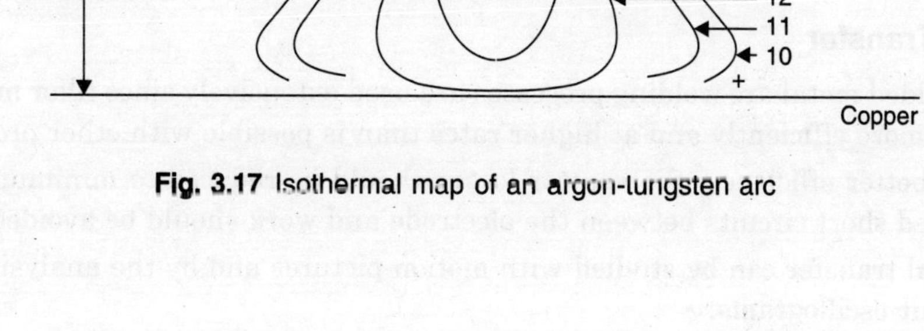

Fig. 3.17 Isothermal map of an argon-tungsten arc

3.7.4 Radiation Losses

- Radiation loss of energy may be over 20 percent of the total input in the case of argon welding arcs.
- Radiation losses from other gases may be about 10 percent.

3.7.5 Electrical Features

- Every arc offers impedance to the flow of current. The specific impedance is inversely proportional to the density of the charge carriers and their mobility.
- The total impedance also depends upon the radial and axial distribution of the carrier density.

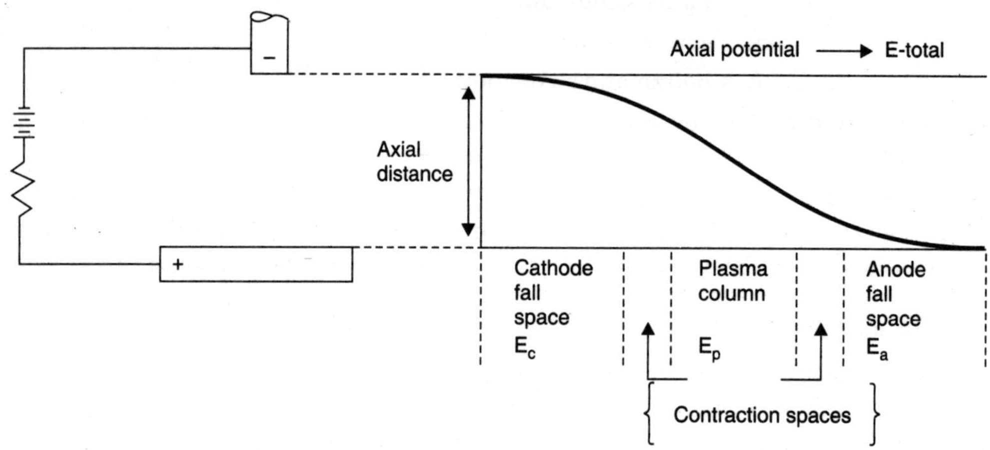

Fig. 3.18 Arc potential distribution between electrode and work.

The current and potential across the cathode fall, Plasma column and Anode Fall regions as shown in Fig. 3.18 are expressed according to

$$\text{Watts} = I\,(E_a + E_c + E_p) \qquad ...(3.10)$$

where E_a = anode voltage drop

E_c = cathode voltage drop

E_p = plasma voltage drop.

3.8 METAL TRANSFER AND MELTING RATES

3.8.1 Metal Transfer

- Shielded metal arc welding processes are used extensively since filler metal is deposited more efficiently and at higher rates than is possible with other processes.
- For better efficiency, the spatter losses should be reduced to minimum and uncontrolled short circuits between the electrode and work should be avoided.
- Metal transfer can be studied with motion pictures and by the analysis of the short circuit oscillograms.

- Metal transfer may be classified as:
  - (*a*) globular (massive drops, short circuiting occurs)
  - (*b*) spray (shower of a large number of small drops).

Generally the metal transfer occurs in some combination of both.

- In GMAW process with argon shielding, when the current is above the transition level, the transfer mechanism can be described as axial spray. With active gases, however, the transfer is globular and some short circuiting is unavoidable.
- Study of metal transfer in arc welding is difficult because the arcs are too small and their temperatures too high and the metal transfers at high rates.
- A combination of the following forces functions to detach the droplet against the force of gravity.
  - (*a*) Pressure generated by the evolution of gas at the electrode tip.
  - (*b*) The electrostatic attraction between the electrodes.
  - (*c*) Gravity.
  - (*d*) The 'pinch effect' caused by a momentary necking of the liquid drop that is, conducting current.
  - (*e*) Explosive evaporation of the necked filament between the drop and electrode due to the very high density of the conducting current.
  - (*f*) Electromagnetic action produced by a divergence of current in the plasma around the drop.
  - (*g*) Friction effect of the plasma jet.

3.8.2 Polarity and Metal Transfer

Electrode Positive

- At low welding currents the size of the droplet in argon develops to a diameter more than the diameter of the electrode.
- The droplet size is roughly inversely proportional to the current and only a few droplets are released per second.
- With long arc length, the droplets are transferred without short circuit, no spatter, and arc is stable.
- Above a critical current level, the characteristics of this transfer change from globular to spray transfer mode.
- In spray transfer, the tip of the electrode becomes pointed and, from it, minute drops are transferred at a rate of about a hundred per second. The current at which this occurs is called transition current. The change is usually abrupt. See Fig. 3.19.
- Axial spray transfer is stable. There is no spatter, the drops are transferred in line with electrode and not through the minimum path. The metal can therefore be directed where needed for making fillet vertical or overhead welds.
- The key to the spray transfer is the 'pinch effect' which automatically squeezes the drops off the electrode; this occurs as a result of the electromagnetic effects of the current.

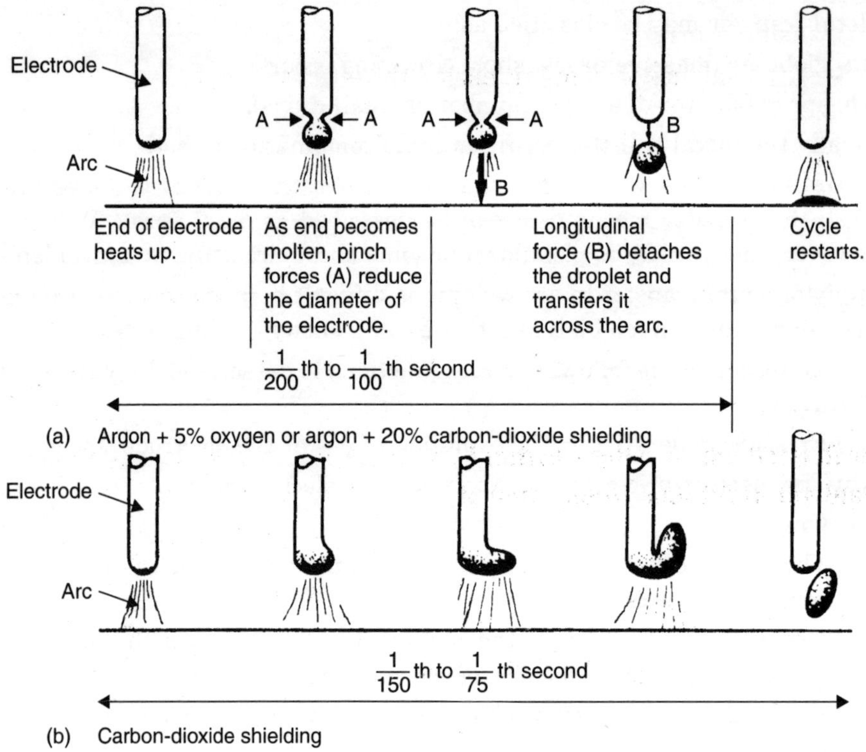

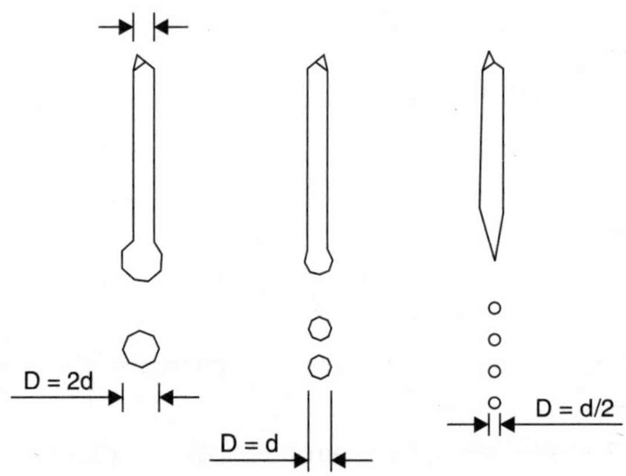

Metal transfer in the spray mode of the pulsed GMAW welding Process

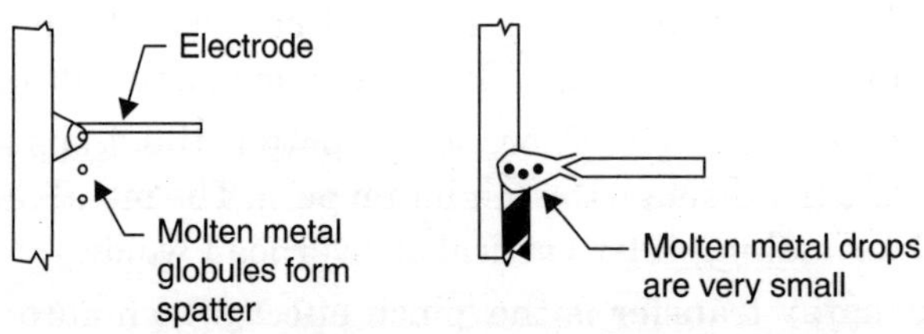

Fig. 3.19 Horizontally held electrode wires are shown producing globular and spray transfer during gas-metal-arc welding

- The transition current depends upon :
  (*a*) electrode diameter, Fig. 3.20 shows the effect.
  (*b*) electrode extension (distance between the point of current pick-up and the arc). As extension increases current for spray transfer decreases (extended wire gets heated).
  (*c*) electrode composition.
  (*d*) metal being welded (less for aluminium and more for steel).
- Spray transfer can be achieved at average current levels below the transition current by using pulsed current. Drops are transferred at the frequency of the current pulses. This technique increases the useful operating range of a given electrode size.
- When useful upper range of the welding current is exceeded a spatter-forming rotation of the arc is initiated on the electrode tip. This is called "Jet rotation".

Electrode Negative

- GMAW arc becomes unstable and spattery when electrode negative is used. The drop size is big and due to arc forces the drops are propelled away from the workpiece as spatter.
- Spray transfer is observed in argon shielded consumable electrode arc only. It appears that argon provides the unique plasma properties with the self-magnetic force to develop axial spray transfer through the arc.

A.C. Arcs

- Arc is extinguished during each half cycle and is reignited as the voltage rises again, current increases and the electrodes get heated again, arc path gets ionised.
- As arc length increases, the arc gas gets less heated and a higher reignition potential is required.

3.8.3 Effect of Other Gases on Metal Transfer

- Helium, although inert gas, does not produce axial spray transfer. The transfer is globular with both polarities at all current levels.
- Helium arcs are useful, nevertheless, because they provide deep penetration.
- Spray transfer can be obtained by mixing small quantities of Argon (about 20 percent). With helium, the deep penetration is still maintained. Normal commercial mixtures contain 25 percent argon as a safety factor.
- Active gases like carbon-di-oxide and nitrogen do not produce spray transfer, spatter on the other hand is increased.
- Spatter can be minimised by burying the arc below the plate surface to trap the spatter in the deep arc crater. This technique is used when:
  (*a*) carbon dioxide is used to shield arcs in mild steel.
  (*b*) nitrogen is used mixed with argon to shield aluminium alloys.
  (*c*) nitrogen is used to shield copper.

- The amount of spatter, massiveness of the drops and instability of transfer generally are greater when electrode is negative.
- Spray transfer can be achieved by painting cesium and sodium on steel wire surface with CO_2 shield using direct current electrode negative polarity.

3.8.4 Short Circuiting Transfer (Dip Transfer)

- Metal is tansferred from the electrodes (consumable) to the work through short circuits. It operates at low currents and low voltage (21 V, 200 A or less), the electrode end melts slowly. As the electrode is fed, arc gap shortens, until the tip touches the weld pool (Fig. 3.19 *c*)

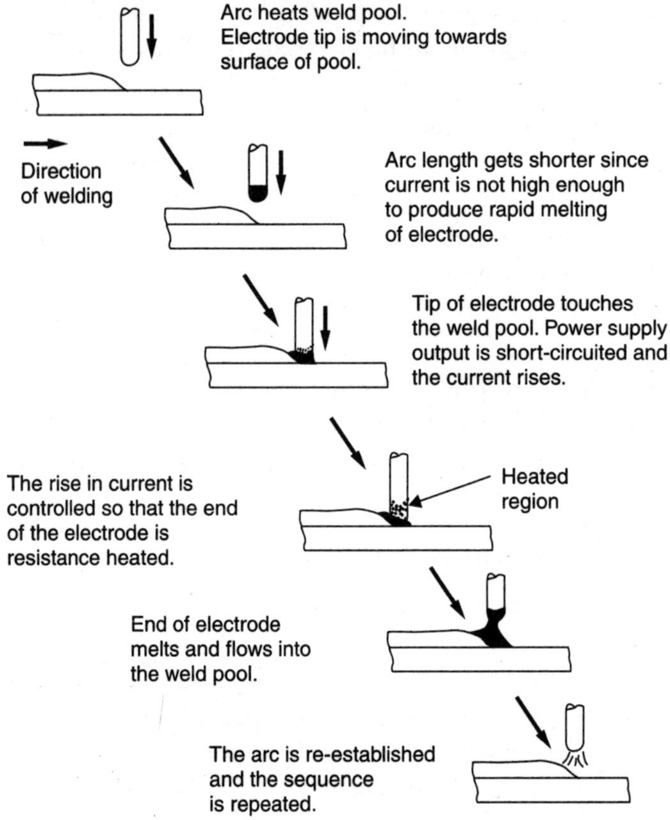

Time for complete sequence = $\frac{1}{200}$th to $\frac{1}{50}$th second

Fig. 3.19 (*c*) Dip transfer in MAGS welding

- Metal transferred in this way is less fluid and less penetrating, free of spatter and easy to handle.
- It is specially useful for joining thin sheets.
- Electrical reactance is used to **control** the rate of **current** rise when the wire and pool are in contact.

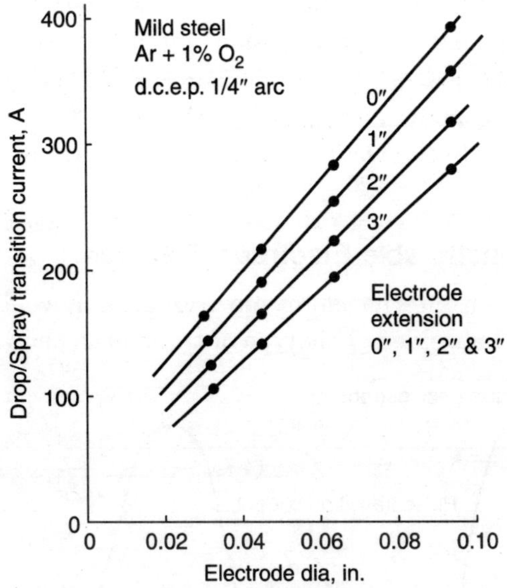

Fig. 3.20 (*a*) Influence of electrode diameter and extension on drop-to-spray transition currents

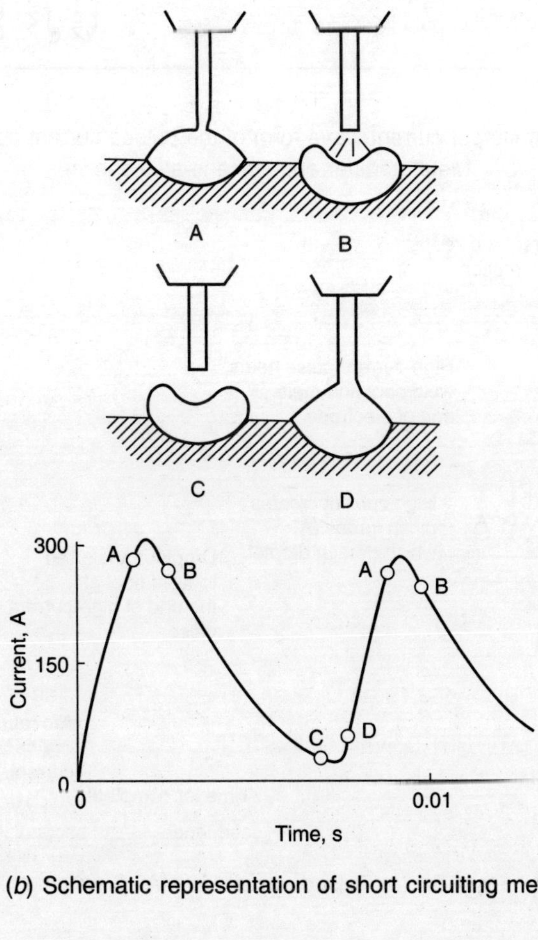

Fig. 3.20 (*b*) Schematic representation of short circuiting metal transfer

- The average **current is also kept** low by using relatively small diameter electrodes.
- With proper equipment adjustment short circuits of the order of hundreds of drops per second are obtained.
- Since **little time** is available to fuse the electrode, the drops formed are very small, and are transferred to the weld by surface tension when electrode tip and weld pool come in contact.

3.8.5 Pulsed Current Consumable Electrode Transfer

- This technique is an alternative of dip transfer for welding in positions and when thin plates are to be welded. This type of transfer is shown in Fig. 3.21 (*a*) and (*b*).

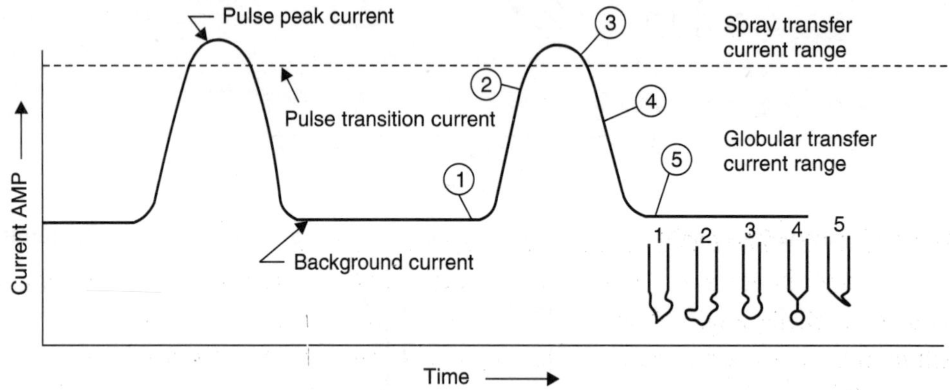

Fig. 3.21 (*a*) Output current wave form of the pulsed current power supply; Metal transfer sequence is also shown

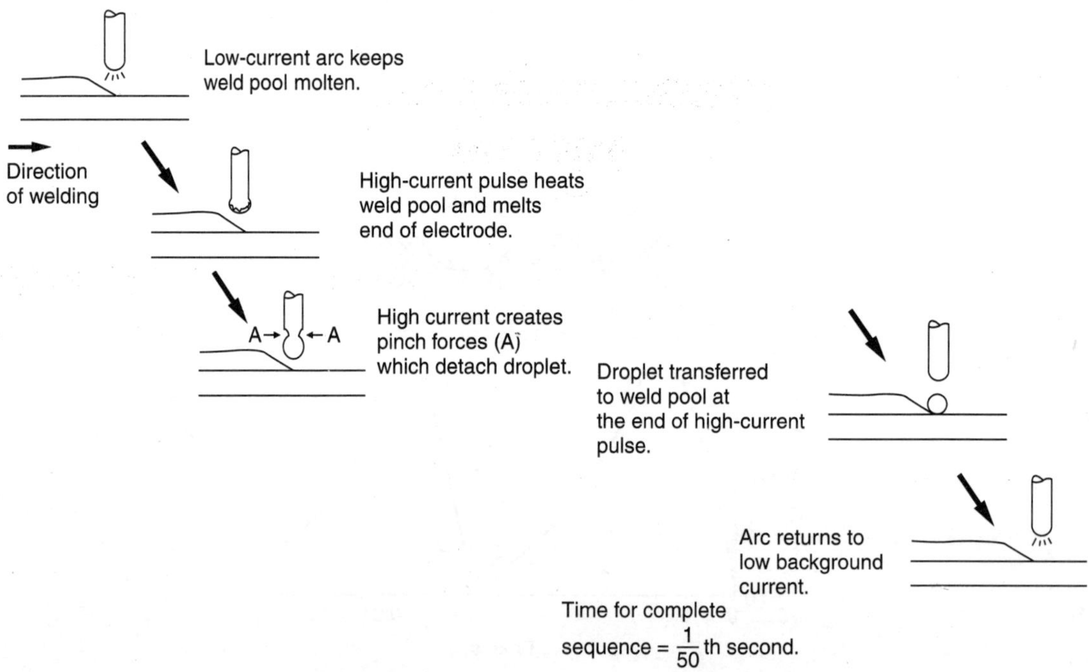

Fig. 3.21 (*b*) Pulsed transfer in MAGS welding

- Current pulses back and forth between the globular and spray transfer are superimposed on the normal background current.
- Time duration between consecutive pulses must be less than that required for globular transfer.
- Droplets are ejected from the electrode tip at regular intervals corresponding to the frequency of current pulses.
- Currents and deposition rates can be decreased so that welding speed can be reduced to cope more easily with thicknesses down to 1.0 mm or even thinner.

3.8.6 Covered Electrode Transfer

- In general the metal transfer is globular on one extreme and spray type on the other.
- Showery spray transfer is desirable. In some cases, however, spray transfer is not used because of spatter associated with it.
- Most of the electrodes contain cellulose or metal carbonates that burn in the arc forming a gas shield to protect the weld from atmospheric contamination. This shield contains mainly active gases like carbon dioxide, carbon monoxide, hydrogen and oxygen. Those gases do not develop a highly conductive arc plasma, the current distribution is such that the liquid metal is forced out of the arc and weld pool as massive drops and spatter.
- These reactions are more intense when electrode is negative, Reverse polarity is, therefore, used with electrodes that do not contain cathode stabilizers (cellulosic electrodes).
- Coverings can be made thermionic by adding rutile, lime and iron-oxide in combination. Such electrodes produce more stable arc, less spatter and form smaller drops with **direct current electrode negative.**
- With AC, current reduces to zero when polarity changes. The binders for such electrodes is changed from sodium silicate to potassium silicate. Potassium has lower ionisation potential, it also increases cathode emissivity to permit an easy reignition.
- Electrodes containing rutile or lime in sufficient quantities are also thermionic and do not require substitution of potassium binders to make them suitable for AC welding.

3.8.7 Melting Rates

General Controlling Parameters

- Most structural metals and their alloys form a cold cathode, its area is small but large quantities of energy are generated to release the electrons needed to support an arc.
- High m.p. materials like carbon, tungsten and molybdenum easily supply electrons to sustain the arc due to their temperature. These metals are called **thermionic.**
- Change from cold cathode to thermionic emission is accompanied by a lowering of the heating energy and, therefore reduction in melting rate.
- Also any improvement to arc stability in a.c. or metal transfer mode in dc en is associated with a reduction in melting rate.

- Electrical resistance heating of the electrode by welding current affects the electrodes melting rate.
- Electrode melting rate can be expressed as :

$$\text{M.R.} = aI + bLI^2 \qquad ...(3.11)$$

where a = anode or cathode constant of proportionality for heating. It depends upon polarity, composition and with dc en, the emissivity of the cathode.

b = constant of proportionality for electrical resistance heating and includes the electrode resistivity.

L = electrode extension or stick out.

I = welding current.

Table 3.3. Relative magnitude of heating coefficients in the melting rate of 1.6 mm diameter wire electrode

Metal	a	b
	Kg/h-A	*Kg/h.A²·mm*
Aluminium (dcep)	5.4×10^{-3}	4.4×10^{-6}
Mild steel (dcep)	8.6×10^{-3}	2.5×10^{-5}
Mild steel (dcen)	1.8×10^{-2}	2.5×10^{-5}

a = Kg/hour. Amp. b = Kg/hour Amp.^{2} mm.

3.8.8 Melting Rates with GMAW

- Melting rate is controlled by:
  (*a*) electrode diameter
  (*b*) electrode extension
  (*c*) cathode or anode heating (current polarity)
  (*d*) current mangnitude
  (*e*) Factors like shielding gas, arc length (arc voltage).
- Equation (3.11) for melting rate can be used to calculate melting rates for electrode positive. Problems develop with dc en, because the cathode heating value becomes quite sensitive to the presence of oxides alkali and alkaline earth compounds.
- The first term of the equation is more significant at low currents and with short electrode extension. The influence of second term becomes pregressively greater as the electrode diameter is reduced and its extension (resistivity) is increased and the current is raised. The relative magnitude of the heating coefficients with 1.6 mm diameter is shown in Table 3.3.

The values of the terms of the equation (3.11) depend upon the material (or alloy) being welded. First term is important for aluminium since its resistivity is low. It gains greater importance when the electrode is negative since the use of any additive that affects cathode emissivity will reduce the value of 'a' and thus reduce melting rate. Fig. (Fig. 2.20) shows that the electrode can be made so much thermionic as to reduce the heating effect represented by the term 'a' for electrode negative below that of electrode positive. Direct current electrode negative arcs have greater significance as they give very high melting rates (Fig. 2.20), but (unfortunately) the transfer is globular and spattery. When a.c. is used the values of '*a*' are an average between the values obtained for dc ep and dc en.

When argon shields are used the upper limit of melting rates is determined by the formation of 'jet-rotation' which needs higher currents and consequently higher diameter electrodes to sustain higher currents. The extent of these ranges is shown in Fig. for steel. This is not true for aluminium. The upper current for aluminium is limited by the formation of a very rough weld surface.

With active gas welding, metal transfer is always globular for all current levels. At lower level of current there is random short circuiting, absence of wetting and power weld quantity. At upper limits of current, there is spatter, poor bead appearance and porosity. When very low melting rates are necessary, the short circuit technique is frequently used.

Melting Rates with SAW

In general the above discussion for GMAW applies to SAW also. The melting rate increases with current. Cathode or anode voltage changes due to change of flux.

3.8.9 Melting Rates with SMAW

- The SMAW is least efficient in converting electrical energy to useful weld heat.
- Current controls the melting rate to some extent, but as the current increases the electrode diameter must be increased proportionately.
- Lower limit of current is defined by incomplete fusion, high viscosity of flux. Upper limit causes excessive resistance heating of the electrode that damages the electrode flux covering and the flux constituents breakdown before reaching to the arc where products of combustion arc needed for shielding.
- Cellulose coating on E6010 electrode of 6 mm diameter is useful in the range between 200-300 A while for the same diameter, the rutile-base E6012 that does not rely on gas formers has a useful range between 200 and 400 A.

3.9 WELDING PARAMETERS AND THEIR EFFECTS

Weld quality, and weld deposition rate both are influenced by various welding parameters and joint geometry. These parameters are the process variables as given below :

1. Welding current
2. Arc Voltage
3. Welding speed.
4. Electrode Feed rate
5. Electrode extension (stick-out)
6. Electrode diameter
7. Joint geometry.

Each of the above parameters affects, to varying extent, the following:

1. Deposition rate
2. Weld-bead shape

3. Depth of penetration
4. Cooling rate
5. Weld induced distortion.

Hence, a proper understanding of the effects of welding parameters (or process variables is important to obtain a sound welded joint with adequate metal deposition rate and minimum distortion. General effect of these variables will be discussed in the following paragraphs.

3.9.1 Welding Current

Melting rate is directly proportional to the energy (current and voltage) used for a given electrode and polarity used in DC welding. Part of this energy Q is used to melt the base metal (q_b), part goes to melt electrode and flux (q_f) rest is dissipated as conduction (q_{ep} + q_{ce}), convection (q_v) and radiation (q_r)

$$Q = q_b + q_f + (q_{cp} + q_{ce}) + q_v + q_r)$$

Also,
$$Q = IV.\ \text{J/S}$$
$$= I^2\ Ra\ \text{J/S}$$

where Q = electrical energy consumed
I = welding current
V = arc voltage
Ra = arc resistance

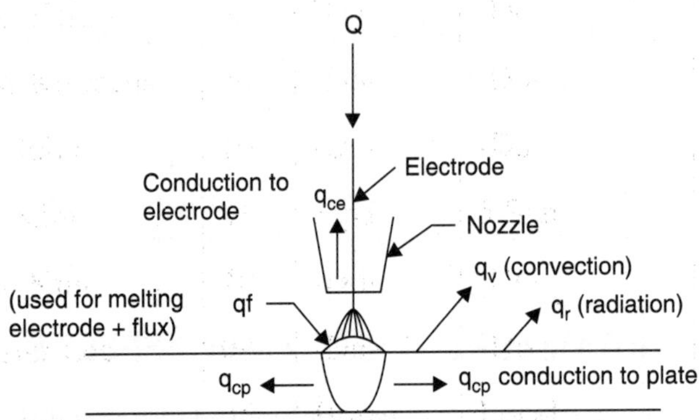

Fig. 3.22 Heat balance in SAW

Welding current is most important variable affecting melting rate, the deposition rate, the depth of penetration and the amount of base metal melted.

If the **current** (for a given welding speed) is too high, it will result in:

- excessive **penetration**
  (thinner plates will melt through)
- Excessive melting of electrode—**excessive reinforcement**
- More heat input to plates being joined **increased distortions**

If the welding current is too low, it will result in:

- inadequate penetration

- lack of fusion

Current could be DC or AC. DC provides steady arc and smooth metal transfer, good wetting action, uniform weld bead size, specially suited to thin section welding, give better quality welds in vertical and overhead welding positions.

3.9.2 Arc Voltage

Arc voltage is the voltage between the job and the electrode during welding. For a given electrode it depends upon the arc length. Open circuit voltage on the other hand is the voltage generated by the power source when no welding is done.

Open circuit voltage varies between 50–100 V whereas arc-voltages are between 17 V to 40 V. When the arc is struck, the open circuit voltage drops to arc voltage and welding load comes on power supply.

The arc voltage depends on arc length and type of electrode.

As arc length increases, arc resistance increases, (resulting in higher voltage drop (*i.e.,* arc-voltage increases and arc current decreases. This decrease in current depends upon the slope of volt-ampere curve explained earlier.

Arc length is the distance between the molten electrode tip to the surface of molten weld pool. Proper arc length is important in obtaining a sound joint. As the metal droplet transfers through the arc there is a variation in instantaneous arc voltage. Welding will be quite smooth if the arc voltage variation and hence the arc length is maintained consistant. As a general rule arc length should not be more than the electrode diameter.

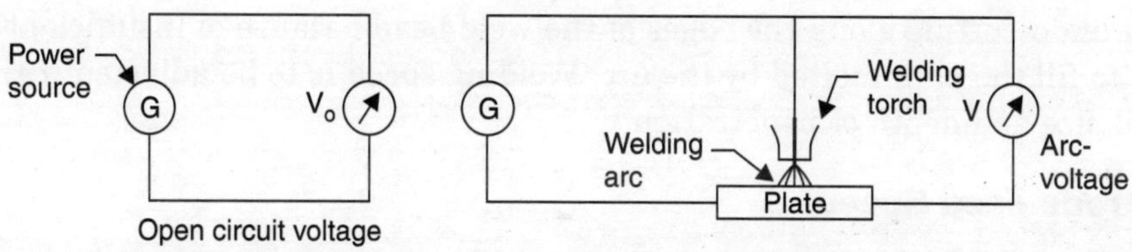

Fig. 3.23 Concept of open circuit voltage and arc-voltage

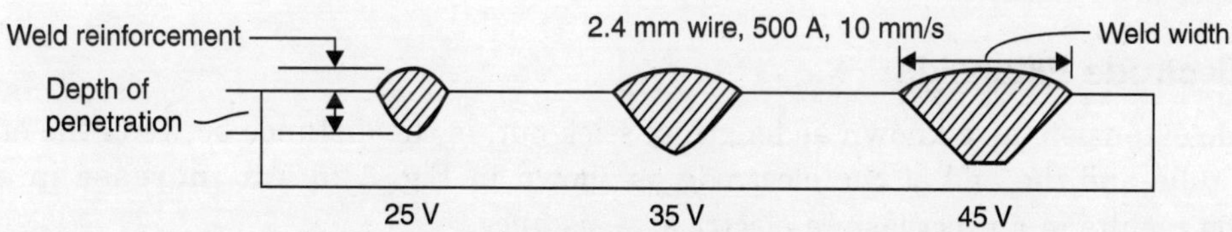

Fig. 3.24 Effect of arc-voltage variations on weld bead shape

Short arc: causes short circuits during metal transfer

Long arc—lacks direction and intensity, gives heavy spatter, low deposition rate and formation of undercuts.

Though arc length needs to be controlled in order to obtain a quality welding, it is much easier to monitor and control arc voltage.

Weld-bead appearance depends on arc-voltage. Increase in arc-voltage tends to cause porosity, spatter flatten the weld bead and increase weld width. Reduction in arc-voltage leads to : narrower weld-bead, higher crown, deeper penetration. Trials are, therefore, made to obtain optimum arc voltage.

3.9.3 Welding Speed

Welding speed is the linear rate at which the arc moves with respect to plate along the weld joint. Welding speed generally conforms to a given combination of welding current and arc voltage.

If welding speed is more than required

- Heat input to the joint decreases.
- Less filler metal is deposited than requires, less weld reinforcement height
- Undercut, arc blow, porosity and uneven bead shape may result.

If welding speed is slow

- Filler metal deposition rate increases, more weld reinforcement
- Heat input rate increases
- Weld width increases and reinforcement height also increases more convexity.
- Penetration decreases beyond a certain decrease in speed.
- A large weld pool, rough bead and possible slag inclusion.

With all variables held constant, weld penetration depth attains a maximum at a certain intermediate welding speed. At excessively low welding speeds the arc strikes a large molten pool, the penetrating force gets cushioned by the molten pool. With excessively high welding speeds, there is substantial drop in thermal energy per unit length of welded joint resulting in undercutting along the edges of the weld bead because of insufficient backflow of filler metal to fill the path melted by the arc. Welding speed is to be adjusted within limits to control weld size and depth of penetration.

3.9.4 Electrode Feed Speed

Electrode feed rate determines the amount of metal deposited per unit length or per unit time. In most welding machines the welding current adjusts itself with electrode feed speed to maintain proper arc length.

3.9.5 Electrode Extension

Electrode extension, also known as length of stick out, is the distance between the end of the contact tube and the end of the electrode as shown in Fig. 3.25. An **increase** in electrode extension results in an increase in electrical resistance.

This causes resistance heating of electrode extended length, resulting in additional heat generation and increase of electrode melting rate. But the energy so consumed reduces the power delivered to the arc. This reduces arc voltage and thus decreases bead width and penetration depth.

To maintain proper head geometry alongwith a desired penetration and higher melting rate (*i.e.,* large electrode extension), the machine voltage setting must be increased to maintain proper arc length. At current densities above 125 A/mm^2, electrode extension becomes important. An increase of upto 50% in deposition rate can be achieved by using long electrode extensions without increasing welding current. This increase in deposition rate is accompanied with decrease in penetration.

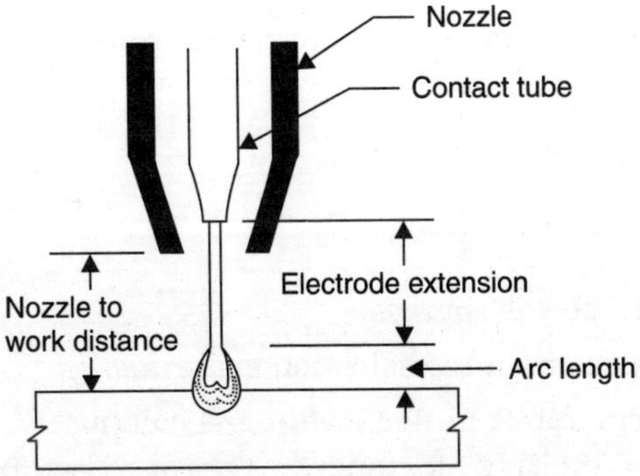

Fig. 3.25 GMA welding terminology

Thus when deep penetration is desired long electrode extension is not desirable. On the other hand, for thinner plates, to avoid the possibility of **melting through,** a longer electrode extension becomes beneficial. It is also important to note that the increase in arc extension make it more difficult to maintain correct position of electrode tip with respect weld centreline.

3.9.6 Electrode Diameter

Electrode affects bead configuration, affecting penetration and deposition rate. (Fig. 3.26). At any given current, a smaller diameter electrode will give higher current density causing a higher deposition rate compared to large diameter electrode. A larger diameter electrode, however requires a higher minimum current to achieve the same metal transfer characteristics. Thus larger electrode will produce higher deposition rate at higher current. If a desired feed rate is higher than the feed-moter can deliver changing to larger size electrode will permit desired deposition rate and vice versa. In case of poor fit-up or thick plates welding larger electrode size is better to bridge large root openings then smaller ones.

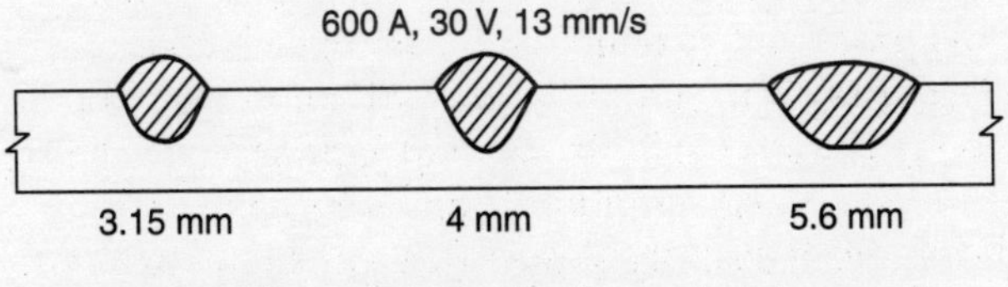

Fig. 3.26 Effect of electrode size on bead geometry

■ QUESTIONS

3.1 What characteristics are desired in a welding heat source?

3.2 Regarding welding power sources discuss

(*a*) Arc volt-amp. characteristic and compare it with Ohm's Law

(*b*) Arc-length in regard to Arc voltage, V-I characteristics for different arc-lengths.

(*c*) V.I. Characteristics of power supply used in

(*i*) Manual GTA welding (drooping).

(*ii*) Automatic Welding (constant potential).

3.3 Discuss the arc welding power supply equipment commonly used such as:

(*a*) Reactors

(*b*) Transformers

(*c*) Generators

(*d*) Rectifiers

(*e*) Solid-state welding power sources.

3.4 Discuss the welding power source selection criteria.

3.5 Discuss how the energy input in Arc welding is computed. What do you mean by heat transfer efficiency and melting efficiency in regard to net arc-energy calculation?

3.6 During submerged arc welding of mild steel, with an arc voltage of 20 V and current of 200 A, a welding speed of 5 mm/s was used. The cross-sectional area of the joint is 20 mm^2. Heat required to melt steel may be taken as 10 J/mm^3 and the heat transfer efficiency is 0.85. Calculate the volume of base metal melted in mm^3/s and the melting efficiency.

Chapter 4

Shielded Metal Arc (SMA) Welding

- Shielded Metal Arc Welding (SMAW) is a welding process in which coalescence of metals is produced by heat from an electric arc maintained between the tip of a consumable electrode and the surface of the base metal in the joint being welded.
- This is the most commonly used arc welding process, the equipment is cheap, welder has more freedom of movements, and it is possible to weld a wide variety of metals by changing only the electrode type.

4.1 PRINCIPLE OF OPERATION

- The electrode and the work are part of an electric circuit. Two cables come out from the power source. One is connected to the workpiece and the other to the electrode holder. Welding commences as an arc is struck between the tip of a consumable electrode and the workpiece region where welding is needed. Arc temperature is of the order of 5000°C. Melting of the workpiece and electrode tip occurs instantaneously. Process requires sufficient electrical energy to melt the electrode and proper amount of base metal. Metal droplets from the electrode are transferred to the weld pool and the electrode moves along the line of welding and is fed to the pool at a rate at which it is consumed to maintain a consistent arc length. Electrode melting rate depends upon the welding parameters used, electrode size, covering ingredients, polarity used etc. Shielded metal arc welding operating variables will now be discussed.

4.2 WELDING CURRENT (A.c. Vs. D.c.)

Electrode size and type and thickness of coating on it determine the arc voltage requirement (overall range 16–40 V) and current requirement (within an overall range of 20–550 A). The current could be direct of alternating depending upon the electrode being used. Almost all electrodes work well on d.c. but only a few flux compositions give stable arc operation with a.c.

- Transformers, on the other hand, are easier to maintain and are more robust as compared to d.c. generators or rectifiers. During d.c. welding, direct current flows between the electrode and the opposite terminal clamped to the workpiece. This current flow leads to the formation of a magnetic field which deflects the arc from the joint causing problems. This phenomenon is called arc-blow. It does not occur with a.c. as no stable magnetic fields are produced with a.c. (Fig. 4.1).

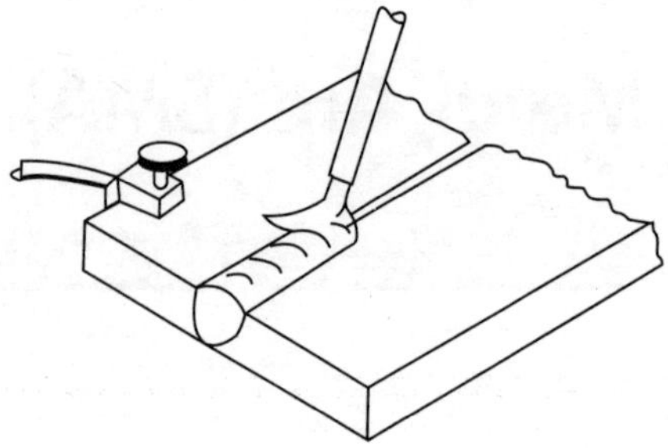

Fig. 4.1 Arc blow in SMA welding with direct current

- A.c. has another problem. The arc is extinguished each time the current pulse is reversed (*i.e.,* for 50Hz power supply, every one-hundredth of a second) To maintain a stable arc, the arc must be instantaneously re-ignited. A voltage in excess of 80 V must be supplied each time the current falls to zero. These high voltages are safety hazard and d.c. with an o.c.v. of about 60 V is preferred from this point of view.

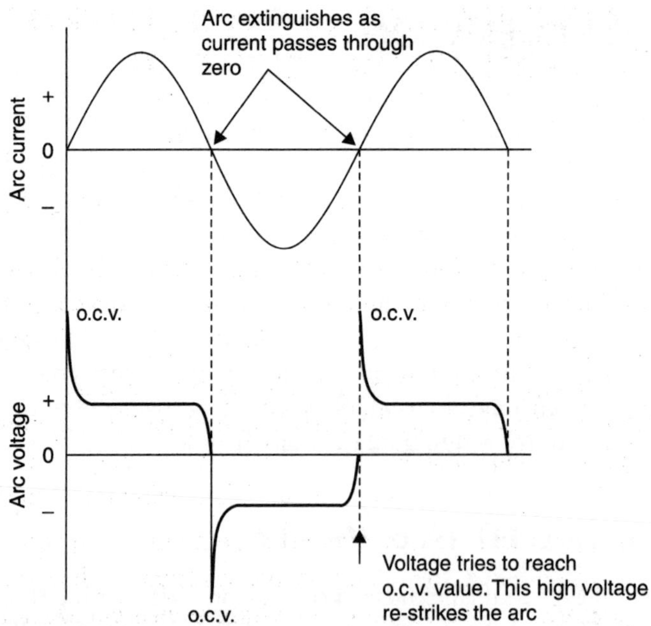

Fig. 4.2 Current and voltage waveforms in a.c. welding

4.3 COVERED ELECTRODES

In addition to establishing the arc and delivering filler metal to the weld, the electrode introduces other materials into and around the arc and weld pool through its covering. The main purpose of using a flux covering is to protect the molten metal from atmospheric contamination, the flux performs the following functions leading to the formation of a successful weld.

- weld-metal protection
- arc-stabilisation
- provides scavengers, de-oxidants, and fluxing agents to cleanse the weld and prevent excessive grain growth in the weld deposit.
- provides a slag blanket to protect hot metal from air, enhance mechanical strength, bead profile and surface cleanliness of weld bead.
- coating melts slower than the core wire, forming a cup the electrode end which additionally protects droplets of molten metal and makes touch welding possible and spatter loss is reduced.
- provides a means of adding alloying elements to enhance weld metal properties or adding iron powder to increase deposition efficiency.

In the following paragraph these factors will be briefly discussed.

4.3.1 Weld-Metal Protection

- Flux melts with the core wire and covers the surface of the molten metal drops and the weld-pool (see Fig. 4.3), excluding oxygen and nitrogen to come in their contact. As the weld-pool progressively solidifies, the flux forms a slag blanket over the weld-bead and continues to protect it from oxidation till it cools to room temperature.

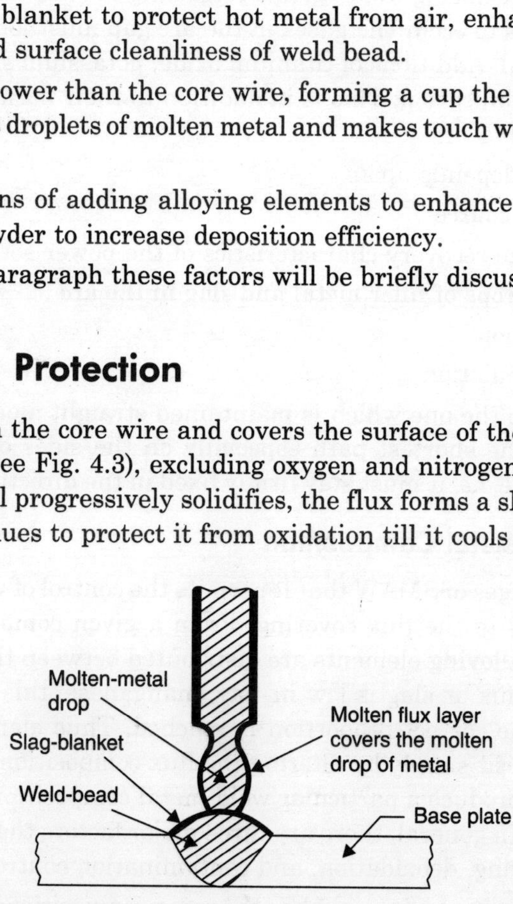

Fig. 4.3 Molten flux covers molten metal droplet and forms a slag blanket over the weld bead excluding oxygen and nitrogen to come in their contact

- The flux must also be completely **detachable.** This is very important especially when multiplc layers are to be deposited. Ideally we require a slag which automatically detaches itself off the weld deposit. This requirement is difficult to reconcile with the need to adhere to the weld-metal during the cooling period. Slag detachability is also influenced by compounds added to the flux to achieve other objectives. A compromise between the antagonistic effects of the compounds added to achieve different objectives is the only solution.

- Additional protection from atmospheric oxygen and nitrogen contamination is provided by adding compounds in the coating which decompose by the heat of the arc and form an additional gaseous shield around the arc and weld-pool. They may be carbonates (giving carbon dioxide) or cellulose (giving hydrogen and carbon monoxide).

4.3.2 Arc Stability

- There are two major aspects of arc stability. It is the ease of initiating and maintaining an electric arc during welding, and reigniting the arc during each half cycle in a.c. welding. For this to occur the gases in the arc gap must ionise rapidly and at lowest possible potential. Additions of titanium oxide, potassium silicate, calcium carbonate facilitate arc stabilisation. This is in addition to their normal purpose of acting as a flux.

Thus arc stability depends upon:

- O.C.V. of power source
- Transient voltage recovery characteristics of the power source
- Size of molten drops of filler metal and slag in the arc
- Arc path ionisation
- Electrode manipulation

A stable arc is also the one which is maintained straight along the electrode, axis and does not waver to find the shortest path especially on the sides of a vee edge preparation during welding in a groove, *i.e.* it must stay firmly fixed in the direction dictated by the welder.

4.3.3 Control of Weld-Metal Composition

This is one of the advantages of SMAW that it permits the control of weld metal composition by adding alloying elements to the flux covering. From a given combination of flux and weld-metal compositions, the alloying elements are distributed between the two in more-or-less the same proportion. If the flux or slag is low in, say, manganese, this metal transfers from the weld to the slag until the correct proportion is reached. Thus elements can be added to or taken from the weld deposit simply by altering the flux composition. The amounts of alloying elements to be added to produce a particular weld-metal composition can be calculated by the electrode manufacturer. In general, there are three major factors that control weld-metal composition. These are: alloying, deoxidation, and contamination control.

Alloying. When the core wire used has the same composition as desired in the weld, we need not add any alloying elements, except to ensure that the elements are not lost during welding. The electrodes used with low carbon, carbon-manganese, and low alloy steels, alloyed core wires turn out to be expensive. Alloying is to be done in the weld pool. Thus low carbon steel core wires could be used and manganese, chromium, molybdenum, etc. could be added through the flux. This helps in producing a large variety of electrodes with the same core wire, especially when small quantities of specific composition are needed.

Deoxidation. During the welding of steel, if the molten weld-metal pool contains excessive oxygen, it gives rise to the formation of carbon monoxide bubbles which get trapped in the solidifying weld metal to form porosity:

$$FeO + C = Fe + CO$$

This also causes loss of carbon which reduces the strength of the weld. This reaction can be supressed by adding deoxidants in the coating. A commonly used deoxidant for steel is silicon (added to the coating as ferro-silicon). Oxygen reacts with silicon in preference to steel as follows:

$$2FeO + Si = 2Fe + SiO_2$$

Silicon oxide formed floats to the weld-pool surface and forms slag. For welding copper the deoxidant used could be phosphorus or zinc to remove the oxygen and could be added to the filler metal and not to flux.

Contamination. The most harmful contaminant entering the molten weld-pool through the flux is hydrogen which leads to the formation of hydrogen cracks. Hydrogen is present in the electrode flux covering both as combined and absorbed moisture. Absorbed moisture can be removed by drying the electrodes before welding. The extent of chemically combined moisture depend upon the compounds used in the coating. Hydrogen has very high solubility in iron at elevated temperature. As the metal solidifies the solubility goes down and hydrogen bubbles are formed and are entrapped. As the metal cools and contracts, the pressure in the bubble exceeds the metal strength at that temperature forming cracks. Oxidising iron-oxide electrodes have been found to give beneficial results in solving the problem of hydrogen cracking.

- Other contaminants could be due to careless handling of the electrodes. Grease, oil, damped sulphurous fumes absorbed from the surroundings etc. may be transferred to the weld pool and cause harm. Careful handling of electrodes is, therefore, necessary.

4.3.4 Flux Covering Ingredients and their Functions

Depending upon the welding situational requirements a number of chemical compounds are used in formulating a flux. In Table 4.1 these compounds are listed with their major functions and secondary benefits for the welding of steels. The electrode flux covering performs the following functions:

1. Provide a protecting atmosphere
2. Forms slag of suitable characteristics to protect molten metal from oxidation
3. Facilitate over head and position welding
4. Stabilise the arc
5. Add alloying elements to the weld metal
6. Refine the metallurgical structure
7. Reduce weld spatter
8. Increase deposition efficiency
9. Remove oxides and impurities
10. Determine the depth of arc penetration

11. Affect weld-bead shape
12. Slow down the weld cooling rate
13. Contributes weld metal from powdered metal in the coating.

Table 4.1 Electrode Covering Ingredients with Functions

Function	*Ingredients*
1. Fluxing agents	Silica, CaO, Flourspar.
2. Slag formers	Rutile, Titania, Potassium titanate, limenite, Asbestos, Alumina, Silica flour, Iron oxide, Calcium fluoride (Flourspar) Feldspar, Manganese dioxide, Wollastonite.
3. Arc stabilisers	Potassium oxalate, Potassium silicate, Zirconium carbonate, Potash, Feldspar, Lithium carbonate, Titania.
4. Gas forming materials	Cellulose, Limestone, Woodflour, Calcium carbonates, other carbonates.
5. Alloying	Ferro-manganese, Ferro-chrome, Ferromolybdenum, Electronickel, Ferro-titanium, Metal powders.
6. Deoxidisers	Ferrosilicon, Ferromanganese.
7. Binders	Sodium silicate, Dextrin, Potassium silicate, Gum arabic, Sugar, Asbestos.
8. Slipping agents (for easy extrusion)	Glycerine, China clay, Kaolin clay, Talc, Bentonite clay, Mica.

Modern coated electrodes were first developed by Oscar Kjellberg of Sweden in 1907. Since that time considerable research has been done on electrode coating to obtain:

- good tensile and impact properties matching the base metal.
- most satisfactory electrode running characteristics.
- low cost formulation.

All this research has led to the development of a few standard covering types which have been coded and classified in the international specifications for electrodes as follows:

- Cellulosic,
- Rutile,
- Oxidising Iron-oxide and
- Basic

Table 4.2 compares the characteristics of these electrodes.

Cellulosic coverings. These coatings contain large quantities of organic materials. Cellulose exceeds 30% by weight. Other organic materials like wood flour, charcoal, cotton, starches and gums are also used to partially replace cellulose. It produces gaseous atmosphere of approximately the following composition,

$$55\%\ CO, + 42\%\ H_2 + 1.5\%\ H_2O + 1.0\%\ CO_2$$

The presence of hydrogen increases the voltage across the arc column making it more penetrating. For a given current cellulosic electrodes give 70% more deeper penetration than other electrodes. As most of the covering decomposes, the slag layer formed is thin and is easily removed. Hydrogen content of the weld is high. It is not recommended for welding high

Table 4.2. Characteristics of different types of electrodes

S.No.	Type	Classification AWS/ASTM	Coating Ingredients	Gas shield	Gas content of weld deposite ml/00 g: Diffusible* hydrogen	Gas content of weld deposite ml/00 g: Residual hydrogen	Applications
1.	Cellulosic	E–6010	Typically 40% cellulose 25% TiO_2 ; 20% $MgSiO_3$; 15% Fe-Mn bonded with sodium or potassium silicate.	Approximately 40% H_2 : 40% CO + CO_2 and 20% H_2O	15–30	1–5	General purpose electrode for carbon steel. Most commonly used type in U.S.A. Pipe welds. More heavily coated rods are used for deep penetration. Most heavily coated arc cutting electrodes.
2.	Rutile	E–6012 and E–6013	Typically 4% cellulose 50% TiO_2 ; 10% $CaCO_3$; 6% SiO_2 ; 20% Mica ; 10% Fe-Mn bonded with sodium or potassium silicate.		10–30	0.5–4.0	General purpose welding of carbon steel ; most generally used type in U.K. and other countries.
3.	Iron oxide (Deoxidized)	E–6020	Oxides and carbonate of iron and manganese with mineral silicates and ferro-manganese.		10–20	0.5–4.0	Give sound deposit with satis factory mechanical properties. Easy slag removal and good appearance of weld bead. Declining use.
4.	Basic low hydrogen	E–7015 and E–7016	Typically 60% $CaCO_3$; 30% CaF_2 ; 2.5% Fe-Mn ; 4% Fe-Si ; 2.5% Fe-Ti bonded with sodium or potassium silicate.	Approximately 80% CO and 20% CO_2	0.5–7.5 (dried immediately before use at 150°C)	0.0–2.0	Lowest hydrogen content. Good notch-ductility. Used for carbon steel where notch-ductility must be optimum: critical ship structures and sub-zero temperature applications. Low alloy steel electrodes: stainless steel electrodes.

*Electrodes giving upto 10 ml diffusible hydrogen per 100 gm deposited metal are called hydrogen controlled eletrodes.

strength steels. Because the coating does not contain much of ionisation compounds, they work well on d.c. To make them suitable for working on a.c. potassium, silicate is added to the coating.

Rutile coverings. Here the main ingredient is titanium-oxide. This compound is a good slag former and arc stabiliser. These electrodes are general purpose. By varying the amount of fluxing agents, viscosity and surface tension can be adjusted to give electrodes either for flat position only or for all position welding. Mechanical properties are adequate. Flux requires combined moisture to retain binding strength. The moisture, if excessively driven off, binding of the flux will suffer. It is retained and, therefore, hydrogen content of the weld deposit is high (25–30 ml/100 g.). This is higher than the quantity allowable (10 ml/100 g) for high strength steel welds.

Oxidising type covering. This covering contains mainly iron-oxide and silicates with or without manganese oxides. During welding it forms heavy solid slag with oxidising properties giving rise to welds which are low in carbon and manganese. The resultant deposit is soft and low in strength. Its use is limited to sheet metal fabrication.

Basic coverings. These coverings contain calcium carbonate and calcium fluoride (fluorspar) as bonding agents, and deoxidants. This results in a basic slag which is fairly fluid. The solidified slag is heavy, friable glassy brown. They are mainly used for welding high strength steels. Use of compounds containing combined moisture is avoided. They are baked at 400-450°C temperature which is high enough to drive-off nearly all the combined moisture. With the arc heat calcium carbonate forms carbon-dioxide and carbon monoxide gases. The gas evolution rate is substantially lower. It is, therefore, necessary to maintain a short arc to avoid oxygen and nitrogen contamination.

The arc characteristics can be modified by using easily ionisable metals in the coating. The use of potassium silicate as a binder instead of sodium silicate makes the electrode suitable for a.c. welding also. But for high quality welding d.c. is preferred.

Flux covering thickness. This varies with each class and brand of electrode, and is usually expressed as coating factor, which is the ratio of coating diameter to the core wire diameter (see Fig 4.4)

$$C.F. = \frac{D}{d}$$

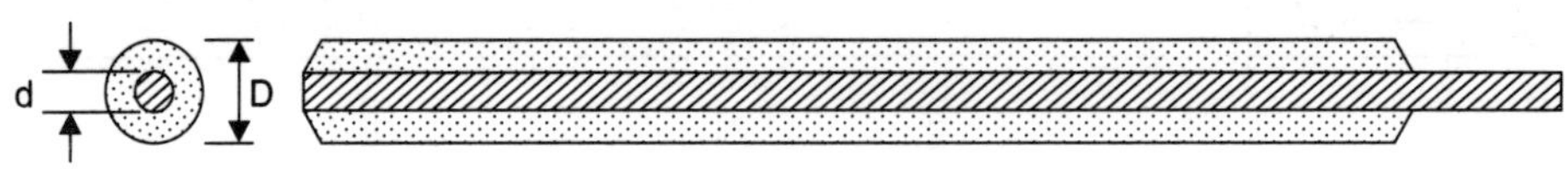

Fig. 4.4 SMAW electrode

These electrodes are often classified as light coated, medium coated and heavy coated depending on their coating factor as given below

Light coated	1.2 – 1.35
Medium coated	1.4 – 1.70
Heavy coated	1.8 – 2.20

As the coating thickness increases the weldpool becomes deeper and narrower, and the electrode is said to have **deep penetration** characteristics. Electrodes with very thick coatings are used for cutting metals.

Alloying elements and iron powder. Subtantial amounts of alloying elements are sometimes added to the coating so as to obtain a desired composition of the weld deposit. Iron powders can be added to the coatings in amounts from 10–50% of the coating weight to increase weld deposition rates.

4.3.5 Current Ranges for SMAW Electrodes

These ranges are given in Table 4.3.

Table 4.3. Current ranges for SMAW electrodes

Core-wire diameter mm	*Lengths of electrode*	*Welding Current (Amperes)*		
		Light work	*Normal work*	*Heavy work*
2.5	250/300/350	55	70	85
3.2	350/450	90	110	130
4.0	350/450	140	165	180
5.0	350/450	180	210	240
6.0	350/450	200	255	315
6.3	350/450	220	260	320

4.3.6 Electrode Core-wire Composition

According to AWS A5.1–81, the core wire for the electrodes in this specification is usually a rimmed or capped steel having a typical composition of 0.1% C, 0.45% Mn, 0.03% S, 0.02% P, and 0.01% Si. IS : 2879-1975 recommends rimming quality steel with the following composition (maximum percent) 0.1% C, 0.38–0.62% Mn, 0.03% S, 0.03% P, 0.03% Si, 0.15% Cu.

4.3.7 Factors Affecting Electrode Selection

Each situation needs a number of factors to be considered before specifying a particular electrode. These factors are:

(*a*) composition of metal to be welded

(*b*) mechanical properties desired in the joint

(*c*) weldability problems – any risk of weld metal cracking

(*d*) heat input limitations

(*e*) welding power source available a.c. or d.c.

(*f*) welding position

(*g*) type of joint

(*h*) parent metal thickness

4.4 MILD STEEL AND LOW-ALLOY STEEL ELECTRODES

Having answered these and other questions relevant to the specific situation an electrode type and size is selected which gives desired performance at minimum cost. The electrodes are marketed by different manufacturers in different brand names. They also give a standard code number based on international or national standards. These code numbers are useful in comparing the electrodes from different manufacturers and in knowing the characteristics of the electrodes completely regarding the mechanical properties of the weld deposit, type of covering, type of current (a.c./d.c.) and welding positions in which the electrode can be used. These standards are explained further in the following paragraphs.

4.5 WELDING ELECTRODES SPECIFICATION SYTEMS

Various systems of electrode specifications are used in different countries.

Most important ones are from:

1. International Organisation for Standardisation (ISO)
2. American Welding Society
3. Indian Standards Institution
4. British Standards Institution
5. Deutsches Institut Für Normung (DIN).

They cover some or all of the following groups of electrodes

1. Mild steel electrodes
2. Low alloy steel electrodes
3. Stainless steel electrodes
4. Surfacing electrodes
5. Cast iron electrodes
6. Copper and copper alloy electrodes
7. Nickel and nickel alloy electrodes
8. Aluminium and aluminium-alloy electrodes.

As mild steel and low alloy steel electrodes are most commonly used, the important welding electrode specification systems for these electrodes will be discussed in the following paragraphs.

4.5.1 International Standards Organisation System of Coding

ISO-2560-1973(E): Covered Electrodes for Manual Arc Welding of Mild Steel and Low-alloy Steel. Code for Identification.

- Prefix E: indicates covered electrodes for manual arc welding. (See Fig. 4.10)
- Next symbols: 43 or 51 indicate that all weld metal tensile strength is in the range of 430–510 MPa or 510–610 MPa respectively. Upper limits may exceed by 40 MPa.
- For each range of tensile strength, there are six sub-groups based on elongation (on L = 5d) and temperature for minimum impact value of 28 J (see Table 4.4).

– Next come one or two letters symbol for covering type

A = Acid (iron-oxide) ; AR = Acid rutile; B = Basic ; C = Cellulosic ;

O = Oxidising ; R = Rutile ; RR = heavy coated rutile ; S = other type

Symbols up to this stage are compulsory, beyond this the symbols indicate :

– Weld deposition efficiency in increments of 10 (110, 120, 130, etc.)

– Next digit indicates welding position

1. all positions;
2. all positions except vertical down
3. flat butt ; flat fillet ; horizontal/vertical fillet weld
4. flat butt, flat fillet
5. as 3 plus vertical down.

– Next comes the symbol for electrical characteristics *i.e.,* whether the electrode operates with a.c. as well as d.c. or d.c. alone, the polarity of d.c. and minimum open circuit voltage for a.c. necessary for sustaining the arc. It is given in Table 4.5.

– The last symbol H is used only when the electrode is hydrogen controlled *i.e.* the weld deposit contains diffusible hydrogen content of less than 15 ml. per 100 g of deposited metal (determined by a standard method).

Table 4.4. Electrode designation according to ISO-2560

Electrode designation	*Tensile strength MPa*	*Min. elongation on L = 5 d %*	*Temp. for minimum impact value of 28 J °C*
E 430	434–510	—	—
E431	434–510	20	+ 20
E432	434–510	22	0
E433	434–510	24	– 20
E434	434–510	24	– 30
E435	434–510	24	– 40
E510	510–610	–	–
E511	510–610	18	+ 20
E512	510–610	18	0
E513	510–610	20	– 20
E514	510–610	20	– 30
E515	510–610	20	– 40

Tolerance + 40 MPa, 1 J = 0.102 Kgf.m.

Table 4.5. Symbols for electrical characteristics in ISO-2560

Symbol	*Electrode polarity with direct current*	*Nominal O.C.V. with alternating current volts*
0	+	not used
1	+ or –	50
2	–	50
3	+	50
4	+ or –	70
5	–	70
6	+	70
7	+ or –	90
8	–	90
9	+	90

Example (*a*) ISO 2560

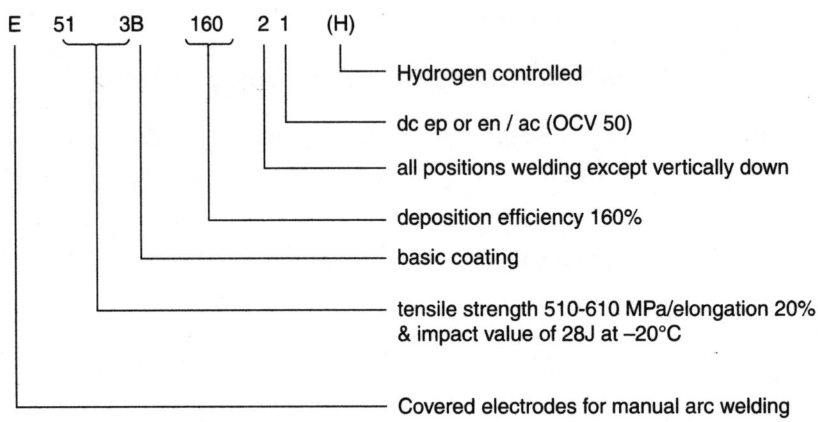

Fig. 4.15 Example of electrode designation according to ISO-2560

4.5.2 British Standards Institute Coding Systems

B.S : 639 : 1976 Covered Electrodes for Manual Metal Arc Welding of Carbon Manganese Steels.

This is based on ISO 2560 except that E is followed by 4 digits instead of 3 digits in ISO. This fourth digit gives more information on elongation and impact value.

In this system minimum yield stress is also specified as also in DIN. This system will be explained with an example (see Fig. 4.5).

Example (*b*)

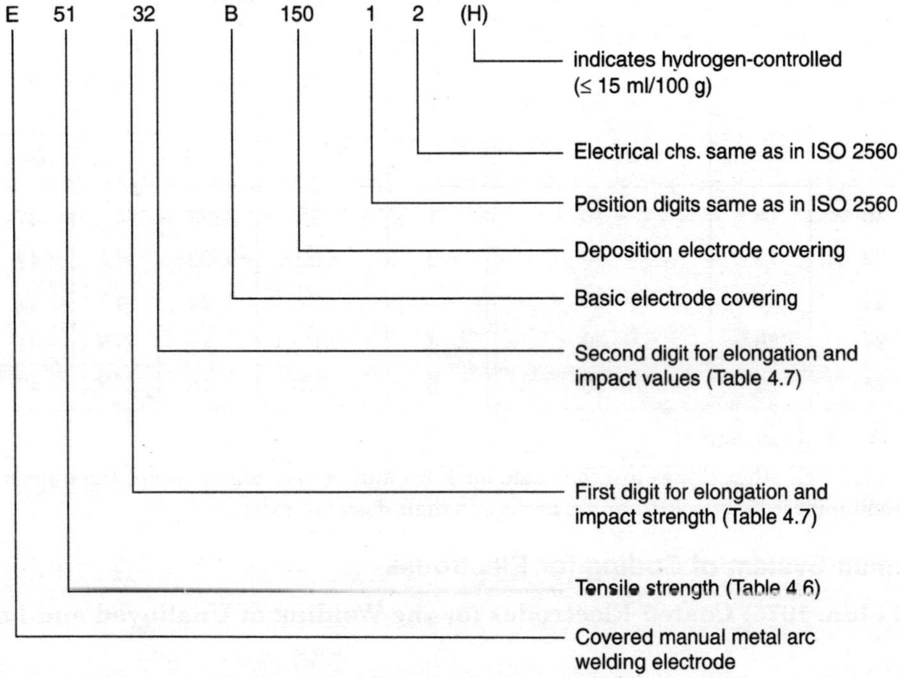

Fig. 4.6 Electrode designation according to BS : 639 : 1976

Table 4.6 Tensile strength BS 639 (1976) and DIN 1913 (1976)

Electrode designation	*Tensile strength, MPa*	*Minimum Yield Stress, MPa*	
		BS : 639 : 1976	*DIN : 1913 : 1976*
E43	430–550	360	330
E51	510–650	380	360

Table 4.7. First and Second digits elongation and impact strength

First Digit	*Min. elongation % L = 5D*		*Temp. for impact value of 28 J (°C)*	*Second Digit*	*Min. elongation % L = 5D*		*Impact prop.*		
	E43	*E51*			*E43*	*E51*	*Impact value J*		*Temp. °C*
							E43	*E51*	
1	20	18	+ 20	1	22	22	47	47	+ 20
2	22	18	0	2	22	22	47	47	0
3	24	20	– 20	3	22	22	47	47	– 20
4	24	20	– 30	4	NR(*a*)	18	NR	41	– 30
5	24	20	– 40	6	NR	18	NR	47	– 50(*b*)

(*a*) NR = Not relevant

(*b*) In DIN all other things are the same for First and Second digits except the impact temperature for second digit if "5" = – 40°C and 6 as second digit does not exist.

4.5.3 German System of Coding for Electrodes

DIN 1913 (Jan. 1976) Coated Electrodes for the Welding of Unalloyed and Low-alloy Steels

The German coding system is also based on ISO : 2560 with some modifications as in BS 639. It starts with prefix E followed by two digits 43 or 51 indicating the range of tensile strengths as in ISO, with the addition that minimum yield strength is specified as 360 MPa and 380 MPa respectively (see Table 4.6).

These two digits are followed by another two digits indicating elongation and impact strength as given in Table 4.8. After this DIN has a departure from ISO 2560 and BS 639. It provides a classification based upon :

(*a*) coating type

(*b*) welding position

(*c*) welding current condition

and then uses the classification number to designate each type of electrode. The details are as follows:

(*a*) **Coating type** is indicated by letter or letters as follows

A–acidic B–Basic C–Cellulosic

A–rutile (thin/medium) RR–rutile (heavy coating)

AR–acid-rutile (mixed) R(c) rutile–cellulose (medium coated)

RR(c)–rutile-cellulose (heavy coated)

B(R)–basic coated with non-basic components

RR(B)–rutile-basic (heavy coated)

They define :

Thin coated, having a coating factor (CF) of 120% ; medium coated, having a CF of 120–155% and heavy coated having a CF of over 155%.

(*b*) Welding position

1. all position.

2. all positions except vertical down.

3. butt-weld flat, fillet-weld flat, fillet-weld horizontal.

4. butt-weld flat, fillet weld flat.

(*c*) **Welding current conditions** are same as in ISO 2560 and BS 639 except that in case of 0 (zero)

0 means dc only electrode positive or negative polarity

0^+ means dc only with electrode positive polarity

0^- means dc only with electrode negative polarity

Combining (*a*), (*b*) and (*c*) twelve classifications of electrodes are given in Table 4.9.

This electrode class coding is followed by a three digit number indicating the deposition efficiency, which is to be used only if it is more than 105%. This is identical to ISO 2560 and BS 639.

Table 4.8 First and second digit for elongation and impact strength in DIN 1913

First digit	*Min. elongation L = 5d (%)*	*Temp. for min impact value of 28 J (°C)*	*Second digit*	*Temp. for impact value of 47 J (°C)*
0	Nil	Nil	0	Nil
1	22	+ 20	1	+ 20
2	22	0	2	0
3	24	– 20	3	– 20
4	24	– 30	4	– 30
5	24	– 40	5	– 40

Table 4.9. Classification numbers of electrodes in DIN 1913

Electrode type	*Welding position code* see (b) above*	*Current condition***	*Coating see (a) above*	*Classification number*
A1	1	5	thin coated A	1
A2	1	5	thin coated A	2
R2	1	5	thin coated R	
R3	2(1)	2	medium coated R	3
R(C)3	1	2	medium coated R(C)	
C4	1 +	0^{+}(6)	medium coated C	4
A5	2	5	heavy coated A	5
RR6	2	2	heavy coated RR	6
RR(C)6	1	2	heavy coated RR(C)	
AR7	2	5	heavy coated AR	7
RR(B)7	2	5	heavy coated RR(C)	
RR8	2	2	heavy coated RR	8
RR(B)8	2	5	heavy coated RR(B)	
B9	1†	0^{+}(6)	heavy coated B	9
B(R)9	1†	6	heavy coated B(R)	
B10	2	0^{+}(6)	heavy coated B	10
B(R)10	2	6	heavy coated B(R)	
RR11	4(3)	5	RR with dep. eff. > 105%	11
AR11	4(3)	5	AR with dep. eff. > 105%	
B12	4(3)	0^{+}(6)	B with dep. eff. > 120%	12
B(R)12	4(3)	0^{+}(6)	B(R) with dep. eff. >120%	

*Bracketed code numbers for welding positions apply only to a smaller sizes and/or low levels of deposition efficiency.

**Bracketed code numbers for current conditions mean conditional qualification.

†Favoured for vertical down.

4.5.4 Indian Standards System

IS : 815-1974 classification and coding of covered electrodes for metal arc welding of structural steels.

The code starts with a prefix E or R meaning thereby

E–electrode produced by solid extrusion

R–extruded with reinforcement

Next come digits

First digit indicates the type of covering

Table 4.10. First digit for type of covering in IS : 815

First digit	*Type of Covering*	*ISO : 2560 Equivalent*
1	High cellulose content	C
2	High titania giving viscous slag	R
3	Appreciable titania, giving fluid slag	RR
4	High oxides or silicates of iron or both and manganese giving inflated slag	A
5	High iron oxides or silicates or both giving heavy solid slag	O
6	High calcium carbonate and fluoride	B
9	Any other covering not specified	S

Second digit indicates welding position and third digit indicates welding current condition as shown in Table 4.11.

Table 4.11. Second and third digit for welding position and current condition in IS : 815

Second digit	*Welding position*	*Third digit*	*Welding current condition*
0	F, H, V, D, O	0	D +
1	F, H, V, O	1	D +, A90
2	F, H	2	D –, A70
3	F	3	D – , A50
4	F, Hf (horizontal fillet)	4	D +, A70
9	Any other welding position not classified above	5	D ±, A70
		6	D ±, A70
		7	D ±, A50
		9	other conditions not classified.

Fourth and Fifth digits are 41 or 51 indicating tensile strength range in combination with yield stress.

Sixth digit indicates percentage with impact strength as given in Table 4.12.

Table 4.12. Digits indicating mechanical properties in IS : 815

Fourth, fifth and sixth	**Tensile strength N./mm²*	*Min. yield stress N/mm²*	*Min. elongation %*	*Temp. for min. impact value of 47 J, °C*
410	410–510	330	–	–
411	410–510	330	20	+ 27
412	410–510	330	22	0
413	410–510	330	24	– 20
414	410–510	330	24	– 30
415	410–510	330	24	– 40
510	510–610	360	–	–
511	510–610	360	18	+ 27
512	510–610	360	18	0
513	510–610	360	20	– 20
514	510–610	360	20	– 30
515	510–610	360	20	– 40

*Upper limit of tensile strength may be exceeded by + 40N/mm².

The coding terminates with one or more of the following suffixes to be used when appropriate.

Suffix letter	*Special property*
H	Hydrogen controlled electrode
J	Iron powder covering deposition efficiency 110-130%.
K	As J with deposition efficiency 130 – 150.
L	As J with deposition efficiency of 150%.
P	Deep penetration.

A hydrogen controlled electrode gives a weld deposit that gives not more than 10 ml of diffusible hydrogen/100 g weld deposit. Appendix A gives types of flux coverings according to DIN, 1913, IS : 815 and AWS.

Types of Flux Covering

IS : 815 describes the standard flux coverings as follows :

Type 1: Electrode with covering having a high cellulose content.

The covering contains at least 15% of material having a high cellulose content and up to 30% of titania (as rutile or titanium white). This type of electrode is characterised by a deep penetrating arc and rapid burn-off rate. Spatter loss is somewhat higher than that with electrodes having the mineral type of covering. A voluminous gas shield is formed as a result of the decomposition of the cellulosic material in the arc region. The weld finish is somewhat coarser than usual, the ripples being rather more pronounced and less evenly spaced. The deposit has a thin cover of slag, which is friable and thus easy to remove. Because of its arc characteristics and the small volume of slag produced, the electrode is particularly easy to use in any welding position. With current values near to the maximum of the range, the electrode may be used in the flat position for deep-penetration welding. The electrode is suitable for all types of mild steel welding and is of particular value for applications involving changes in position of welding, for example, in pipe welding, storage tanks, bridges and ship building. Generally, this type of electrode is suitable for use with DC with the electrode connected to the positive pole. Some types are available which contain arc stabilising materials and are suitable for use with AC.

Type 2: Electrode with covering having a high content of titania and producing a fairly viscous slag.

The covering contains a high proportion of titania (as rutile, titanium white or ilmenite) and the high content of ionisers provides excellent welding properties. An electrode of this type is suitable for butt and fillet welds in all positions and is particularly easy to use for fillet welds in the horizontal-vertical position. Sizes larger than 5 mm are not normally used for vertical and overhead welding. Fillet welds tend to be convex in profile and have medium root penetration. The electrode has smooth arc characteristics and normally produces very little spatter. The slag is dense and completely covers the deposit and is easily detached, except from the first run in a dc ep V-groove. The electrode is particularly suitable for use with AC, and on DC it may be used with the electrode connected to either pole.

Type 3: Electrode with covering containing an appreciable amount of titania and producing a fluid slag.

The covering contains an appreciable amount of titania (as rutile, titanium white or ilmenite), but the addition of basic materials yields a much more fluid slag than produced by electrodes of Type 2. Welding in the overhead and vertical (upwards) position is far easier with this type of electrode than with any other type of mild steel electrode, but its use is not confined to these positions. The electrode has smooth arc characteristics, medium penetration, and normally produces very little spatter. The slag is generally easy to detach, even from the first run in a deep V-groove. The deposit produced by this type of electrode will usually meet normal radiographic tests more readily than the one made with electrodes of Type 2. The electrode is suitable for use with AC and DC and may be used with the electrode connected to either pole.

Type 4: Electrode with covering producing an inflated slag and having high content of oxides and/or silicates of iron and manganese.

The covering consists principally of oxides or carbonates of iron and manganese, together with silicates. The electrode is generally produced with a thick covering and is used for welding in the flat position only. Certain varieties have a thinner covering, and these may be used for welding in all positions but have generally been superseded by other types of electrodes. Both the forms of covering produce a fluid, voluminous slag which freezes with a characteristic internal honeycomb of holes, the so-called inflated slag, which is very easily detached. The weld finish is smooth, the ripples being much less pronounced than on deposits produced by the other types of electrodes. In grooves and fillet welds, the weld profile is concave. The principal application for this type of electrode with a thick covering is for deep groove welding in thick plates, particularly where such welds are subject to strict radiographic acceptance standards. Certain varieties of this type of electrodes are suitable for deep penetration welding. The electrode is suitable for use with DC, usually with the electrode connected to the positive pole, and may be used on AC.

Type 5. Electrode with covering having a high content of iron oxides and/or silicates producing a heavy solid slag.

This type of electrode has a thick covering, consisting principally of iron oxides with or without oxides of manganese. An electrode of this type is used principally for single run fillet welds, where appearance is of primary importance. The covering melts with a pronounced cupped effect at the electrode tip, enabling the electrode to be used touching the work, this procedure being known as touch welding. The degree of penetration is low. A heavy solid slag is produced which is sometimes self-detaching, and in fillet welds, gives a smooth, concave weld metal has low carbon content and a particularly low manganese content. This type of electrode has been used with some success for the welding of certain high tensile steels and also steels having a higher content of sulphur than those used for structural welding, but on such steels the weld profile may be more irregular. Weld metal deposited by this type of electrodes usually has low mechanical properties, the reduction of area and Izod impact values being generally less than the values normally specified. The electrode is particularly suitable for use with AC and DC and may be used with the electrode connected to either pole.

Type 6: Electrode with covering having a high content of calcium carbonate and fluoride.

The covering of this electrode contains appreciable quantities of calcium carbonate and fluoride. The slag is fairly fluid and the deposit is usually convex to flat in profile. This class of electrode is generally suitable for welding in all positions. Electrodes of this class are also known as basic coated, and have the advantage of being particularly suitable for welding medium and high tensile structural steels and other applications, where high mechanical properties and maximum resistance to cracking are required. They are also used for welding steels having higher carbon and sulphur contents than normal structural steels. During manufacture, these electrodes are baked at a high temperature and to obtain the best results they should be properly stored, and if necessary, thoroughly dried to the manufacturer's recommendations before use. In welding with these electrodes, it is necessary to use a short arc and the correct electrode angle to achieve maximum soundness in the weld deposit. Properly used in this way, the electrode will produce welds to high radiographic acceptance standards. Most of the electrodes recently developed can be used with AC but with some types DC is preferred, in which case the electrode should be connected to the pole recommended by the manufacturer. Coatings of this type are commonly used for electrodes dopositing high tensile and alloy weld metals.

Note: The addition of metal powder to any of the above types of covering may affect the characteristics described above.

4.5.5 American Coding System

AWS-A5.1 – 81 Specification for Carbon Steel Covered Arc Welding Electrodes

The American Coding System starts with a prefix E which means an electrode. Then comes a two digit number 60 or 70 designating tensile strength in ksi (60 ksi or 70 ksi). The actual stipulated minimum tensile strength values and the associated yield strength values vary according to the type of covering as given in Table 4.13. The impact strength requirements are given in Table 4.14.

The third digit indicates the welding positions in which the electrode can be used satisfactorily, as follows:

1. F, H.V. OH
2. F, H-fillet
3. F, H, V-down, OH.

The last two digits together indicate current conditions and the type of covering. Table 4.15 gives complete classification and their significance.

Table 4.13. Strength and elongation requirements for all-weld-metal tension test in the as-weld condition (AWS.A-5.1)

AWS Code	*Min. tensile strength*		*Min. yield strength*		*Min. elongation on L = 4d*
	Ksi	*MPa*	*Ksi*	*MPa*	*%*
E6010	62	430	50	340	22
E6011	62	430	50	340	22
E6012	67	460	55	380	17
E6013	67	460	55	380	17
E6020	62	430	50	340	22
E6022	67	460	Not required		Not required
E6027	62	430	50	340	22
E7014	72	500	60	420	17
E7015	72	500	60	420	22
E7016	72	500	60	420	22
E7018	72	500	60	420	22
E7024	72	500	60	420	17
E7027	72	500	60	420	22
E7028	72	500	60	420	22
E7048	72	500	60	420	22

For each increase of 1% in elongation, the tensile strength or yield strength or both may decrease by 7 MPa to a minimum of 420 MPa for tensile strength and 330 MPa for yield strength for E60 series and to a minimum of 480 MPa for tensile and 400 MPa for yield strength for E70 series, except for E6012, E6013 tensile and yield strength may reduce to a minimum of 450 and 365 MPa respectively. Since E-6022 electrodes are for single-pass welding, the elongation and yield measurement is not necessary.

Table 4.14. Impact requirements as per AWS-A5.1

AWS classification	*Charpy–V notch impact requirement, min*
E6010, E6011 E6027, E7015 E7016, E7018* E7027, E7048	27 J at – 29°C
E7028	27 J at – 18°C
E6012, E6013 E6020, E6022 E6014, E7024	Not required

*Upon agreement between the supplier and the purchaser classified as E7018 may be supplied to a minimum Charpy-V notch impact requirement of 27 J at – 46°C. Such electrodes shall be identified as E7018-1.

Table 4.15. Type of covering, welding position and type of current as per AWS-A5.1

AWS classification	*Type of covering*	*Welding positions*	*Type of current***
	E60 series electrodes		
E6010	High cellulose sodium (C)	F, V, OH, H	D^+
E6011	High cullulose potassium (C)	F, V, OH, H	D^+, A
E6012	High titania sodium (R)	F, V, OH, H	D^-, A
E6013	High titania potassium (RR)	F, V, OH, H	$D^\pm$, A
E6020		H-fillets	D^-, A
E6022†	High iron oxide (A)	F	$D^\pm$, A
E6027	High iron oxide, iron powder (A)	H-fillets, F	D^-, A
	E70 series electrodes		
E7014	Iron powder, titania (RR)	F, V, OH, H	$D^\pm$, A
E7015	Low hydrogen sodium (B)	F, V, OH, H	D^+
E7016	Low hydrogen potassium (B)	F, V, OH, H	D^+, A
E7018	Low hydrogen potassium iron powder (B)	F, V, OH, H	$D^\pm$, A
E7024	Iron powder, titania (RR)	H-fillets, F	$D^\pm$, A
E7027	High iron oxide, iron powder (A)	H-fillets, F	D^-, A
E7028	Low hydrogen potassium, iron powder (B)	H-fillets, F	D^+, A
E7048	Low hydrogen potassium iron powder (B)	F, OH, V, V-down	D^+, A

*Letters in brackets indicate equivalent ISO 2560 symbols for types of covering.

** The standard refers to D + as reverse polarity and D – as straight polarity and A as a.c.

† Electrodes of the E6022 classification are for single-pass welds.

Chemical composition limits for weld-metal as per AWS-A5.1

– For electrodes E6010, E6011, E6012, E6013, E6020, E6022, E6027, no specific chemical limits are given.

AWS classification	*Chemical composition* Mn	Si	Ni	Cr	Mo	V
E7018, E7027	1.6	0.75	0.3	0.2	0.3	0.08
E7014, E7015						
E7016, E7024	1.25	0.9	0.3	0.2	0.3	0.08
E7028, E7048						

Note: For obtaining above chemical composition dc en should be used.

The total of all elements for E7018, E7027 shall not exceed 1.75 except for silicon and in the case of other six electrodes it shall not exceed 1.5 except for silicon.

Apparently, ISO 2560 and the various national standards based on it have put forward a universal coding system, in which all possible electrodes could fit. The AWS standard has, on the other hand, considered the types which are in general industrial usage in the U.S.A. and then brought out a system to fit them.

AWS A5.1 has provided description of electrode classification in the Appendix. Following are the extracts:

E6010–high cellulose sodium

E6010 electrodes are characterised by a deeply penetrating, forceful, spray type arc and readily removable, thin friable slag, which may not seem to completely cover the deposit. Fillet welds are usually relatively flat in profile and have a rather coarse, unevenly spaced ripples. The coverings are high in cellulose, usually exceeding 30% by weight. The other materials generally used in the covering include titanium dioxide, metallic deoxidisers such as ferromanganese, various types of magnesium or aluminium silicates, and liquid sodium silicate as a binder. These electrodes are recommended for all-position work, particularly on multiple pass applications in the vertical and overhead positions and where weld of radiographic soundness are required. These electrodes have been designed for use with direct current, reverse polarity. The maximum amperage that can generally be used with the larger sizes of these electrodes is limited in comparison to that for other classification due to the high spatter loss that occurs with high amperage.

E6011–high cellulose potassium

E6011 electrodes are designed to duplicate the usability characteristics and mechanical properties of the E6010 classification, using AC. Although also usable with DC, reverse polarity, a slight decrease in penetration will be noted when compared to the E6010 electrodes. Penetration, arc action, slag, and fillet weld appearance are similar to those of the E6010 electrodes. The coverings are also high in cellulose content and are designed as the high-cellulose potassium type. In addition to the other ingredients normally found in E6010 coverings, small quantities of calcium and potassium compounds are usually present. High amperage results in high spatter loss.

E6012–high titania sodium

E6012 electrodes are characterised by medium penetration and dense slag which completely covers the bead. The coverings are high in rutile content, usually exceeding 35% by weight. The coverings generally also contain small amounts of cellulose and ferromanganese, and various siliceous materials such as feldspar and clay with sodium silicate as a binder. Also, small amounts of certain calcium compounds may be used to produce satisfactory arc characteristics on direct current, straight polarity. Fillet welds tend to be convex in profile with a smooth, even ripple in the horizontal position, and a widely spaced convex ripple in the vertical position, which becomes smoother and more uniform as the size of the weld is increased. The E6012 electrodes are all-position electrodes. Their ease of handling, good fillet weld profile, and ability to bridge gaps under conditions of poor fitup and to withstand high amperages make them very suited to this type of work. Weld metal from these electrodes is generally lower in ductility and may be high in yield strength.

E6013–high titania potassium

E6013 electrodes, although very similar to the E6012 electrodes, have distinct differences. Their slag system promotes better slag removal and a smoother arc transfer than E6012 electrodes. E6013 electrodes were designed specifically for light sheet-metal work. However, the larger diameters are used on many of the same applications as E6012 electrodes and provide similar penetration. Coverings of E6013 electrodes contain rutile, cellulose, ferro-manganese, potassium silicate as a binder, and other siliceous materials. The potassium compounds permit the electrodes to operate with alternating current at low amperage and low open-circuit voltages. E6013 electrodes are all-position electrodes and are similar to the E6012 electrodes in operating characteristics and bead appearance. The arc action tends to be quieter and the bead surface smoother with a finer ripple. In addition, the weld metal is definitely freer of slag and oxide inclusions than E6012 weld metal and gives better radiographic soundness.

E7014–iron powder, titania

E7014 electrode coverings are similar to those of E6012 and E6013 electrodes, but with the addition of iron powder for obtaining higher deposition rates. The covering thickness and the amount of iron powder in it are less than for E7024 electrodes. The iron powder also permits the use of higher amperage than are used for E6012 and E6013 electrodes. The amount and character of the slag permit E7014 electrodes to be used in all positions. Typical weld beads are smooth with fine ripples. Penetration is approximately the same as that obtained with E6012 electrodes which is advantageous when welding over gaps due to poor fit-up. The profile of fillet-welds tends to be flat to slightly convex. The slag is easily removed. In many cases it removes itself.

E7015–low-hydrogen sodium

E7015 electrodes are low-hydrogen electrodes to be used with direct current, reverse polarity. Their slag is chemically basic. E7015 electrodes are commonly used for making small welds on heavy sections, since they are less susceptible to cracking. They are also used for welding high sulphur and enameling steels. The arc of E7015 electrodes is moderately penetrating. The slag is heavy, friable, and easy to remove. The weld beads are convex, although fillet welds may be flat. E7015 electrodes are used in all positions up to 4 mm size. Larger electrodes are used for groove welds in the flat position and fillet welds in the horizontal and flat positions. Amperage for E7015 electrodes are higher than those used with E6010 electrodes of the same diameter. The shortest possible arc should be maintained for best results with E7015 electrodes. This reduces the risk of porosity. The necessity for preheat is reduced; therefore, better welding conditions are provided.

E7016–low-hydrogen potassium

E7016 electrodes have all the characteristics of E7015 electrodes plus the ability to operate on AC. The core wire and coverings are very similar to those of E7015, except for the use of a potassium silicate binder or other potassium salts in the coverings to facilitate their use with AC. Most of the preceeding discussion of E7015 electrodes applies equally well to the E7016 electrodes.

E7018–low-hydrogen potassium, iron powder

E7018 electrode coverings are similar to E7015 coverings except for the addition of a high percentage of iron powder. The coverings on these electrodes are slightly thicker than those of the E7015 and E7016 electrodes. The iron powder in the coverings usually amounts to between 25 and 40% of the covering weight. E7018 low-hydrogen electrodes can be used with either AC or DC, reverse polarity. They are designed for the same applications as the E7015 electrodes. As is common with all low-hydrogen electrodes, a short arc should be maintained at all times. In addition to their use on carbon steel, the E7018 electrodes are also used for dissimilar joints involving highstrength, high carbon, or alloy steels. The fillet welds made in the horizontal and flat positions are slightly convex in profile, with a smooth and finely rippled surface. The electrodes are characterised by a smooth, quiet arc, very low spatter, adequate penetration, and can be used at high travel speeds. Electrodes identified as E7018-1 have the same usability and design characteristics as E7018 electrodes, except that their manganese content is set at the high end of the range. They are intended for use in situations requiring a lower transition temperature than is normally available from E7018 electrodes when used out of position or with high-heat input.

E7048–low-hydrogen potassium, iron powder

Electrodes of the E7048 classification have the same usability, composition, and design characteristics as E7018 electrodes, except that E7048 electrodes are specifically designed for exceptionally good vertical-down welding.

E6020-E6022–high iron oxide

E6020 electrodes have a high iron oxide covering. They produce flat or slightly concave, horizontal fillet and groove welds with either AC or DC, straight polarity. They are characterised by a spray type arc and a heavy slag, well honeycombed on the underside, which completely covers the deposit and can be readily removed. Medium penetration will be obtained with normal amperages. However, these electrodes are capable of operating at high amperages and in that case will penetrate deeply. The E6020 electrodes are generally considered better than all other classifications for deep penetration fillet welds. E6020 electrodes contain manganese compounds and silica in their covering, along with large amounts of iron oxide and sufficient deoxidisers. The slag coverage is so extensive and the slag-metal reaction of such a nature that the electrodes do not normally depend on gaseous protection. Fillet welds tend to have a flat or concave profile and a smooth, even ripple. In many cases the surface of the deposit is dimpled. E6020 electrodes are recommended for horizontal fillet and flat welds, where radiographic soundness is important. Radiographic quality welds can be obtained even with high deposition rates in heavy plates. These electrodes are not usually used on thin sections, because of the higher amperages that are generally used. Electrodes of the E6022 classification are recommended for single pass, high-speed, high current flat and horizontal lap and fillet welds in sheet metal. The weld bead profile tends to be more convex and less uniform, especially since the welding speeds are higher.

E7024–iron powder, titania

E7024 electrode coverings contain large amounts of iron powder in combination with ingredients similar to those used in E6012 and E6013 electrodes. The coverings on E7024 electrodes are very heavy and usually amount to about 50% of the weight of the electrode. The E7024 electrodes are well suited for making fillet welds. The welds are slightly convex to flat in profile, with a very smooth surface and an extremely fine ripple. These electrodes are characterised by a smooth, quiet arc, very low spatter, and low penetration. They can be used with high travel speeds. Electrodes of this classification can be operated on AC or DC, either polarity.

E6027–high iron oxide, iron powder

E6027 electrode coverings contain large amounts of iron powder in combination with ingredients similar to those found in E6020 electrodes. The coverings on E 6027 electrodes are also very heavy and usually amount to about 50% of the weight of the electrode. The E6027 electrodes are designed for fillet or groove welds in the flat position with AC or DC, either polarity, and will produce flat or slightly concave, horizontal fillets with either AC or DC, straight polarity. E6027 electrodes have a spray-type arc. They will operate at high travel speeds. Penetration is medium and spatter loss is very low. They produce a heavy slag, which is honeycombed on the underside. The slag is friable and easy to remove. Welds produced with E6027 electrodes have a flat to slightly concave profile with a smooth, fine, even ripple and good wash up the sides of the joint. The weld metal may be slightly inferior in radiographic soundness to that from E6020 electrodes. High amperages can be used, since a considerable portion of the electrical energy passing through the electrode is used to melt the covering and the iron powder it contains. These electrodes are well suited for fairly heavy sections.

E7027–high iron oxide, iron powder

E7027 electrodes have the same usability and design characteristics as E6027 electrodes, except that they are intended for use in situations requiring slightly higher tensile and yield strengths than are obtained with E6027 electrodes. In other respects, all previous discussion for E6027 electrodes also apply to E7027 electrodes.

E7028–low-hydrogen potassium, iron powder

E7028 electrodes are very much like the E7018 electrodes. They differ as follows: the slag system of E7028 electrodes is similar to that of E7016 electrodes, rather than E7018 electrodes. E7028 electrodes are suitable for horizontal fillet and flat welding only, whereas E7018 electrodes are suitable for all positions. The E7028 electrode coverings are much thicker. They make up approximately 50% of the weight of the electrodes. The iron content of E7028 electrodes is higher (approximately 50% of the weight of the coverings). Consequently, on horizontal fillet and flat position welds, E7028 electrodes give a higher deposition rate than the E7018 electrodes for any given size of electrode.

4.5.6 Testing of Electrodes

All electrode standards describe in great detail the procedures for executing all-weld tensile and impact test. Some of them also describe methods for determining weld deposition efficiency and hydrogen in the weld deposit. The tensile strength, yield strength and elongation values obtained in the tensile test, and the values obtained in the other two tests provide the symbols for the coding of an electrode.

All-weld metal means weld deposit which is not diluted by the base metal. In all-weld tensile and impact tests, the test specimens are so prepared that the area which is subjected to test is pure, undiluted weld metal.

The electrode standards also prescribe supplementary tests which are not related to the code symbols, but are meant to evaluate the performance of an electrode and its suitability for welding certain grades of steel. These tests in various combinations are used for the quality control of production batches and their acceptance by consumers as indicated in the standards. The various tests included (✓) in each standard are indicated in Table 4.16.

Table 4.16. Standard tests for electrodes

Type of test	*ISO 2560*	*BS 639*	*DIN 1913*	*IS:814/815*	*AWS A5.1*
All-weld tensile and impact	✓	✓	✓	✓	✓
Transverse bend	×	✓	✓	✓	✓
Transverse tensile	×	×	✓	✓	✓
Deposition efficiency	×[b]	✓	×	✓	×
Diffusible hydrogen	×[c]	✓	×[e]	✓	×[d]
Chemical composition of weld metal	×	×	×	×	✓
Weld soundness test (radiography)	×	×	×	×	✓
Fillet weld	×	×	×	×	✓
Deep penetration	×	×	×	✓	×

While IS : 815 deals with classification and coding, IS : 814 covers specification and testing. Hence the tests are distributed among them.

ISO 2401 describes this test.

ISO 3690 describes the method.

AWS describes coating moisture test as a substitute for diffusible hydrogen test.

DIN 8572 describes the method.

■ QUESTIONS

4.1 What do you mean by shielded metal arc welding? Briefly discuss its principle of operation, currents (d.c. and a.c.) used. Covered electrodes used. What is arc blow? How can it be minimised.

4.2 What do you mean by weld-bead geometry? On a sketch of a weld-cross-section show weld width, reinforcement height, depth of penetration. How do you calculate percentage weld-metal?

4.3 How the welding arc, molten droplets and newly deposited weld bead is protected from the oxygen and nitrogen present in the open air atmosphere? How weld-metal composition is controlled.

4.4 Briefly discuss the electrode flux covering ingredients and their functions. What do you mean by hydrogen controlled electrodes?

4.5 What are the internationally recognised types of electrode flux covering. How cellulosic coverings differ from rutile in their behaviour and in applications. What are the basic ingredients of Iron-oxide and basic low hydrogen electrodes, list their special applications?

4.6 What is coating factor? What factors affect electrode selection ? Briefly discuss the International Standards Organisation System of coding of mild and low-alloy steel electrodes. How does it differ from Indian standard system.

4.7 Discuss AWS Specification for carbon steel covered electrodes. Why is it very commonly used system throughout the world?

Chapter 5

Thermal And Metallurgical Considerations in Welding

A welding engineer needs the knowledge of welding metallurgy in order to control :

– the chemistry and soundness of weldmetal.

– the micro-structure of the weldmetal and heat-affected-zones (HAZs).

Metallurgy consists of two parts:

– Process metallurgy (*e.g.*) convertion of ore to metals, refining and alloying, shaping through casting, forging and rolling etc.).

– Physical metallurgy (deals with heat-treatment, testing, metallographic studies related to design and application).

Welding involves both:

– Process metallurgy-electrode covering and SAW fluxes formulation.

– Physical metallurgy–control of cooling rates and controlling the microstructure of weldmetal and HAZs (through welding heat input control and pre-and post-heating). The ultimate aim is to obtain the desired mechanical properties.

5.1 GENERAL METALLURGY

5.1.1 Structure of Metals

The pattern of solidification of metals is shown in Fig. 5.1. As the liquid metal cools and solidification temperature approaches initial crystals are formed. The crystals then grow into large solid grains. At the end of solidification the large solid grains meet each other at grain boundaries. Each grain has a crystalline structure with the atoms in the crystals arranged in a specific geometric pattern (F.C.C., B.C.C., HCP. Fig. 5.2). The orientation of grain lattice in each grain is different as each grain has developed independently. This orderly arrangement is disrupted at the grain boundaries, and has its repercussions on the metal properties. This applies to pure metals. Metals are commonly used in the industries as alloys (in combination with other metals or non metals).

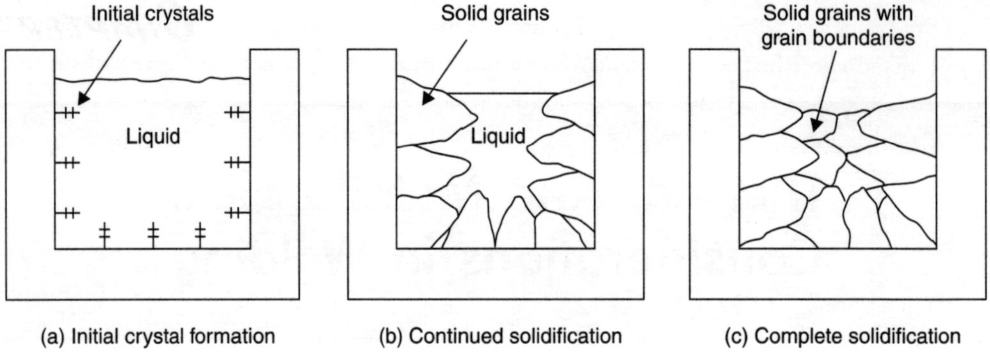

Fig. 5.1 Pattern of solidification of metals

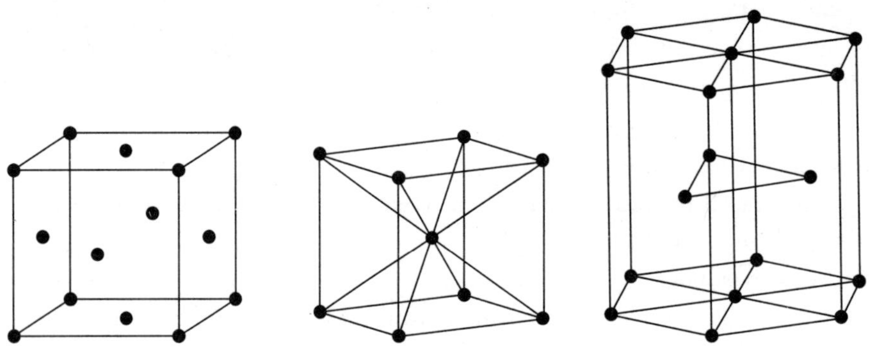

Fig. 5.2 The three most common crystal structures in metals and alloys. Left: face centred cubic (FCC) Centre: Body centred cubic (BCC) and right: hexagonal close packed (HCP).

Alloying elements dissolve in parent metal as follows:

(*a*) Substitutional solid solution in which alloying atom replaces the parent metal atom in the lattice (Fig. 5.3 (*b*)). This occurs when the solute and solvent atoms are similar in size and chemical behaviour.

(*b*) Interstitial solid solution in which alloying atom places itself in the space between the parant metal atoms without displacing any of them. See Fig. 5.3 (*a*). Example of this is carbon in iron (mild steel).

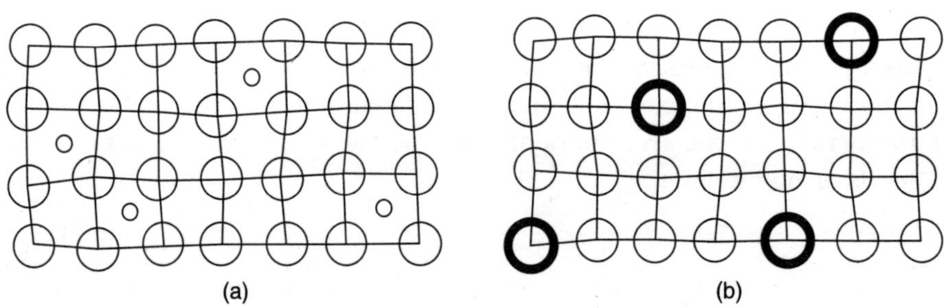

Fig. 5.3 Solution. Left: interstitial alloying; Right: Substitutional solid solution

(*c*) **Multiphase alloys.** In many alloys, several alloying elements are used which do not completely dissolve either way. They produce multiphase alloys in which several phases having their own crystalline structure exist side-by-side.

A suitably polished and etched specimen of an alloy when observed under a microscope at high magnification shows grains, grain boundaries and phases in the microstructure. This microstructure depends upon the alloy chemistry and its thermal history.

(*d*) **Grain boundaries.** Since the atomic arrangement here is in disarray, the interatomic space may be larger than normal, movement of individual atoms of elements, through the solvent structure may occur resulting in a phenomenon called **segregation.**

(*e*) **Grain size.** The grain boundaries also **resist deformation of individual grains,** thus **improving the strength** of an alloy at normal temperatures. At elevated temperatures the atoms at the grain boundaries slide **more easily.** Thus, for better strength at lower temperatures coarse-grained structures are desireable. Metals could be coarse-grained or fine-grained depending upon the solidification rate. Grain-size control is more important in the case of weld-metal.

5.1.2 Phase Tranformation

Multiphases can coexist in an alloy as discussed earlier. Phase change occurs on melting. In some metals phase change occurs in solid state due to heating or cooling–called allotropic transformation. Iron, titanium, zirconium and cobalt show allotropic transformation.

5.1.3 Iron Carbon Phase Diagram

Iron-carbon phase diagram is shown in Fig. 5.4. Steel undergoes definite internal changes when subjected to temperatures above its critical range. If the steel cools naturally from this temperature it returns to its normal condition similar to that found after normalizing. Time needs to be allowed during cooling cycle so that the internal changes that occurred during heating have time to reverse.

- If the time needed to modify the internal changes is not allowed, the properties of steel change on cooling.
- Critical points are designated as Ac_1 , Ac_2 and Ac_3 for heating and Ar_3 Ar_2 and Ar_1 for cooling. These letters were taken from French language.

  A = Arrent (stop).

  C = Chauffage (heating)

  r = Refroidissment (cooling)

Thus, Ac_1 = stop heating at the number 1 critical point

Ar_1 = stop cooling at the lower critical point.

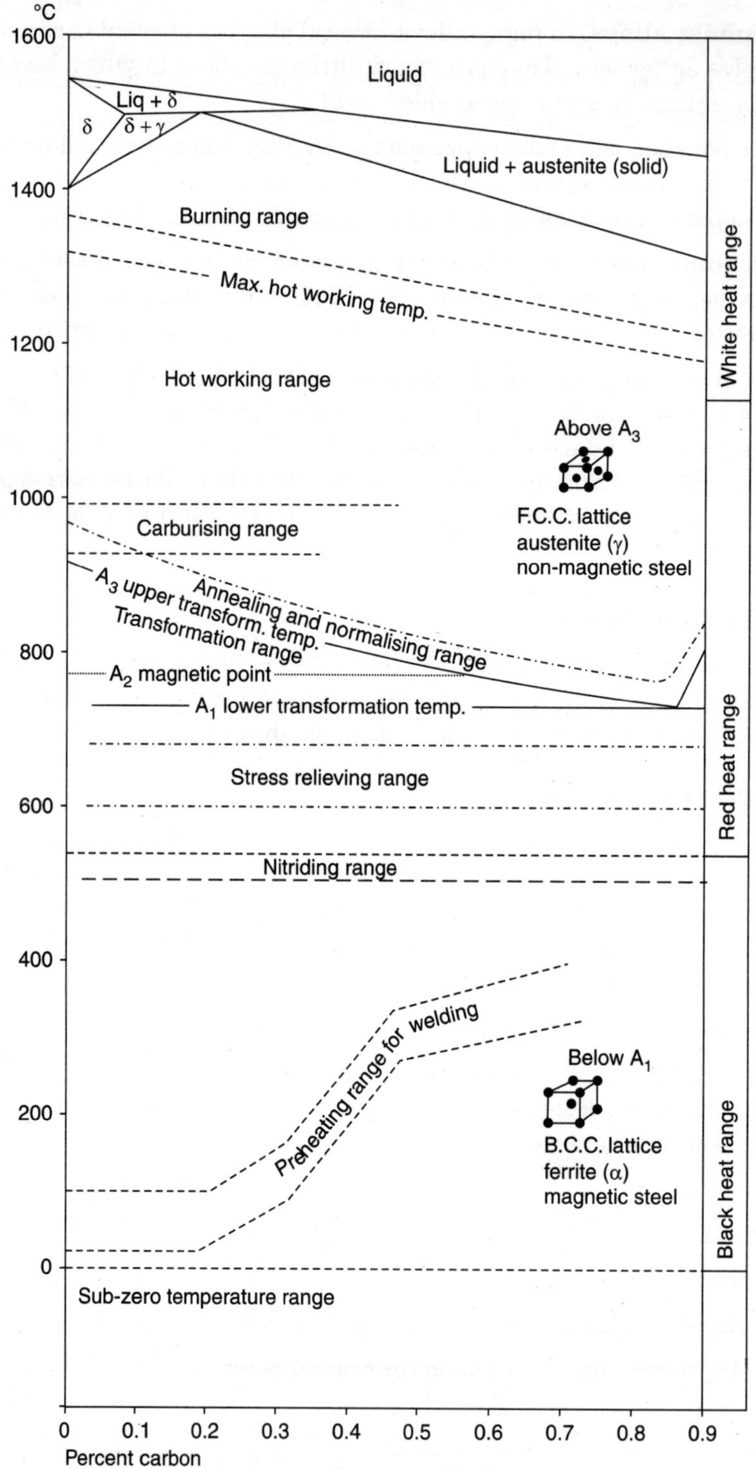

Fig. 5.4 Iron-carbon phase diagram

5.1.4 Critical Range

If a piece of SAE-1030 steel is heated its colour will change though the temper colours up into red range becoming more and more brighter as the temperature increases. At 723°C, the colour will remain constant for a short time even though the heat is being supplied. Upto this point the metal will expand at a uniform rate proportionate to the temperature.

At the Ac_1 point the expansion stops and the material begins ro shrink until to Ac_3 point (813°C) is reached. At this point on, the material will start expanding again to its normal expansion rate.

When the steel is heated to or beyond Ac_3 point it becomes nonmagnetic. The critical point Ac_3 falls as the carbon content increases.

5.1.5 Micro-structural Changes

When SAE 1030 steel is examined under a microscope, it is found to contain mostly ferrite and cementite (alternate layers). Cementite is one of the iron carbides, a hard chemical compound of iron and carbon. When this steel is heated, no change is seen upto Ac_1 temperature. At this temperature, ferrite begins to act as a solvent in which all the carbide goes into solution in the solid condition. This solution is, therefore, called solid solution. This combination is known as austenite.

When steel from Ac_3 temperature is cooled rapidly (quenched), the austinite changes to martensite, the hardest and most brittle iron. This happens because no time has been allowed for the austenite to change back to ferrite and cementite.

5.1.6 Carbon Steels

Table 5.1 shows the weldability of different types of plain carbon steels.

Table 5.1 weldability of steel

Name	*Carborn Content %*	*Application*	*Weldability*
Ingot Iron	0.03 (max)	Deep drawing sheet and strip	Excellent
Low carbon steel	0.15 (max)	Welding electrodes special plates and shapes, sheet, strip	Excellent
Mild Steel	0.15 – 0.30	Structure shapes, plates and bars	Excellent
Medium carbon steel	0.3 – 0.50	Machinery parts	Fair (pre-heat and post heat freq. reqd.)
High carbon steel	0.5 – 1.00	Springs, dies, railroad rails	Poor (pre-heat and post heat necessary)

5.1.7 Low Alloy Steels

These steels contains usually less than 0.25% carbon and frequently less than 0.15% carbon. Ni, Cr, Mn and Si are added to increase strength at room and elevated temperatures, to improve notch toughness at lower temperatures, to improve their corrosion resistance and response to heat treatment. These additions, sometimes reduce their weldability. Proper choice of filler metal and welding procedures will develop comparable properties in welded joints in these steels. Some of these steels can give upto 690 MPa (100,000 psi) yield strength and still retain better notch toughness than ordinary Plain carbon steels.

These steels find their applications in high temperature service in welded structures such as boilers, oil refinery towers, and chemical processing plants.

5.1.8 High Alloy Steels

- These are high quality expensive steels with outstanding mechanical properties, corrosion and oxidation resistance and elevated temperature strength and ductility. They are used in dies, punches and shears.
- Most of the high alloy steels are stainless steels *i.e.,* they resist attack by many corrosive media at atmospheric or elevated temperatures. They contain at least 12% Cr and many have substantial amount of nickel. Other elements are added to impart special properties. There are three basic types of stainless steels: austenitic, ferritic and martensitic. Some of these steels are precipitation hardenable.
- The martensitic stainless steels contain the smallest amount of chromium and they can be quite hardenable. They need special care during welding since martensite tends to be produced in the HAZ and be very hard. Preheating and post heat treatment are necessary to prevent cracking.
- The ferritic stainless steels contain 12–27% Cr and no austenite–forming elements. The ferrite phase is present upto the melting temperature of these steels and the steels develop little or no austenite upon heating. They are essentially non-hardenable.
- Austenitic stainless steels contain elements that stabilize the austenite at all temperatures and thus eliminate the austenite–to–ferrite or–martensite transformation. Nickel is frequently used to achieve this objective. As these alloys do not undergo austenite–ferrite transformation, they cannot be hardened by heat-treatment. Thus, there are no hardened areas in the HAZ of welds produced. These steels, therefore, have excellent weldability. Carbon contributes to elevated temperature strength but it reduces corrosion resistance by forming a chemical compound with chromium.

5.1.9 Isothermal Transformation and Time Temperature Transformation Diagrams.

Iron–carbon equilibrium diagrams, as discussed before, do not give information regarding the transformation of austenite to any structure other than equilibrium structures. It also does not give details on cooling rates required to produce other structures. A more practical diagram in this regard is the Time–Temperature–Transformation (T.T.T.) Diagram. It graphically shows the cooling rates required for the transformation of astenite to pearlite, bainite or martensite and the temperatures at which such changes take place are also given as shown in Fig 5.5 for 0.8 percent plain carbon steel (every composition of steel has its own TTT diagram).

To produce this diagram samples of 0.8% carbon steel were heated to austenitizing temperature (845°C) and then placed in environments in which they could abruptly fall to a series of temperatures starting from 705°C to room temperature. This could be done by plunging the samples into various solutions of brine, oil or water at the desired temperature and then holding each specimen for a specified length of time. After this time that specimen will be cooled quickly and examined under a microscope.

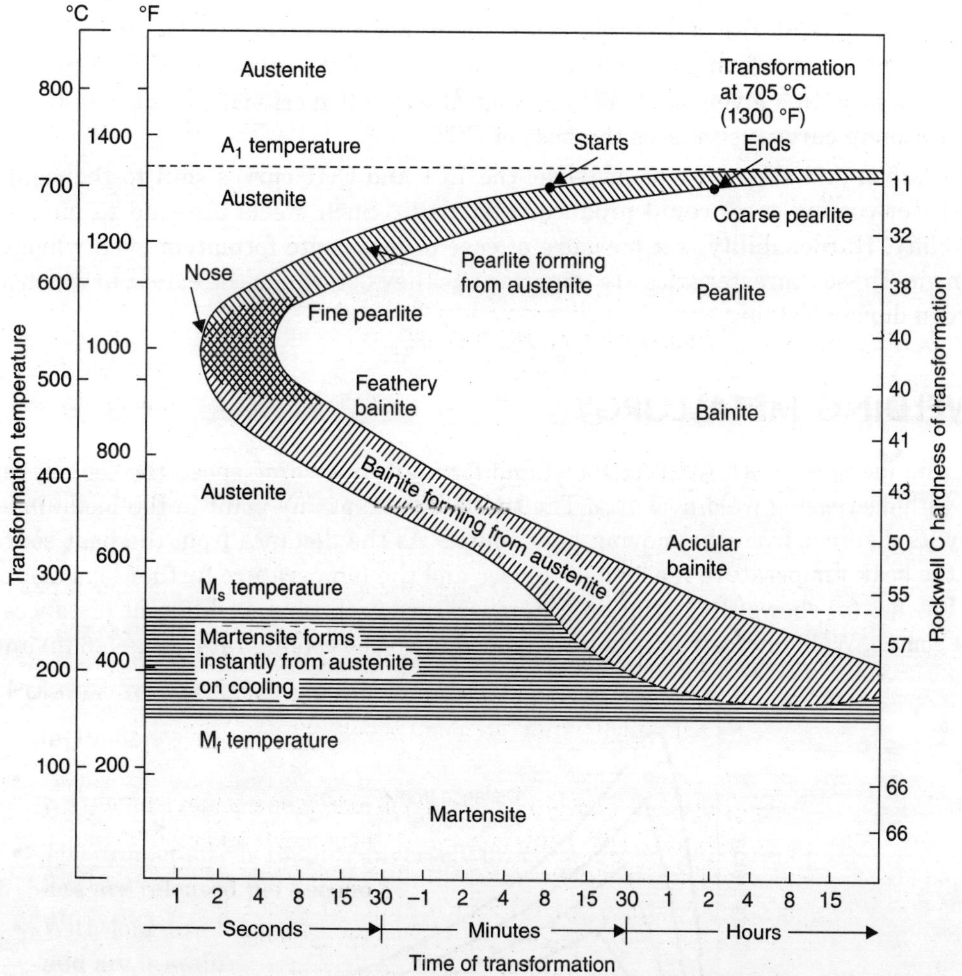

Fig. 5.5. The TTT diagram for the transformation of austenite in a euctectoid (0.8% carbon) plain carbon steel.
M_s = Martensite start temperature
M_f = Martensite finish temperature

The sample held at 705°C did not begin to transform for about 8 minutes and did not finish transfoming untill about 60 minutes are elapsed. The structure formed was coarse pearlite and tho eample was fairly soft (hardness Rc 15).

The transformation was quicker for the specimens held at 565°C. It started in one second and completed in 5 seconds. Transformation took the shortest length of time at this temperature and, therefore, the nose of the curve is located at 565°C (for 0.8%C plain carbon steel). The microstructure obtained is fine pearlite (hardness Rc 41). As temperature decreased further, the transformation start time again increased and structure was bainite. The specimens cooled to room temperature rapidly enough just to miss the nose of the curve had an entirely different microstructure (martensite). Martensite forms by a transformation which occurs only on cooling. It starts at 230°C and completes at 120°C for 0.8% C steel.

In case the cooling is not isothermal but continuous, these curves do not apply. Therefore, continuous cooling transformation (CCT) diagrams have also been developed for steels. These diagrams give information about the slowest cooling rates which will allow 100% martensite to form in a given steel. This cooling rate is called critical cooling rate the rate at which the cooling curve just misses the nose of CCT.

As carbon and alloy content increase, the TTT and CCT curves shift to the right, This means slower cooling rates could produce martensite. Such steels are said to have higher hardenability. Hardenability is a measure of ease of matensite formation even when cooled slowly in air. These characteristics are important as they determine the extent to which a steel will harden during welding.

5.2 WELDING METALLURGY

Cooling rate increases with welding speed and for a given welding speed the cooling rate increases with decreasing weld-pool size. The thermal cycle at any point in the medium is governed by its distance from the moving heat source. As the distance from the heat source increases the peak temperature reached decreases and the temperature further lags behind the source. Fig. 5.6 (*a*) shows the variation of temperature with time at different distances from the heat source. Weld microstructures will depend upon the cooling rates [Fig. 5.6 (*b*) and (*c*)].

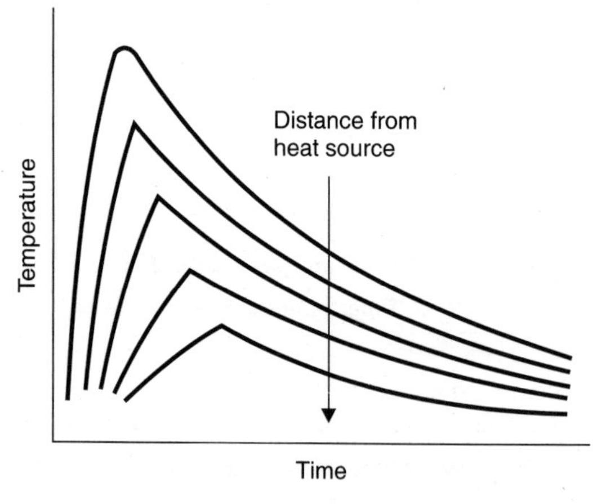

Fig. 5.6 (a) Temperature variation with time at various distances from heat source

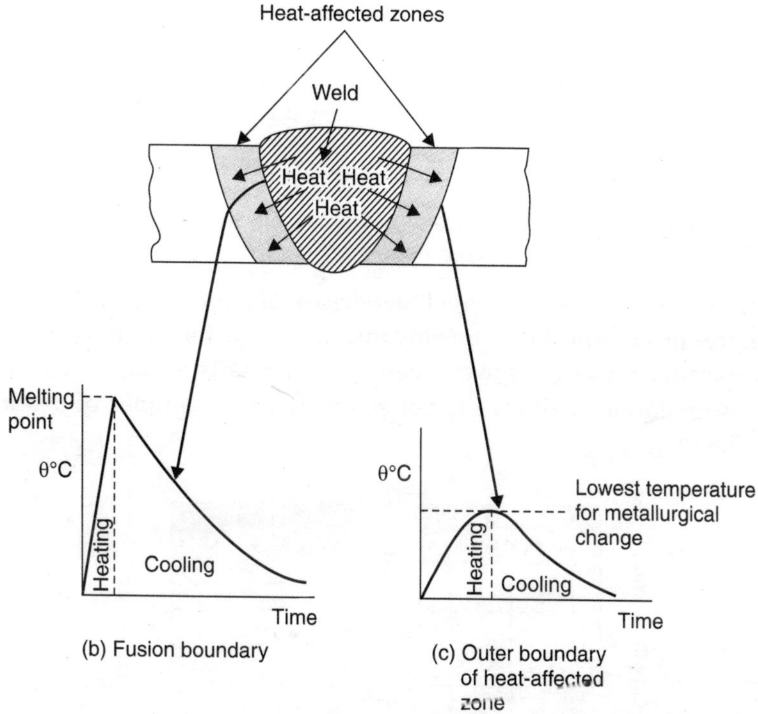

Fig. 5.6 Variation of temperature with time at different distances from the heat source (*b*) fusion boundary (*c*) outer boundary of HAZ

5.2.1 Weld-Metal and Solidification

Welded joints contain a melted zone, which on solidification comparises the weld-metal. It is composed of varying mixtures of filler metal and base metal melted in the process. Its chemical composition can be tailored by the composition of the filler metal used but its micro-structure and the attendent mechanical properties are a direct result of the sequence of events that occur just before and during the period of solidification. These events include gas metal reactions in the vicinity of the weld, reactions with non-metallic liquid phases (slag or flux) during welding and solid-state reactions occuring in the weld after solidification. Let us first consider solidification.

Solidification. In arc-welding the molten weld pool is contained in a surrounding solid metal. Thus a liquid-solid interface, present at the fusion boundary provides an ideal nucleation site (heterogeneous nucleation). There is no homogeneous nucleation and thus the supercooling is negligible. Since the heat flow in welding is highly directional towards the cold metal, hence the weld acquires a columnar structure having long grains parallel to the direction of heat flow (Fig. 5.7).

In the case of pear-shaped growth shown on the right, the columnar grains growing from apposite sides meet at the middle of the weld. This midplane solidifies last and often contains impurities and porosity. It is prone to fracture at low strains. This defect is called ingotism and can be corrected by adjusting the joint gap configuration and weld procedure.

There is a unique dependence by the dendrite arm spacing on energy input. The more rapid the solidification, the more closely spaced are the dendrites.

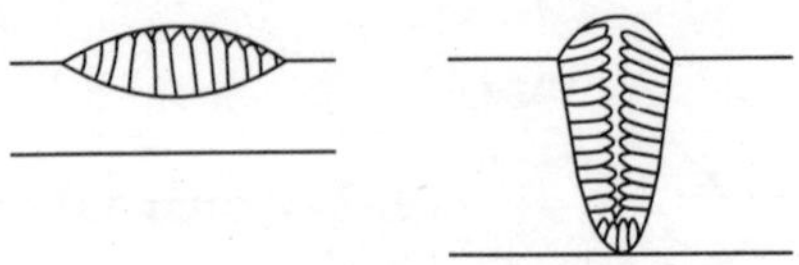

Fig. 5.7 Columnar structure of welds Left: Shallow weld; Right: Deep pear-shaped weld.

When solidification is extremely rapid, dendrites do not develop fully, under these conditions a much shorter projection of the freezing interface into the liquid weldpool occurs which is called a cell structure. Spacing between cells are normally smaller than those between dendrites and the segregation of solutes is not so extensive. Examples of dendrites and cells are shown in Fig. 5.8.

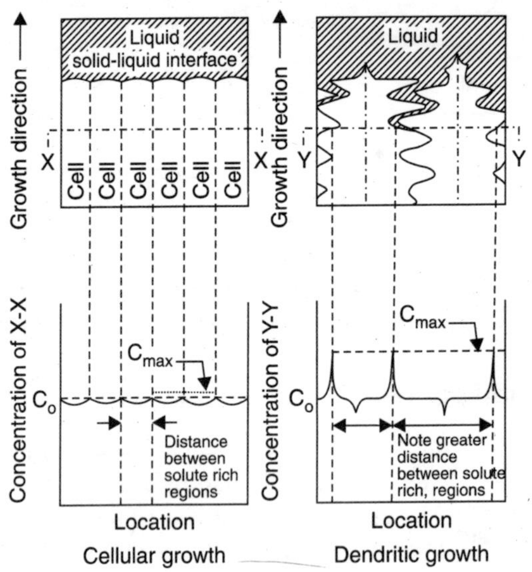

Fig. 5.8 Schematic of solute distribution for cellular and dendritic growth patterns.

5.2.2 Gas-Metal reaction

The absorption of gas from the arc or flame into the weld-pool causes gas-metal reaction (since both the metal and the gas are at higher temperatures). There are two types of such reactions. In the first type the gas may be just dissolved in the liquid metal. In the second type, the gas and liquid metal may chemically react to form stable chemical compounds. In case this chemical compound is soluble it may cause embrittlement of the welded joint.

An insoluble reaction product may produce surface scale or slags and thus physically interferes with the formation of the weld pool. In this case the excess gas is either prevented or a flux is used to dissolve or disperse the reaction product.

When the gas is dissolved in the liquid weld pool, the gas evolves during cooling as its solubility decreases with fall of temperature. Gas bubles are formed. If these bubles are trapped, the weld becomes porous and of low quality. This defect is common in metals whose oxides are easily reducible by hydrogen, and can be avoided by the addition of a suitable deoxidant in the filler metal.

Another important gas-metal reaction is the diffusion of the gas into the parent metal from the weld pool. When the temperature of the thermal cycle is high, this diffusion process may be quite fast. The diffusion of hydrogen into the HAZ may again cause an embrittlement of the welded joint.

5.2.3 Liquid-Metal Reactions

During welding, non-metallic liquid phases are produced that interact with the weld metal. These may be slag layers formed by the melting of flux in SMAW, SAW, etc. They may also be produced as a result of reactions occuring in the molten weld-pool and remain in or on top of the weld metal after welding.

The flux layers used in SMAW or SAW etc. processes are designed to absorb deoxidation products produced in the arc and molten metal. They usually float to the surface of the weldpool forming part of the slag, but some may remain in the metal as inclusions.

Another important effect of liquid solid interaction is hot cracking, which occurs during solidification. The interdendritic liquid, the last region to freeze, has a substantially lower freezing temperature than the bulk dendrite. The shrinkage stresses produced during solidification act upon this small liquid region and produce interdendritic cracks. These cracks occur at temperatures close to bulk solidification temperature, therefore, they are called hot cracks.

5.2.4 Solid State Reactions

Among the solid state reactions, the most important phenomenon is the formation of cold cracks or delayed cracks. This type of cracking is confined to steels that can be hardened. These steel contain a hard phase called martensite.

The cracks occur after the weld completely cools down, sometimes hours after or even weeks after welding. This is always associated with the presence of hydrogen in the weld metal.

At high temperature the steel is F.C.C. austenite, a form in which hydrogen is quite soluble. On cooling the austenite changes to pearlite or martensite, and there is drastic reduction of hydrogen solubility. In plain carbon steels this transformation takes place at a relatively high temperature (about 700°C), even if cooling is rapid, there is sufficient mobility so that much of the rejected hydrogen diffuses out of the metal. Moreover the transformation product (forrite plus carbide) formed in the HAZ are relatively ductile and crack resistant.

A rapidly cooled hardenable steel transfoms at a much lower temperature (generally below 400°C) and often room temperature, so the hydrogen is locked into the structure which may also be hard and brittle. It is this combination that induces cracking. This has led to the development of low hydrogen electrodes. These electrodes have to be protected from moisture.

5.2.5 Macro and Microstructure of Weld, Heat–Affected Zone (HAZ) and Parent Metal

The metallurgical changes that takes place in weld and HAZ significantly affect the weld quality. The wide variety of changes that may take place depend on various factors, *e.g.*,

(*a*) the nature of the material (*i.e.* single-phase, two-phase)

(*b*) the nature of the prior heat-treatment

(*c*) the nature of the prior cold working

We now consider typical examples of these changes.

Let us consider the fusion welding of two pieces of a single-phase material, which have been cold worked to yield a desired orientation. These cold worked grains result in a high strength and low ductility. However, on fusion welding, a random grain growth again takes place within the melt boundary, which, in turn, results in a low strength. Within the heat affected zone, the grains become coarse due to heat input (annealing), and a partial recrystallization also occurs. In either case, the strength falls much below that of the parent material. With increasing distance from the melt boundary, the grains become finer until the heat unaffected zone with elongated grains is reached. All these changes are shown in Fig. 5.9.

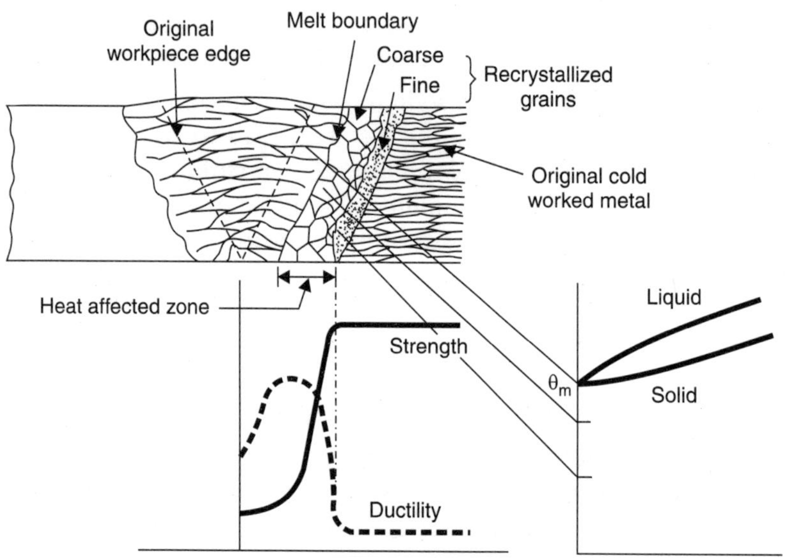

Fig. 5.9 Characteristics of welded joints in pure metals.

Let us now consider a two-phase material which derives its strength mostly from precipitation hardening. In this case, the strength within the melt boundary is again too low. But, in the immediately adjacent heat affected zone, the thermal cycle results in heating and quenching followed by further aging. This aging process recovers some of the strength. The material beyond this zone is only overaged due to the heat of welding and becomes harder with the loss of strength. Hence, the strength and ductility variation near the joint are as shown in Fig. 5.10.

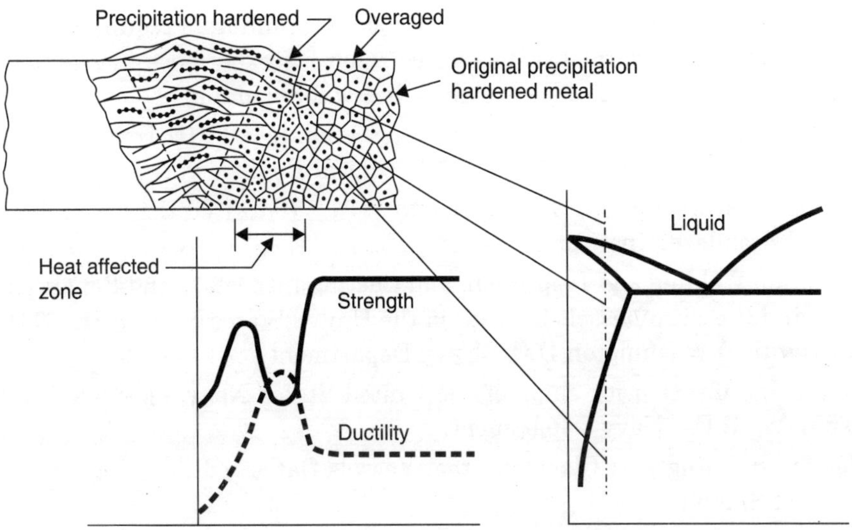

Fig. 5.10 Characteristics of welded joints in precipitation hardened alloy

The two examples we have considered clearly demonstrate that various types of metallurgical changes are possible during welding, particularly for complex alloys. These changes are governed by the non-equilibrium metallurgy of such alloys, and must be clearly understood to yield a satisfactory fusion weld. Also, a decision on the postwelding heat treatment to be given, must be taken to restore the desirable characteristics of the joint.

5.3 THERMAL AND MECHANICAL TREATMENT OF WELDS

Various thermal and mechanical treatments are often performed on welds to reduce the residual stresses and distortion. They include preheat, postweld thermal treatments, peening, and so forth. These treatments also change the metallurgical properties of weldments.

5.3.1 Reasons for Treatment

- To restore the base properties affected by the welding heat.
- To modify weld-deposit properties.
- To relieve stresses and produce desired micro-structure in base material, HAZ and weld metal.
- The extent of harm the weld has caused determines the subsequent treatment.
- Improve weldability (for example preheat improves weldability).
- To reduce "metallurgical notch" effect resulting from abrupt changes in hardness etc.
- To improve resistance to crack propagation.

5.3.2 Code Requirements

Some welded constructions are required to be in accordance with the recommendations of a code such as the ASME Boilers and Pressure Vessels Code, thermal treatments are specified for certain types of weldments. These recommendations are based upon the existing evidence necessitating the thermal treatment. These are codes for minimum requirements. The fabricator should employ other treatments also based upon his experience in addition to the code requirements. Some important codes are given below for example :

1. ASME Boiler and Pressure Vessels Code, Section I, III, VIII Divs. 1 and 2 (latest edition). New Yorlk: American Society of mecanical Engineers.
2. Code for Pressure Piping, Ansi B 31.1 to B 31.8 (latest edition) New York: American National Standards Institute.
3. Fabrication Welding and Inspection, and Casting Inspection and Repair for Machinery, Piping and Pressure Vessels in Ships of the United States Navy, MIL–STD–278 (Ships) (latest edition) Washington D.C. : Navy Department.
4. General Specification for ships of the United States Navy, spec. 59-1 (latest edition) Washington D.C. : Navy Department.
5. Rules for Building and Classing Steel Vessels (latest edition) New York : American Bureau of Shipping.
6. Structure Welding Code AWS D 1.1 (latest edition as revised). Miami : American Welding Society.
7. United States Coast Guard Marine Engineering Regulations and Materials, spec. CG - 115 (latest edition). Washington D.C. : United States Coast Guard.

As these documents are constantly revised, the latest available versions should be obtained and followed.

5.3.3 Common Thermal Treatments

Preheat. Preheat temperatures may be as low as 26°C for out door welding in winter to 650°C when welding ductile cast iron and 315°C when welding highly hardenable steels. In many situations the temperature of preheat must be carefully controlled. The best way is to heat the part in a furnace and held at the desired temperature.

- Preheating is very effective means of reducing weld metal and base metal cracking. **It retards the cooling rates and reduces the magnitude of shrinkage stresses.**
- Also the thermal conductivity reduces as temperature increases (for iron thermal conductivity at 595c is 50% of its value at room temperature). This also reduces the cooling rate resulting in favourable metallurgical structure, HAZ also remains at the transformation temperature for a longer period of time permitting the formation of ferrite and pearlite or bainite instead of martensite.
- When an area being welded is under severe restraint, localized preheat may increase the amount of shrinking and cause cracking.

Thus preheat must be used with caution, since detrimental effects may result under certain conditions.

Electrical strip heaters are commonly used on site for preheating.

These must be properly insulated to avoid danger of shock to welders.

Induction heating, using **60 Hz (or 50 Hz)** transformers of suitable capacities built for this purpose, is a common method of preheating pipe joints for welding.

5.3.4 Postweld Thermal Treatment

- Stress relief heat-treatment is defined as the uniform heating of a structure to a suitable temperature, holding at this temperature for a predetermined period of time, followed by uniform cooling (uneven cooling may result in additional stresses).
- Stress relief heat treatment is usually performed below the critical range so as not to affect the metallurgical structure of the work.
- The percentage relief of internal stress depends upon the type of steel (its yield strength). The effects of varying time and temperature are shown in Fig. 5.11.

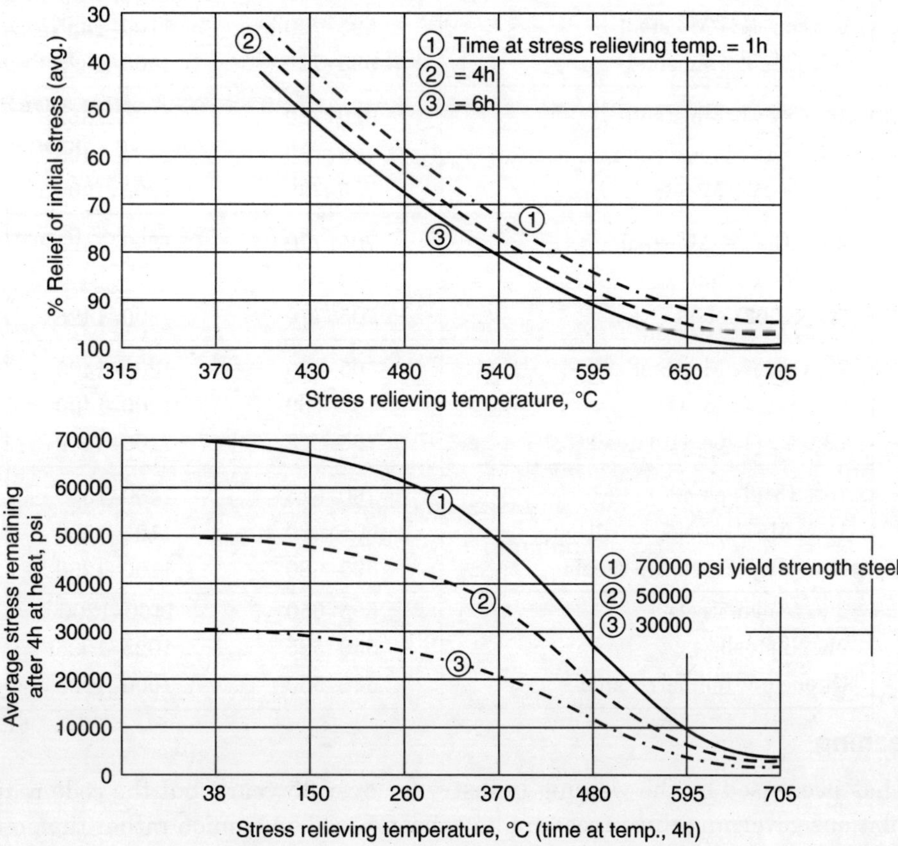

Fig. 5.11. Effect of temperature and time or stress-relief

- The temperature reached is more effective than the time at that temperature in stress relieving. Temperatures closer to recrystallisation temperature are more effective.
- Microstructure, tensile and impact strength values are affected by stress relief treatment. Temperature for stress relief should be so chosen as to develop or retain the desirable properties while at the same time provide the maximum stress relief (Table 5.2).
- Controlled low temperature stress relief treatment could be done when the structures are big enough to be stress relieved in a furnace. The material on either side of

the weld bead is heated to 175°-205°C while the weld itself is relatively cool. This causes thermal expansion in the base metal and a reciprocal tensile stress in the weld beyond the yield. When the metal cools and contracts, the stress falls below the yield. When the process is used properly a partial reduction in the longitudinal stresses of butt welds is achieved.

Table 5.2. Typical thermal treatments for weldments

Material	*Soaking temperature*	
	°C	*°F*
Carbon steel	595–680	1100–1250
Carbon–½% Mo steel	595–720	1100–1325
½% Cr–½% Mo steel	595–720	1100–1325
1% Cr–½% Mo steel	620–730	1150–1350
1¼% Cr–½% Mo steel	705–760	1300–1400
2% Cr–½% Mo steel	705–760	1300–1400
2¼% Cr–1% Mo steel	705–770	1300–1425
5% Cr–½% Mo (Type 502) steel	705–770	1300–1425
7% Cr–½% Mo steel	705–760	1300–1400
9%Cr–1% Mo steel	705–760	1300–1400
12% Cr (Type 410) steel	760–815	1400–1500
16% Cr (Type 430) steel	760–815	1400–1500
1¼% Mn–½% No	605–680	1125–1200
Low-alloy Cr-Ni-Mo steels	595–680	1100–1250
2 to 5% Ni steels	595–650	1100–1200.
9% Ni steels	550–585	1025–1085
Quench & tempered steels	540–550	1000–1025

5.3.5 Peening

Peening has been used by the welding industry for over 35 years, but the code requirements and regulations governing this procedure have been based on opinion rather than on scientific data because there has been no practical method for measuring the effect of peening.

Various specifications and codes require that the first and last layers of a weld should not be peened.

The results of laboratory tests conducted by American Bureau of Shipping and explosion tests by the Naval Research Laboratory confirm the requirement prohibiting the peening of the first and the last layers.

In conducting peening, the following special precautions may be necessary:

(1) Work hardening should be considered when certain AISI 300 series steels are involved.

(2) Hot shortness may preclude hot peening of certain bronze alloys.

(3) AISI 400 series steels have relatively poor notch ductility in the as-welded condition. Utmost care should be exercised if peening is attempted.

(4) The relative elongation values for ductility of welds and metals should be considered before employing the peening process.

Peening equipment should be selected with care The hammer, pneumatic tools, and so forth should be sufficiently heavy for striking force to be effective without producing excessive work hardening, but not so heavy as to involve bending moments or produce cracks in the weld.

5.4 RESIDUAL STRESS AND DISTORTION IN WELDS

As the weldment is locally heated, the weldmetal and HAZ adjacent to it are at a temperature substantially above that of the unaffected base metal. As the molten pool solidifies and shrinks it causes shrinkage stresses on the surrounding weld metal and HAZ area. In the beginning, the contraction the weld metal applies is small, the metal is hot and weak. As it solidifies, the weld metal applies increasing stresses on the weld area, the base metal may yield.

The sequence of thermal events in welding is far from simple and is not easily amenable to mathematical analysis. It is possible to describe qualitatively the contraction of a weld and to ascribe to the different stages empirical data established by observations made over a period of many years.

5.4.1 Thermal Expansion and Contraction

To understand residual stresses and distortion let us consider the shrinkage that occurs during welding when the source of heat has already passed. This is made up of three components or stages

(*a*) Liquid contraction (liquid to liquid)

(*b*) Solidification shrinkage (liquid to solid)

(*c*) Solid metal contraction (solid to solid)

From Fig. 5.12 we can see that as the solification front proceeds to the weld centre line, the solid metal occupies a smaller space than the liquid metal it replaces (*i.e.,* its density increases). The molten metal also contracts.

- The surface of weld pool should recede below the original level (formation of weld crater at the end of the weld bead, when the heat source is suddenly removed). However, at the same time further molten metal from the leading edge of the weldpool is fed into the area, the actual shrinkage is thus **not** shown up.

5.4.2 Contraction of Solid Metal

Contraction of weld metal is volumetric. It could be estimated along the length and across it.

Longitudinal contraction is given by

$$l_1 = l_0\,(1 - \alpha\,\Delta\,\theta) = l_0 - l_0\,\alpha\,\Delta\,\theta$$

where l_0 = original length, α = coefficient of linear expansion = 14.3×10^{-6}/°C

l_1 = length after cooling through temperature change $\Delta\theta$

For 1 meter length of weld, the shrinkage along length

$$l_0\,\alpha\,\Delta\,\theta = 1000 \text{ mm} \times 14.3 \times 10^{-6}/°\text{C} \times (1500 - 20)°\text{C}$$
$$= 1000 \times 14.3 \times 10^{-6} \times 1480 \text{ mm}$$
$$= 21.2 \text{ mm/meter length}$$

The value 21 .2 is based on α which does not remain constant over the range of temperature, but it indicates that the contraction is appreciable.

In practice, the measured contraction is significantly less.

- The practical observation shows 1 mm/m. This is because of the restraint provided by the adjoining cold plates.
- When the weld metal tries to contract, its contraction is restrained, so it is plastically deformed.
- Tensile forces ultimately set-up in the weld region and corresponding compressive forces are set in the plate by reaction (Fig. 5.13).
- If the cold plates are perfectly rigid, the welded joint will be of the same length as the original plates. The compressive stresses are of considerable magnitude exceeding the yield stress of the parent plate. The result is that the plates get deformed so reducing the overall length of the joint and thus resulting in 1 mm/meter contraction shrinkage quoted above. A compressive force of about 150–170 N/mm^2 is required to achieve a compressive strain of about 1 mm/meter.

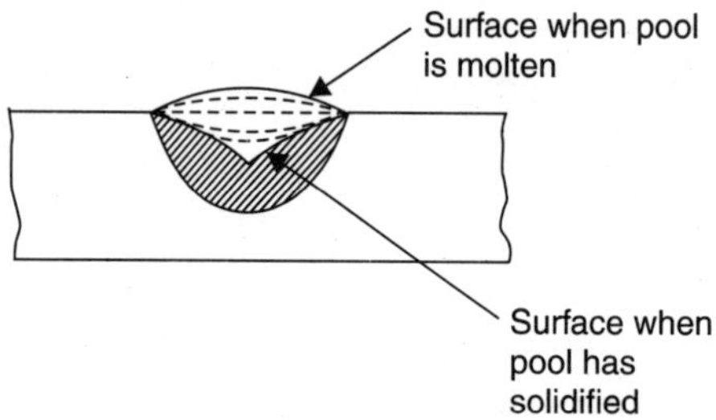

Fig. 5.12 Shrinkage during solidification

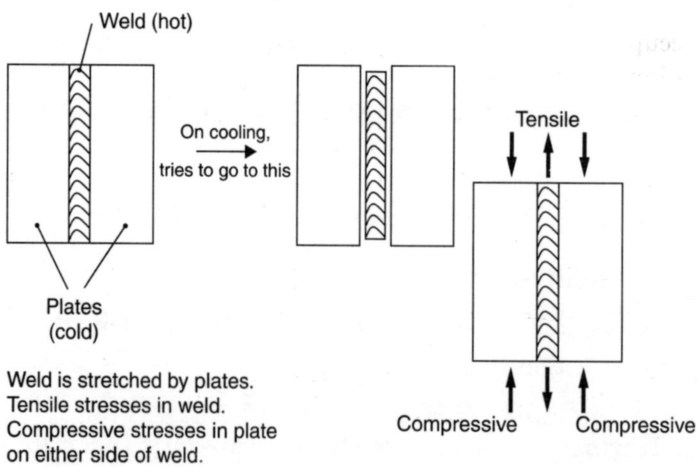

Fig. 5.13 Deformation of a weld metal element during cooling.

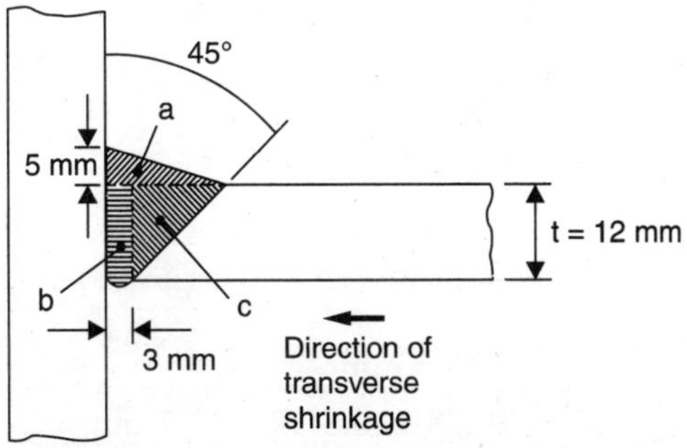

Fig. 5.14 Estimation of transverse shrinkage in '*T*' butt joint

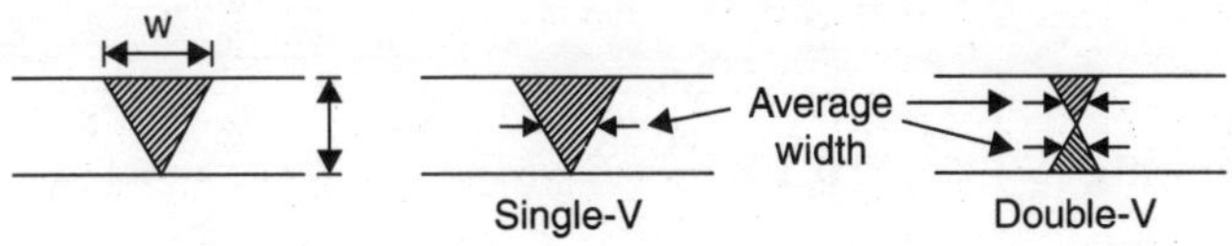

Fig. 5.15 Transverse shrinkage in '*V*' butt welds.

5.4.3 Transvers Shrinkage

Similar conditions apply when look at shrinkage to the weld, where the contracting weld metal tries to pull the plates towards the centre-line of the joint and as a result the whole joint area is in transverse tension. Again we have a situation where, because the hot weld metal has a lower yield stress than the cold plates, deformation first takes place in the weld but, at a later stage of cooling, as the relative yield stresses become more equal, some yielding of the parent material occurs and the overall width of the welded plates is reduced.

Strictly, the amount of transverse shrinkage which takes place depends on the total volume of weld metal, but' as a general rule, for a given plate thickness, the overall reduction in width transverse to the joint at any point is related directly to the cross-sectional area of the wéld. Similarly, as we would expect, the total shrinkage increases with the thickness of the plate, since the weld area is greater. It is possible to state this relationship in a general way:

$$\text{transverse shrinkage} = k\,\frac{A}{t}$$

where k = an empirical factor with a value between 0.1 and 1.17

A = cross-sectional area of weld

t = thickness of plate

This formula can be used to predict the shrinkage that will occur in a butt joint (Fig. 5.14) and has been found to give good correlation with practical observations. In the case of a single-V butt joint the calculation can be simplified, since the ratio A/t is equal to the average width and the formula is reduced to

Transverse shrinkage = k × average width of weld

It should be noted that for a double-V weld the average width is not zero but is the value for one of the *V*'s.

Estimation of Transverse shrinkage in a 'T' butt joint (Fig. 5.14)

$$\text{Transverse shrinkage} = 0.1 \times \frac{A}{t}$$

$$A = a + b + c$$

$$= \frac{1}{2} \times 5 \times (12 + 3) + (3 \times 12) + 1/2 \times 12 \times 12)$$

$$= 145.5 \text{ mm}^2$$

Transverse shrinkage $= 0.1 \times 145.5/12 = 1.21$ mm.

Estimation of Transverse shrinkage in 'V' butt welds, (Fig. 5.15).

Area of weld, $a = \frac{1}{2} \times w \times t$

Transverse shrinking $= 0.1 \times \frac{A}{t}$

$$= 0.1 \times \frac{\frac{1}{2} \times w \times t}{t}$$

$$= 0.1 \times w/2$$

$$= 0.1 \times \text{average width.}$$

5.4.4 Angular Distortion and Longitudinal Bowing

Taking both longitudinal and transverse shrinkage, based on what has been said above the final shape of two plates welded together with a butt joint should be as shown in Fig. 5.6 (*a*). In practice, however, such a simple treatment does not apply, principally because the shrinkage is not distributed uniformly about the neutral axis of the plate and the weld cools progressively, not all at one time.

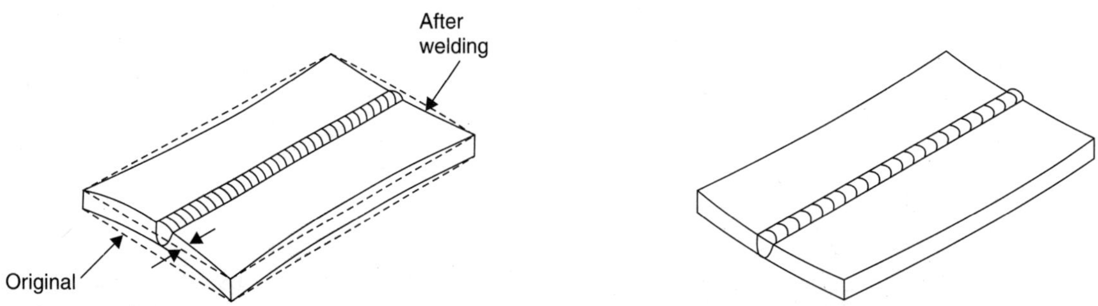

(a) Changes in shape resulting from shrinkage which is uniform throughout the thickness

(b) Asymmetrical shrinkage tends to produce distortion.

Fig. 5.16 Change in shape and dimensions in butt-welded plate.

If we look at a butt made with a 60° included-angle preparation, it is immediately apparent that the weld width at the top of the joint is appreciably greater than at the root.

Since the shrinkage is proportional to the length of metal cooling, there is a greater contraction at the top of the weld. If the plates are free to move, as they mostly are in fabricating operations, they will rotate with respect to each other. This movement is known as angular distortion (Fig. 5.16 *b*) and poses problems for the fabricator since the plates and joint must be flattened if the finished product is to be acceptable. Attempts must be made, therefore, to reduce the amount of angular distortion to a minimum. Clamps can be used to restrain the movement of the plates or sheets making up the joint, but this is frequently not possible and attention has to be turned to devising a suitable weld procedure which aims to balance the amount of shrinkage about the neutral axis. In general, two approaches can be used: weld both sides of the joint or use an edge preparation which gives a more uniform width of weld through the thickness of the plate (Fig. 5.17).

In the direction of welding, asymmetrical shrinkage shows up as longitudinal bowing Fig. 5.18. This is a cumulative effect which builds up as the heating-and-cooling cycle progresses along the joint, and some control can be achieved by welding short lengths on a planned or random distribution basis, Fig. 5.19. Welding both sides of the joint corrects some of the bowing, but can occasionally be accompanied by local buckling.

Angular distortion and longitudinal bowing are observed in joints made with fillet welds (Figs. 5.20 and 5.21), Angular distortion is readily seen, in this case as a reduction of the angle

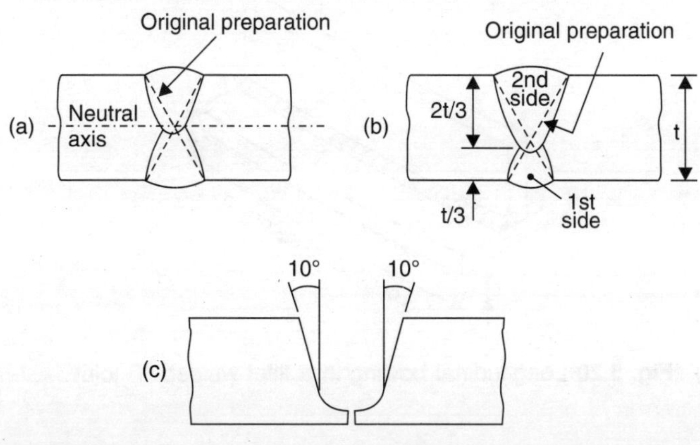

Fig. 5.17 Edge preparation designed to reduce angular distortion

(*a*) Double-*V* joints balance the shrinkage so that more or less equal amounts of contraction occur on each side of the neutral axis. This gives less angular distortion than a single 'V'.

(*b*) It is difficult to get a completely flat joint with a symmetrical double 'V' as the first weld run always produces more angular rotation than subsequent runs; hence an asymmetrical preparation is used so that the larger amount of weld metal on the second side pulls back the distortion which occurred when the first side was welded.

(*c*) Alternatively, a single-*U* preparation with nearly parallel sides can be used. This gives an approach to a uniform weld width through the section.

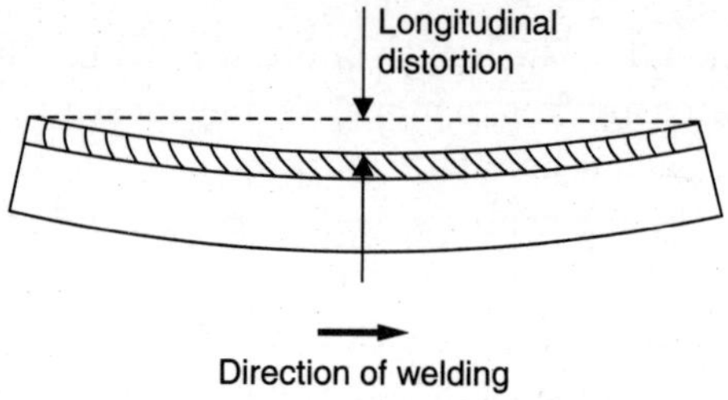

Fig. 5.18 Longitudinal bowing or distortion in a butt joint

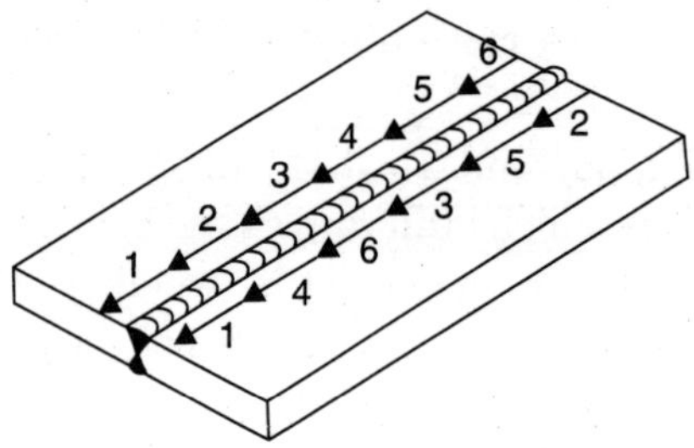

Fig. 5.19 Sequences for welding short lengths of joint to reduce longitudinal bowing

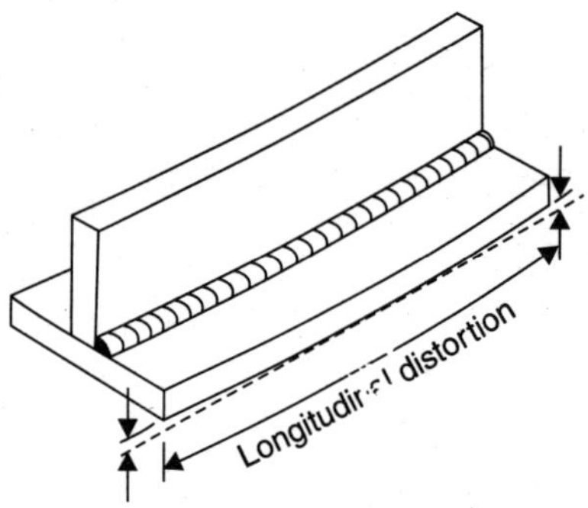

Fig. 5.20 Longitudinal bowing in a fillet-welded 'T' joint

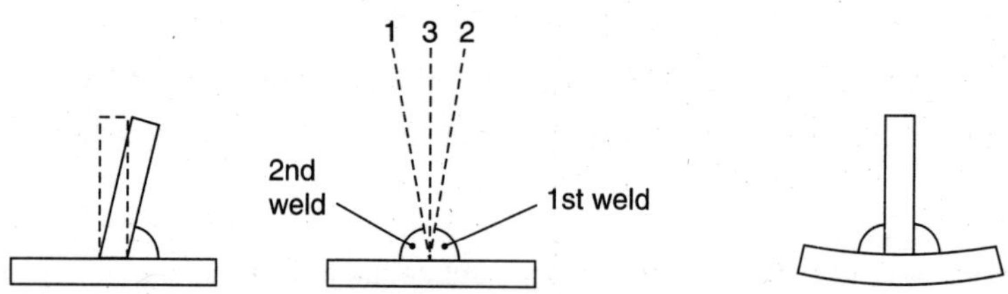

(a) Distortion caused by fillet weld

(b) Use of presetting to correct distortion in fillet welded 'T' joint

1 = plate centre-line before welding

2 = plate centre-line after first weld

3 = plate centre-line after second weld

(c) Distortion of flange

Fig. 5.21 Distortion in fillet welding of 'T' joints

between, the plates and is greatest for the first weld. Although the second weld, placed on the other side of the joint, tends to pull the web plate back into line, the amount of angular rotation will be smaller. With experience, the joint can be set up with the web plate arranged so that the first angle is greater than 90° and thus ends up with the web and flage at right angles. Even so, warping in the flage plate cannot be ignored.

5.4.5 Effect of Heat Distribution

Finally, in our consideration of shrinkage and distortion we must not ignore the importance of heat input. As we have seen in Chapter 2 and 3, the heat from the weld pool during solidification flows into the plate adjacent to the fusion boundary. The width of metal heated to above room temperature is greater than that of the fused zone, and the picture used above of a hot weld-metal element between cold plates is an over-simplification. The heat flowing into the plates establishes a temperature gradient which falls from the melting point at the fusion boundary to ambient temperature at some point remote from the weld.

The heated-band width is directly proportional to the heat input in joules per mm length of weld and is therefore dependent on the process being used. It follows that the amount of distortion and shrinkage will also vary from one welding process to another. If the heat source moves slowly along the joint, the heat spreads into the plate and the width of hot metal which must contract is greater. The effect is less noticeable in thick plate but in sheet material, say 2 mm thick, the differences are most marked. The GMA system, with its fast speed of travel, gives a narrow heat band compared with the spread in oxy-acetylene welding, and it is possible to arrange the manual processes in ascending level of distortion, *i.e.*, GMA, SMAW, GTA and oxy-acetylene welding.

5.4.6 Residual Stresses

Solving the problem of distortion control during welding and determining shrinkage allowances for design purposes are of such importance in fabrication that it is easy to overlook the fact that they are the products of plastic deformation resulting from stresses induced by contraction in the joint. As long as these stresses are above the yield point of the metal at the prevailing temperature, they continue to produce permanent deformation, but in so doing they are relieved and fall to yield-stress level. They then cease to cause further distortion. But, if at this point we could release the weld from the plate by cutting along the joint line, it would shrunk further because, even when distortion has stopped, the weld still contains an elastic strain equivalent to the yield stress. We can visualise the compeleted joint as an element of weld metal being stretched elastically between two plates.

The stresses left in the joint after welding are referred to as residual stresses. From our discussion of shrinkage and distortion, it can be seen that there will be both longitudinal and transverse tension. In the case of the longitudinal stresses, the weld itself and some of the plate which has been heated are at or near yield stress level (Fig. 5.22). Moving out into the plate from the heat-affected zone, the stresses first fall to zero. Beyond this there is a region of compressive stress.

It must be emphasised that all fusion welds which have not been subjected to post-weld treatments-in other words, the vast majority of welded joints contain residual stresses. Procedures developed to minimise distortion may well alter the distribution of the residual stresses but do not eliminate them or even reduce their peak level. Having said this, since we cannot avoid the formation of residual stresses, it is appropriate to ask if we are worried by their presence. As with so many engineering situations the answer is not a simple yes or no. There are numerous applications where the existence of residual stresses would have little or no influence on the service behaviour of the joint-storage tanks, building frames, low-pressure pipework, and domestic equipment all provide examples of situations where the joints can be used in the as welded condition without detriment.

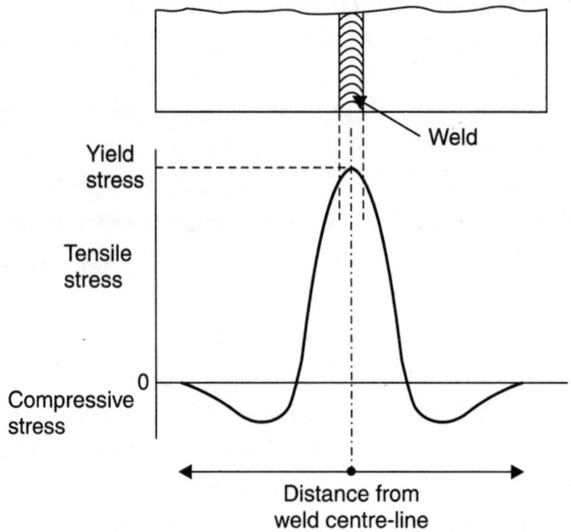

Fig. 5.22 Distribution of residual stresses in a butt-welded joint

If the service requirements do indicate that the residual stresses are undesirable, the designer must take them into account when selecting materials and deciding upon a safe working stress. This approach can be seen in the design of ships, where the combination of low temperatures and residual stress could lead to a type of failure known as brittle fracture. The designer selects a material which is not susceptible to this mode of failure even at the low temperatures which may be experienced during the working life of the ship; the presence of residual stresses is then important. Similarly, in many structures subjected to loads which fluctuate during service–for example, bridges, earth-moving equipment, and cranes–the designer recognises the existence of residual stresses by choosing a working-stress range which takes account of the role these stresses play in the formation and propagation of fatigue cracks.

There are, however, some specific applications where it is essential to reduce the level of residual stresses in the welded joint. With pressure vessels, because of the risk of a catastrophic failure by brittle fracture, stress-relieving is often a statutory or insurance requirement. Again, some metals in certain environments corrode rapidly in the presence of tensile stress, *i.e.*, stress corosion will occur. In these cases, a joint in the as welded condition containing residual stresses suffers excessive attack; this is retarded if the joint is stress-relieved. Finally, when machining welded components, removing layers of metal near the joint may disturb the balance between the tensile and compressive residual stresses and further deformation or warping can occur. This can make it difficult to hold critical machining tolerances and it may be desirable in these circumstances to stress-relieve to achieve dimensional stability.

5.4.7 Stress Relieving

Various methods are available to reduce the level of residual stresses in welded joints. Heat treatment, overloading, and vibratory treatment can all be used, but the most common method is based on a controlled heating-and-cooling cycle, *i.e.*, thermal stress relief. This technique makes use of the fact that the yield stress of a metal decreases as the temperature is raised. If a welded joint is heated to, say, 600°C, the residual tensile stress, which was equivalent to the yield stress at room temperature, is in excess of the yield stress of the metal at 600°C. Localised plastic deformation occurs, and the tensile stresses are reduced. At the same time, the compressive stresses which were in equilibrium with the tensile stresses are also reduced, to restore the equilibrium.

In stress-relieving practice, the temperature is raised until the yield stress has fallen to a low value at which residual stresses can no longer be supported. This clearly depends on the metal being treated, since the relationship between yield stress and temperature is critically influenced by alloy content, and this is reflected in the temperatures recommended in BS 5500: 1976 for the stress-relieving of fusion-welded pressure vessels (Table 5.3).

Table 5.3 Stress-relieving temperature for fusion welded pressure vessels

Type of steel	*Stress-relieving temperature (°C)*
Low-carbon	580–620
Carbon-manganese	600–650
Carbon–1/2% molybdenum	620–660
1 % chromium–1/2% molybdenum	620–660
2¼% ckromium–1% molybednum	660–700
5% chromium–1/2% molybdenum	700–740
3½% nickel	500–620

If thermal treatment is to give a satifactory reduction of residual-stress levels, it is important that differential expansion and contraction must not occur, otherwise new residual stresses will be included. The heating and cooling must be carefully controlled so that the temperature is uniform throughout the component, and special furnaces equipped with comprehensive temperature-control systems have been designed for this purpose. In these furnaces the whole of the component of fabrication is heated, thus easing the problem of avoiding temperature gradients. Localised heating for stress relief is usually not recommended, especially with joints in flat plates, since there is always the risk of creating further stresses. In this connection, pipe welding poses particular problems. Stress relieving might often be desirable to reduce corrosion problems, but it would be impracticable to heat-treat a complete pipework installation. Local stress relief of pipe joints in situ is, therefore, allowed by some authorities, provided that the temperature distribution is controlled. This is usually achieved by specifying the minimum temperature at the joint line and at some specific point remote from the weld a typical example is shown in Fig. 5.23.

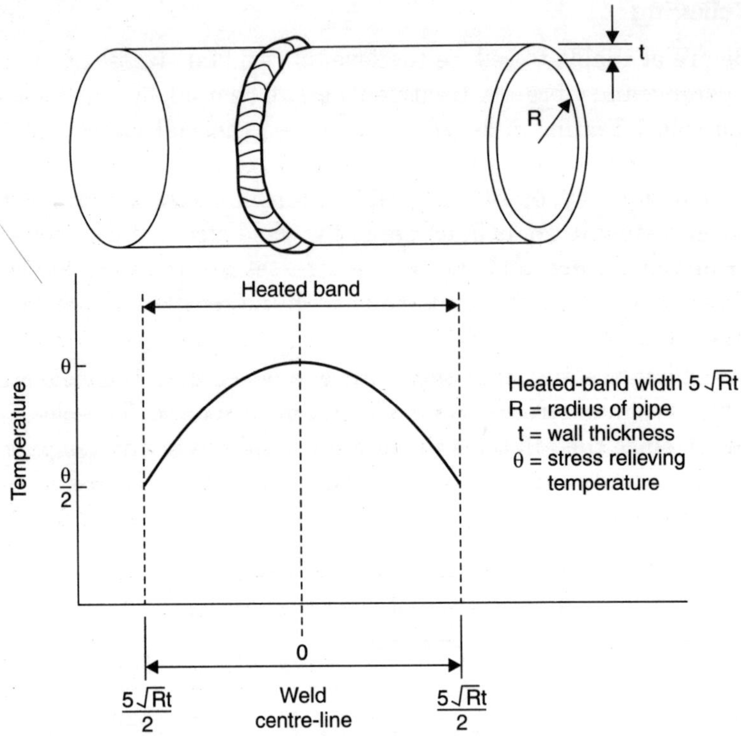

Fig. 5.23 Typical specification for temperature distribution during local stress relief of welded butt joints in pipe

■ QUESTIONS

5.1 Why a welding engineer needs a knowledge of welding? What do you mean by weldability of a metal? What factors affect weldability?

5.2 Briefly discuss the isothermal transformations, Time Temperature Transformations in steel. What is meant by welding metallurgy? Discuss solidification, phenomenon, gas-metal reactions, liquid metal reactions, solid states reactions in regard to welding.

5.3 What is HAZ in welding? Why a weld usually fails in HAZ area?

5.4 Discuss thermal and mechanical treatment of welds. Why heat treatment of welds is necessary for obtaining quality welds? What common thermal treatments are carried out on welds.

5.5 Briefly discuss the welding of 'Cast Irons', Aluminium and its alloys and welding of austenitic stainless steels.

CHAPTER **6**

Analytical and Mathematical Analysis

The amount of heat input to the weld at its rate determines the geometry of the weld bead deposited and the width of the heat affected zone. It also affects the microstructure of the weld and heat affected zone, which in tern affects the mechanical properties of the joints obtained. In the following paragraphs we shall be discussing the factors like the determination of heat input to the weld, maximum heat input rate, in fusion welding of plates and resistance welding of thin sheets.

The discussion will also include the heat flow in welding peak temperatures reached adjascent to the weld and in the HAZ, estimation of the width of HAZ and the effect of pre-heat of this width. Determination of cooling rates has also been included in the discussion as it affects the weld microstructure and consequently the mechanical properties of the welds.

The following sections provide practical working equations for consumable electrode welding applications and other weld processes. The following important quantities can be estimated using the heat flow equations :

1. Peak temperatures
2. Width of HAZ
3. Cooling rates
4. Solidification rates.

Before going into the details of the above equations, let us first concentrate on the heat input to the weld.

6.1. HEAT INPUT TO THE WELD

The heat input, Q in watts, in the case of arc welding is given by,

$$Q = VI \text{ J/S} \qquad \text{...(6.1)}$$

For the melting of the weld at the joint, the exact amount of heat that enters the joint can be calculated (for an electrode moving at a speed of S_w mm/s) using the following relation.

$$H = \frac{Q}{S_w} \text{ J/mm} \qquad \text{...(6.2)}$$

But the actual heat utilized by the joint depends upon how effectively this heat is transferred from electrode tip to the joint. Hence heat transfer efficiency factor f_1 enters the calculations of net heat available at the joint.

$$H_{net} = \frac{f_1 VI}{S_w} \text{ J/mm} \quad ...(6.3)$$

All of this net heat is not used for melting since part of it is conducted away to the base plate. The heat actually used for melting H_m can be obtained by another efficiency factor f_2

$$H_m = \frac{f_1 f_2 VI}{S_w} \quad ...(6.4)$$

where $f_2 = \dfrac{\text{Heat required to melt the joint}}{\text{Net heat suplied.}}$

Ex. 1. Calculate the melting efficiency in the case of arc welding of steel with a current of 200 A at 20 V. The travel speed is 5 mm/s, and the cross-sectional area of the joint is 20 mm^2. Heat required to melt steel may be taken as 10 J/mm^3 and heat transfer efficiency is 0.85.

Volume of base metal melted = 20 × 5 = 100 mm^3/s

Heat required for melting = 100 × 10 = 1000

$$f_2 = \frac{1000}{f_1 VI} = \frac{1000}{0.85 \times 20 \times 200} = 0.2941 = 29.41\%$$

6.2 RELATION BETWEEN WELD CROSS-SECTION AND ENERGY INPUT

There is a simple but important relationship between the weld metal cross-section. A_w, and energy input :

$$A_w = \frac{f_2 H_{net}}{Q} = \frac{f_1 f_2 H}{Q}$$

where $H_{net} = \dfrac{f_1 EI}{v}$ J/s.

Q = Heat required for melting in Joules/mm^3

E = voltage supplied in volts.

I = current consumed in Amp.

v = welding speed in mm/s

$A_w = (A_m + A_r)$ in mm^2

$A_w = A_m$ if no filler metal is added

$$A_w = \frac{f_1 f_2 EI}{vQ}$$

where $A_w = (A_m + A_r)$

f_1 = heat transfer efficiency from electrode to plate

f_2 = melting efficiency

$= \dfrac{\text{Heat required to melt the joint}}{\text{Net heat supplied}}$

H_{net} = Net heat available at the weld joint (J/s)

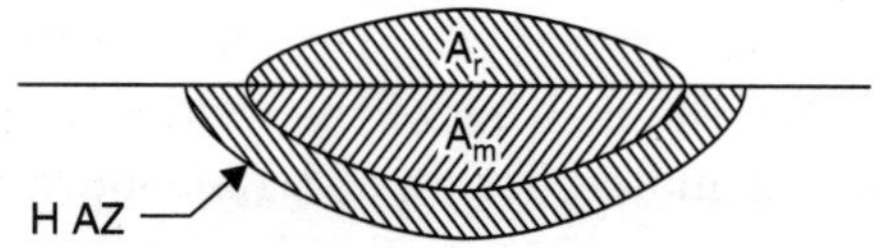

Heat source	f_1
MMA/GMA	0.8 – 0.66
SAW	0.9 – 0.99
GTAW	0.21 – 0.48

Example 1. *An arc weld pass is made on steel under the following conditions :*

E = 20 V *I = 200 Av = 5* mm/s

$f_1 = 0.9$ $f_2 = 0.3$ $Q = 10$ J/mm^3.

Estimate the cross-sectional area of the weld pass.

Solution.

$$A_w = \frac{(0.9)(0.3)(20)(200)}{(5)(10)} = \mathbf{21.6\ mm^2.\ \ Ans.}$$

6.3 THE HEAT INPUT RATE

In many situations, in practice, we are interested in determining the minimum heat input rate '*Q*' in watts required to from a weld of a given width '*w*' in a '*V*' grove as shown in the Fig. 6.1. It can be calculated* for two dimensional heat source or a three dimensional heat source using equations (6.1) and (6.2) respectively.

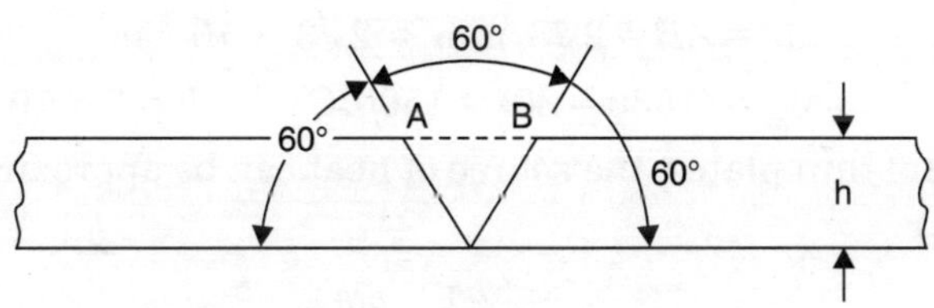

Fig. 6.1 Plate geometry for calculating the heat input rate

The following symbols are used in these equations.

α = thermal diffusivity of the work in (m²/s),

K = thermal conductivity of work material (W/m-°C)

$K_{(steel)}$ = 43.6 W/m –°C $\alpha_{steel} = 1.2 \times 10^{-5}$ m²/s = $\dfrac{K}{PC}$

P = density and C = specific heat

θ_m = M.P. of metal

w = weld width in (m)

h = plate thickness in (m)

v = welding speed (in m/s)

θ_m = MP of steel = 1530°C

θ_0 = room temperature = 30°C (assumed)

$\rho_c = 0.0044$

For two dimensional heat source

$$Q = 8\,K\quad \theta_m h\left(\frac{1}{5} + \frac{vw}{4\alpha}\right) \qquad ...(6.1)$$

and for three dimensional heat source

and $$Q = \frac{5}{4}\,\pi\,\omega\,K\,\theta_m\left(\frac{2}{5} + \frac{vw}{4\alpha}\right) \qquad ...(6.2)$$

It can be observed from these equations that "$v\omega/\alpha$" is the most important parameter Theoretical results fail to accomodate many practical difficulties *e.g.*

1. Inhomogeneous conducting medium (liquid pool + solid)
2. Absorption and rejection of the latent heat at the forward and rear edges, respectively, of the weld-pool.

Still the above two equations provide a good estimate.

In arc welding with short circuit transfer, the heat input is given by

$$Q = CVI \qquad ...(3)$$

where V = arc voltage, I = arc current and

C = fraction of total time for which the arc is on.

If the (actual) Heat input rate given by equation (3) is less than Q

($Q = (CVI) < Q_{given}$ by equations (1) or (2) a lack of side fusion occurs.

In a butt welding process using arc-welding, the arc-power was found to be 2.5 KVA. The process is used to weld 2 plates of steel 3 mm thick, with 60° V-edge preparation angle.

Determine the maximum possible welding speed. The metal transfer is short circuit type and the arc is on for 85% of the total time given.

Solution. The rate of heat input is given as

$$Q = CVI$$

$$= 0.85 \times 2.5 \times 10^3\ w = 2.12 \times 10^3\ w$$

The minimum weld width to be maintained

$$w = AB = 2\sqrt{3}\ \text{mm.} = 2\sqrt{3} \times 10^{-3}\ \text{m.}$$

$$\theta_m = (1530 - 30) = 1500°\text{C} \qquad h = 3 \times 10^{-3}\ \text{m}$$

As in the welding of thin plates, the source of heat can be approximated as a line source. Thus, using equation (1)

$$Q = 8 \times \text{K}\ \theta_m\ h\left(\frac{1}{5} + \frac{vw}{4\alpha}\right)$$

$$2.12 \times 10^3 = 8 \times 43.6 \times 1500 \times 3\left(\frac{1}{5} + \frac{vw}{4\alpha}\right) \times 10^{-3}$$

$$\left(0.2 + \frac{vw}{4\alpha}\right) = 1.35$$

$$v = \frac{1.15 \times 4\alpha}{w}$$

$$w_{min} = 2\sqrt{3} \times 10^{-3}\ m,$$

$$v = \frac{1.15 \times 4 \times 1.2 \times 10^{-5}}{2\sqrt{3} \times 10^{-3}}$$

$$= 0.0158 = 0.016\ \text{m/sec.} = 0.95\ \text{m/min.}$$

6.4 HEAT FLOW EQUATIONS—A PRACTICAL APPLICATION

An important parameter that needs to be calculated is the peak temperature reached at any point in the material during welding. The cooling rate from this peak temperature will determine the metallurgical transformations likely to take place in the *HAZ*.

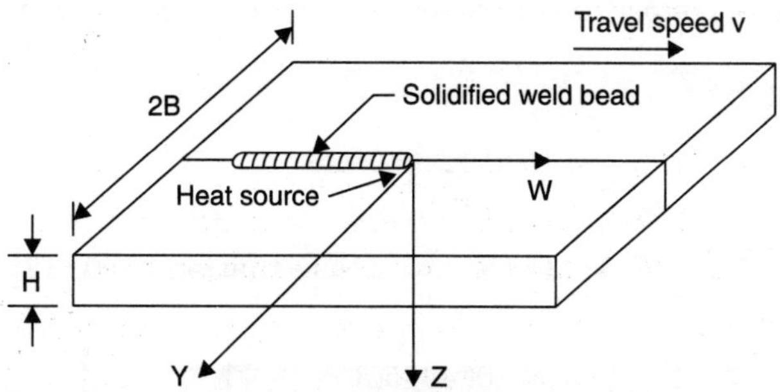

Moving co-ordinate (W, Y, Z).

Fig. 6.2

Peak temperatures can be calculated using the following equations

$$\frac{1}{(T_p - T_0)} = \frac{\sqrt{2\pi e}\ \rho cty}{H_{net}} + \frac{1}{T_m - T_0} \quad ...(1)$$

where e = base of natural logarithm = 2.71828128

Thus $\sqrt{2\pi e} = \sqrt{2 \times 3.14 \times 2.71821828} = 4.13.$

$\rho_c - 0.0044.$

Peak Temperature (T_P)

Peak temperature equation. For a single pass full penetration butt weld in sheet or plate. Peak temperature in the base metal adjacent to the weld T_P in HAZ region

$$\frac{1}{T_p - T_0} = \frac{4.13\ \rho\ CtY}{H_{net}} + \frac{1}{T_m - T_0} \quad ...(2)$$

is given by equation (1) where

T_P = the peak or max. temp. °C, at a distance Y_{mm} from the weld fusion boundary (this eq. doesnot apply for temps. within the weld metal)

t = plate thickness

T_0 = initial plate temperature °C

T_m = melting temperature of base metal

ρ_c = 0.0044

Uses of this equation

1. Determining peak temperature in specific locations in HAZ.
2. Estimating width of HAZ.
3. Effect of preheat on width of HAZ.

Example 1. *A single full penetration weld pass is made on steel using the following parameters:*

$E = 20\ V, I = 200\ A, v = 5\ mm/s, T_0 = 25°C, T_m = 1510°C$

$\rho_C = 0.0044\ J/mm^3.°C,\ \ t = 5\ mm,\ \ f_1 = 0.9\ \ \ \ H_{net} = 720\ J/mm.$

Calculate the peak temperatures at distances of 1.5 and 3.0 mm from the weld fusion boundary.

(*i*) At $Y = 1.5$ mm.

$$\frac{1}{T_p - 25} = \frac{4.13\,(0.0044)\,5(1.5)}{720} + \frac{1}{1510 - 25}$$

$T_P = \mathbf{1184°C.}$ Note that at $Y = 0$, $T_P = T_m$.

(*ii*) At $Y = 3.0$ mm

$$\frac{1}{T_p - 25} = \frac{4.13\,(.0044)\,5(3)}{720} + \frac{1}{1510 - 25}$$

$T_P = \mathbf{976°C.}$

6.5 WIDTH OF HEAT AFFECTED ZONE

For this calculation the outer extremity of the HAZ must be clearly identified with a specific peak temperature. For example for most carbon or alloy steels, there is a distinct etching boundary (as observed on polished and etched weld cross-section), corresponding to a peak temperature of 730°C.

Now the problem reduces to the determination of the distance Y_Z at which $T_P = 730°C$.

$$\frac{1}{730 - 25} = \frac{4.13\,(0.0044)\,5\,Y_Z}{720} + \frac{1}{1510 - 25}$$

$Y_z = 5.9$ mm

Thus a region 5.9 mm wide, adjacent to the fusion boundary will be structurally changed, *i.e.*, it may be affected by the heat of welding.

If the steel plate is preheated to 200°C, its effect will be to widen the HAZ width. This plate was tempered at 430°C. Any temp. above this 430°C will modify its property. Now T_P becomes 430°

$$\frac{1}{430 - 200} = \frac{4.13\,(0.0044)\,(5)\,Y_Z}{720} + \frac{1}{1510 - 200}$$

$Y_z = 28.4$ mm.

Without preheat this width would be

$$\frac{1}{(430 - 25)} = \frac{4.13\,(0.0044)\,(5)\,Y_Z}{720} + \frac{1}{1510 - 25} = \mathbf{14.2\ mm} \quad \textbf{Ans.}$$

Thus preheating has doubled the width of HAZ.

Finally if the net energy input is increased 50% to × 1.5 × 720 = 1080 J/mm

$$\frac{1}{430 - 25} = \frac{4.13\,(0.0044)\,(5)\,Y_Z}{1080} + \frac{1}{1510 - 25}$$

$Y_Z = \mathbf{21.3\ mm.}$ **Ans.**

The weld width is also increased by 50%.

6.6 COOLING RATES

Calculation and comparison of cooling rates require careful specification of conditions, because it varies with position and time. Most useful method is to determine the cooling rate on the center line of the weld at the instant the metal is passing through a particular temperature of interest, T_C. At temperatures well below melting, the cooling rate in the weld and its immediate HAZ is substantially independent of position.

In carbon and low alloy steels the temperature of interest is best taken near the pearlite "nose" temperature on the TTT diagram. The exact temperature is not critical but should be the same for all calculations and comparisons. A value of $T_C = 550$ is quite satisfactory for most steels.

For thickplates requiring several passes (more than six) to complete the joint. The cooling rate (for the first pass or each pass). R is given by :

$$R = \frac{2\pi K(T_C\ T_0)^2}{H_{net}}$$

where R = cooling rate at a point on the weld centerline, °C/s at just that moment when point is cooling past T_C.

K = Thermal conductivity of the metal J/mm-s°C.

T_C = temperature at which cooling rate is calculated

T_0 = initial plate temperature, °C.

The cooling rate is maximum at the weld centreline. The above equation gives this maximum cooling rate. At fusion boundary it is only a few percent lower. Thus this equation applies to the entire weld and the HAZ. If the plates are thin requiring fewer than four passes :

$$R = 2\pi K\rho C\left(\frac{t}{H_{net}}\right)^2 (T_C - T_0)^3 \qquad \text{...(2)}$$

where t = thickness of base metal mm

ρ = density of metal, g/mm^3

C = sp. heat of base metal, J/g. °C

The difference between thick and thin plate.

In thick plates the heat flow is three dimensional. This equation (eq. 2) applies to small boad-on-plate welds on thin plates.

Relative plate thickness factor, τ is defined as follows to distinguish between thick and thin plates.

$$\tau = h\sqrt{\frac{\rho_C(T_C - T_0)}{H_{net}}}$$

$\tau \le 0.75$ **thin** plate equation is valid

$\tau \ge 0.75$ thick plate equation is valid.

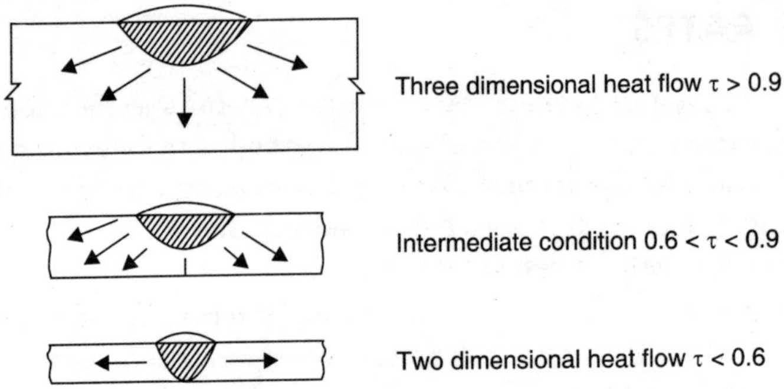

Fig. 6.3 Relative plate thickness factor τ for cooling rate calculations

Example. *Find the best welding speed to be used for the welding of 6 mm steel plates with an ambient temperature of 30°C with the welding transformer set at 25 V and current passing is 300 A. The arc efficiency is 0.9 and possible travel speeds are 6 to 9 mm/s. The limiting cooling rate for satisfactory performance is 6°C/s at a temperature of 550°C.*

Solution. Given $T_0 = 30°C$, $T_C = 550°C$, $K = 0.028$ J/mm-s-°C

$R = 6°C/s$, $V = 25$ V, $I = 300$ A, $h = 6$ mm, $f_1 = 0.9$, $\rho_C = 0.0044$ J./mm^3°C.

1. Assume a travel speed of 9 mm/s

$$\text{Heat input} = H_{net} = \frac{f_1 \, VI}{v} = \frac{0.9 \times 25 \times 300}{9} = 750 \text{ J/mm}$$

To check whether it is a thick or thin plate

$$\tau = h\sqrt{\frac{\rho_C (T_C - T_0)}{H_{net}}} = 6\sqrt{\frac{.0044\,(550 - 30)}{750}} = 0.3314$$

This being less than 0.6, it is thin plate, cooling rate will be calculated by using the thin plate equation

$$R = 2\pi\, K\rho_C \left(\frac{h}{H_{net}}\right)^2 (T_C - T_0)^3 \,.$$

$$= 2\pi \times 0.028 \times 0.0044 \left(\frac{6}{750}\right)^2 (550 - 30)^3 = 6.9659°\text{C/s}.$$

This value is higher than the critical cooling rate required, we may reduce the travel speed to 8 mm/s and recalculate the cooling rate.

This cooling rate is higher than the limiting cooling rate of 6°C/s (given) at a temperature of 550°C : We, therefore, reduce the travel speed to **8 mm/s** and recalculate :

$$v = 8 \text{ mm/s}$$

$$\text{Heat input,} \quad H_{net} = \frac{0.9 \times 25 \times 300}{8} = 843.75 \text{ J/mm}$$

To check whether it is a thick or thin plate :

$$\tau = h\sqrt{\frac{\rho_C (T_C - T_0)}{H_{net}}} = 6\sqrt{\frac{0.0044\,(550 - 30)}{843.75}} = 0.312.$$

This being less than 0.6, it is a thin plate. Using thin plate equation for cooling rate.

$$R = 2\pi K \rho_C \left(\frac{h}{H_{net}}\right)^2 (T_c - T_0)^3$$

$$= 2\pi \times 0.028 \times 0.0044 \left(\frac{6}{843.75}\right)^2 (550 - 30)^3 = 5.504°C/s.$$

This is a satisfactory cooling rate, the welding speed can be finalised at 8 mm/s.

These equations could also be used to calculate the preheat temperature required to avoid martensitic transformation in the weld zone.

6.7 CONTACT-RESISTANCE HEAT SOURCE

The electrical resistance could be used as a source of heat. It could be

(*a*) contact resistance of interfaces or

(*b*) Resistance of molten flux and slag

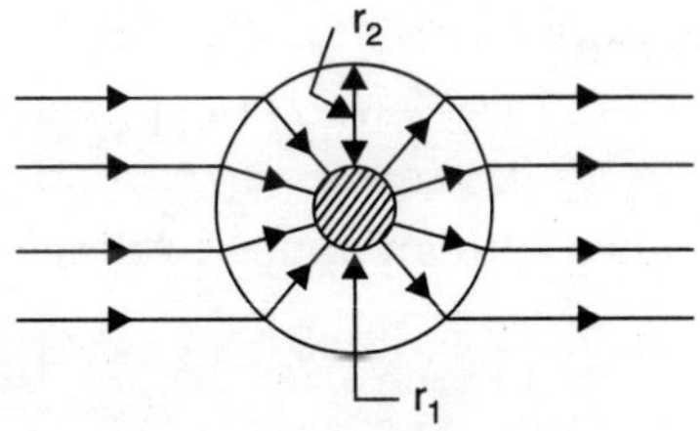

Fig. 6.4

Resistance of each hemispherical constriction

$$R = \rho(r_2 - r_1)/S$$

where ρ = resistivity of material

$(r_2 - r_1)$ = length of current path

S = geometric mean area of the two hemispheres of radii r_1 and r_2 respectively.

$$= \sqrt{(2\pi r_2^2)(2\pi r_1^2)} = 2\pi r_1 r_2$$

$$\text{Now } R = \frac{\rho(r_2 - r_1)}{2\pi r_1 r_2} = \frac{\rho}{2\pi r_1} \text{ as } r_2 >> r_1$$

Total constriction resistance R_c of n such spheres/unit area

$$R_c = \frac{1}{n}\frac{\rho}{\pi r_1} = \frac{\rho}{n\pi r_1}$$

This approximation does not cause an error of more than 15%

Thus $R_c = 0.85\ \rho/n\pi r_1$

Heat generation rate by this contact resistance with an applied voltage of V is $Q = V^2/R_C$ per unit area.

However after a very short time (≈ .001 sec) the contact resistance drops to $\frac{1}{10}$th of its original value. Due to softening of material due to increase in temperature.

Example. *In a resistance welding process*

applied voltage = 5 V

Bridges formed $n = 25/cm^2$

Bridge radius $r_1 = 0.1$ *mm.* $= 0.01$ *cm*

resistivity of material $\rho = 2 \times 10^{-5}$ *ohm-cm.*

$$R_C = \frac{0.85\,\rho}{n\pi r_1} = \frac{0.85 \times 2 \times 10^{-5}}{25 \times 3.14 \times 0.01} = 0.00022 \text{ ohm-cm}^2$$

Rate of heat generated/unit area

$$Q = \frac{V^2}{R_C} = \frac{5 \times 5}{.00022} \text{ W/cm}^2$$

$$= 1.136 \times 10^5 \text{ W/cm}^2.$$

Examples for Revision

Example 1. *Two different pairs of sheets of the same material have to be spot welded. In one pair, there are 25 bridges/cm² and the average radius of each bridge is 0.1 mm. The other pair of sheets contains 50 bridges/cm² with the same average radius of each bridge. Determine the ratio of the voltages to be applied in these two cases to generate the same rate of heating/unit area.*

The rate of heat generated by contact resistance with an applied voltage V is $\frac{V^2}{R_C}$

$$R_C = \frac{0.85\,\rho}{n\pi r_1}$$

ρ = resistivity of the material

V = applied voltage

R_c = constriction resistance

n = number of bridges/cm²

r = radius of bridge (average)

Case 1. Rate of heat generated/unit area

$$= \frac{V_1^2}{R_{C_1}} = \frac{V_1^2 \times 25 \times \pi \times r}{0.85\,\rho}$$

Case 2. Rate of heat generated/unit area

$$= \frac{V_2^2}{R_{C_2}} = \frac{V_2^2 \times 50 \times \pi \times r}{0.85\,\rho}$$

For equal heat to be generated

$$\frac{V_1^2 \times 25 \times \pi \times r}{0.85\,\rho} = \frac{V_2^2 \times 50 \times \pi \times r}{0.85\,\rho}$$

$$\left(\frac{V_1}{V_2}\right)^2 = 2 \quad \frac{V_1}{V_2} = \mathbf{1.414}$$

Example 2. *The voltage-arc length characteristic of a dc arc is given by :*

$$V = (20 + 4l) \text{ volts.}$$

where l is the arc-length in mm. During a welding operation it is expected that the arc length will vary between 4 mm and 6 mm. It is desired that the welding current be limited to the range 450–550 A. Assuming a linear power source characteristic, determine the open circuit voltage and short circuit current of the power source.

D.C. Arc voltage $V = 20 + 41$

Arc length varies between 4 mm and 6 mm

It is desired that welding current should be between 450 to 550 A (difference 100 A)

Assume a linear power source characteristics

Find open circuit voltage and short circuit current voltage variation range :

$$\left.\begin{aligned} V &= 20 + 4 \times 4 = 36 \text{ V to} \\ &20 + 4 \times 6 = 44 \text{ V} \end{aligned}\right\} \mathbf{8\ V}$$

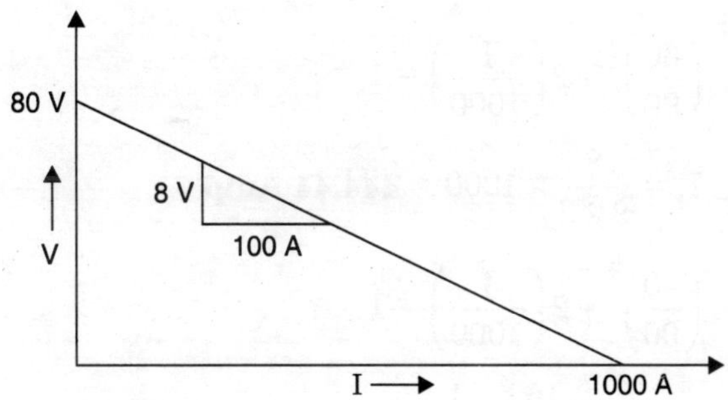

Fig. 6.5

current range (450 – 550) $\simeq$ 100 Amp.

$$\text{Slope} = \frac{8}{100} = 0.08$$

$$V = C - mI = C - \frac{80}{100} I$$

$$36 = C - \frac{80}{100} \times 550$$

$$C = 80 \text{ Thus } V = 80 - 0.08\, I$$

$$V = C - .08\, I$$

$$V = 80 - 0.08\, I$$

When $V = 0$

$$I = \frac{80}{.08} = 1000 \text{ A}$$

Short circuit current = **1000 A**

Open circuit voltage = **80 V**

Example 3. *During an experimental investigation the arc-voltage has been found to be related with arc-length as V = (22 + 4l) volts. The power source characteristics is as follows*

$$\left(\frac{V}{V_0}\right)^2 + 2\left(\frac{I}{I_0}\right) = 1$$

where V_0 *= open circuit voltage and* I_0 *= open circuit current. In one of the observations* $V_0 = 90$ *volts and* $I_0 = 1000$ *Amp. What will be the values of welding currents for arc lengths of 3 mm and 5 mm with corresponding arc voltage of 30 volts and 40 volts.*

Solution. Using the data given

$$\left(\frac{30}{90}\right)^2 + 2\left(\frac{I}{1000}\right) = 1$$

$$I_1 = \frac{8}{9 \times 2} \times 1000 = \mathbf{444.44\ Amp}$$

$$\left(\frac{40}{90}\right)^2 + 2\left(\frac{I}{1000}\right) = 1$$

$$I_2 = \left(1 - \frac{16}{81}\right) \times \frac{1}{2} \times 1000 = \mathbf{400.61\ Amp}$$

The values of welding currents are 444.44 Amp and 400.61 Amp corresponding to arc-voltages of 30 and 40 volts respectively.

■ QUESTIONS

6.1 Briefly discuss how residual stresses and distortions occur in welded structures. How these stress could be minimised and eliminated?

6.2 By means of neat sketches discuss transverse shrinkage in V-butt welds. How can transverse shrinkage be calculated (estimated) in butt welds, fillet welds and T-welds.

6.3 How residual stresses occur in welds? Briefly explain stress-relieving treatment of welds.

CHAPTER 7

Welding of Materials

Some materials are easily weldable while certain others require special procedures to weld them. These materials are called difficult to weld materials. The welding of the following such materials will be discussed in this chapter.

1. Welding of cast irons
2. Welding of aluminium and its alloys
3. Welding of low carbon HY pipe steels
4. Welding of stainless steels

In addition to the above, the welding of dissimilar metals and the hardfacing and cladding will also be discussed.

7.1 WELDING OF CAST IRONS

7.1.1 Composition of Cast Irons

Element	*Gray C.I.*	*Malleable C.I.*	*Nodular C.I.*
Carbon	2.5–3.8	2–3	3.2–4.2
Silicon	1.1–2.8	0.6–1.3	1.1–3.5
Manganese	0.4–1.0	0.2–0.6	0.3–0.8
Sulphur	0.1	0.1	0.02
Phosphorus	0.15	0.15	0.08

7.1.2 Oxy-Acetylene Welding of Gray and Nodular Cast Irons

- Grey cast iron contains much of carbon in flake form. This flake carbon distribution causes it to be brittle and, therefore, the standard set for its welding is not very high.
- Nodular Iron is cast with magnesium, nickel or rare earth addition, the graphite is in the form of spheroids with ferrite or pearlite matrix. This iron has ductility in as cast state upto 4% and on annealing-upto 15–25%. Its weldability is better than that of Grey cast iron as S and P are at low level. Thus the risk of hot tearing in weld metal is reduced. Welding steps are given below.

- A 60 – 90 Vee grove is prepared.
- When repairing a crack a hole should be drilled at each end of the crack to arrest it.
- The job before welding is preheated to 300–650 C in a furnace then covered with asbestos cloth, exposing only the cavity to be welded.
- If furnace is not available the casting is covered with asbestos cloth and locally heated by gas flame. Thick sections should be preheated in a furnace.
- Filler material should have the same composition as the base metal with minimum S and P. Special rods containing Ti and high Si content are also sometimes used.
- Welding rods are square or round cast bars.
- Fluxes for grey iron filler rods are composed of borates, soda ash, and small amounts of ammonium sulphate, iron oxide, etc.
- Torch tip is one size larger than that required for steel of the same thickness.
- Adjust the torch to a neutral flame.
- Move the flame along the groove untill the entire joint is preheated to dull red.
- Concentrate the flame at the bottom of the vee with tip of inner cone about 3.0 to 6.0 mm from the metal surface. As the bottom fuses thoroughly move the flame from side to side to let the liquid metal run down to the pool and rotate the torch to mix the molten metal from side walls to mix with the metal in the pool.
- If metal gets too fluid and runs down raise the flame.
- After the weld pool is formed, heat the filler rod end by outer envelop of the flame, dip the rod into the flux.
- Introduce the Flux coated (dipped) filler rod into the molten pool and apply flame to the tip of the filler rod and the welding is carried out.
- As the weld completes, cover it with asbestos and allow it to cool slowly.
- Post welding stress relieving be carried out for complex shapes. For this purpose keep casting in a furnace at 650°C for one hour per 25 mm thickness and cooled to 260°C or below at a rate not faster than 28°C per hour.

7.2 WELDING OF ALUMINIUM AND ITS ALLOYS

- The most important consideration is the **oxide film.**
- Use of DC reverse polarity (electrode +) is effective for MIG welding while AC is used for TIG welding of Aluminium.
- In AC tungsten inert gas (TIG) welding, when electrode is +ve the oxide of plate is cleaned by ionic bombardment and when it is –ve, the plate gets more heat as it is +ve.
- Because of high thermal conductivity of aluminium.

  (*a*) Nozzle for TIG/MIG welding is larger than that used for steel

  (*b*) Currents used are more than those used for steel, and

  (*c*) Thicker plates are preheated.
- There is no colour change on heating, experience is needed during welding. (Appear-

ance of blusters on surface indicates that welding temperature is reached.

- Shielding gas in MIG welding.

Upto 18 mm plates	100% Argon
18–75 mm plates	75% Argon + 25% Helium
above 75 mm plates	25% Argon + 75% Helium

He and He rich mixtures are never used in AC welding.

7.3 WELDING OF LOW CARBON HY PIPE STEELS

A typical relation for carbon equivalent determination for carbon steels is given as (the elements expressed in wt%)

$$CE = C + (Mn + Si)/6 + (Ni + Cu)/15 + (Cr + Mo + V)/5$$

1. Low carbon HY pipe steels contain less than 0.45% carbon

The mechanical properties and weldability requirements of high strength steel are :

$$Y.S. = 450 \text{ N/mm}^2 \quad UTS = 530 \text{ N/mm}^2,$$

Impat energy > 50 J at – 46°C

HAZ hardness < 22 HR_C (250 VHN)

2. Low carbon content is desirable for high toughness, good weldability and low susceptibility to cold cracking in the HAZ.

3. Niobium and vanadium additions give grain refinement, improve Y.S. and toughness.

4. In X-65 and X 70 low carbon, boron free steels (CE = 0.33), carbon content < 0.04% improves resistance to hydrogen induced cracking, the field weldability and HAZ toughness. Critical material parameter P_{cm} for weld cracking is given by (elements in weight %)

$$P_{cm} = C + \frac{Si}{30} + (Mn + Cu + Cr)/20 + Ni/60 + Mo/15 + V/10 + 5\,B.$$

5. It is necessary to reduce CE and P_{cm} value for high field weldability specially for pipe materials X 65 and X 70. CE < 0.4% and P_{cm} < 0.15% are preferable to obtain HAZ hardness values < 250 VHN

6. Effects of C.E. on UTS and YS of X 65 pipe steel are shown in Fig. 7.1 (*a*) and (*b*).

7. The effect of P_{cm} on HAZ hardness for Low carbon pipe materials is shown in Fig. 7.2.

P_{cm} = critical material parameter.

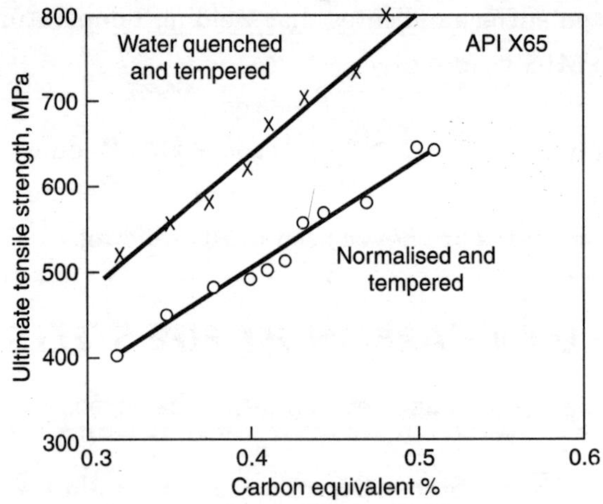

Fig. 7.1 (*a*) Effect of carbon equivalent on UTS of X65 pipe steel.
(R.G. Baker, Proc. Rosenhain Centinary Conf., Royal Society, 1975)

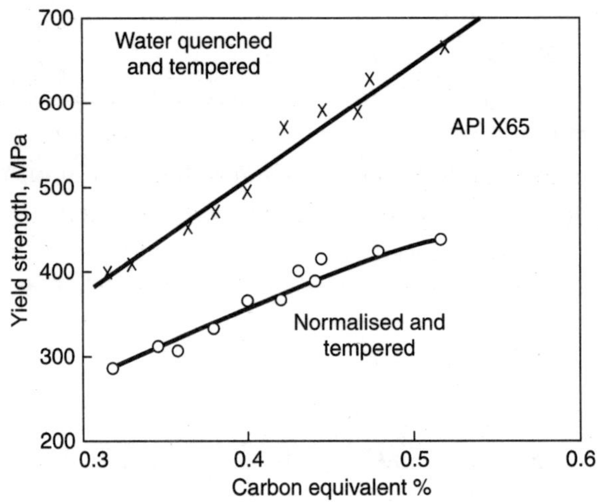

Fig. 7.1 (*b*) Effect of carbon equivalent on YS of X65 pipe steel.
(R.G. Baker, Proc. Rosenhain Centinary Conf., Royal Society, 1975)

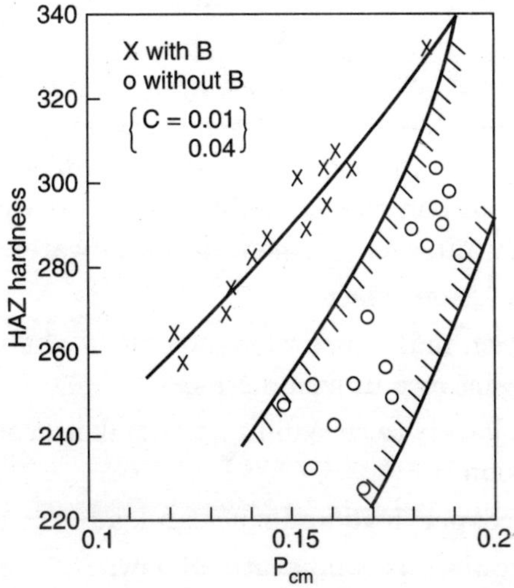

Fig. 7.2 Effect of P_{cm} on HAZ hardness for low carbon pipe steel

7.4 WELDING OF STAINLESS STEELS

Stainless steels are classified according to their matrix structure.

(*a*) austenitic (*b*) ferritic (*c*) martensitic

(*d*) precipitation hardened and (*e*) duplex.

Special features of stainless steels related to welding.

1. Low thermal conductivity (50% of mild steel) results in less heat input for the job and 10% less current is needed for SS electrodes. higher electrode melt. off rates are also obtained. Melting point of stainless steel is 93°C lower.

2. Thermal expansion of Cr-Ni steels is about 50% greater than for mild steel. This increases the chances for warping and buckling. Thus suitable fixture must be used for welding stainless steels.

3. Electrical resistance is 6–12 times higher which causes overheating in the electrodes. Shorter electrodes are, therefore used to reduce electrode heating.

Austenitic stainless steels

1. These steels contain 16–26% chromium
   6–22% Nickel.

2. Type 304 L and 316 L are low carbon grade ($C \leq 0.03\%$).

3. Mo in type 316 improves corrosion resistance and high temperature properties.

4. Types 321 and 347 stainless steels are stabilized against carbide ($Cr_{23}C_6$) precipitation, weld decay and intergranular corrosion by addition of Ti and Nb. The strong carbide formers form TiC and NbC which impart creep resistance. Hence they are also used as creep resisting steels.

5. The 200 series s.s. sin lower Ni which is compensated by Mn and N_2 for austenite formation.

6. Austenitic S.S. (except free machining grades) are easiest to weld and produced welds that are tough.

7. S.S. welding requires 20–30% less heat input than welds in carbon steels, because of low thermal conductivity and high electric resistance. Excess heat will cause distortion, reduce strength and corrosion resistance. Sulpher and Selenium added for free machining, makes the steel unweldable, also high carbon content inhibit weld serviceability. External sources of contamination include carbon nitrogen, oxygen, iron and water.

8. Contaminations and their effects.

- Carbon contamination may cause welds to cracks, change mechanical properties and reduce corrosion resistance in weld areas.
- Iron contamination lowers serviceability, flakes of iron on surface will rust, thus speeding localised corrosion.
- Contamination by copper, lead and zinc can lead to cracking in HAZ of the weld.

9. Welding current required is comparatively low.

10. When stainless steels are heated in the range of 427–870 C or cooled slowly through that range, carbon precipitates at grain boundaries.

11. Formation of these carbides effectively eliminates much of the chromium.

12. It will reduce corrosion resistance especially in HAZ.

13. This carbon precipitation can be minimized by :

(*i*) Reducing the time for which the temperature is between 427°–870°C range.

(*ii*) Selecting low carbon stainless steels to reduce carbide formation.

(*iii*) Addition of Ti, Ta, Columbium which form stable carbide preventing the formation of chromium carbide.

Carbide precipitation

1. Austenitic grades are non-hardening type and welding usually does not adversely affect weld strength and ductility. There is one detrimental effect of heating of Ni-Cr steel *i.e.,* carbide precipitation at the grain boundaries resulting in reduced corrosion resistance. A fine film of Cr-rich carbides containing upto 90% Cr taken from metal layer next to grain boundary gets precipitated along the grain boundary. Precipitation of intergranular chromium carbides is accelerated by an increase in temperature within the sensitized range and by an increase in time at that temperature.

2. Carbide precipitation can be controlled by :

- Using stabilised steels, by adding columbium and titanium which have greater affinity for carbon than does chromium. Columbium is exclusively used for the purpose in welding electrodes as titanium gets lost in transferring across the arc.
- Rapid quenching may minimise carbide precipitation, but this may not always be possible specially in thick sections.
- Limiting carbon content to a maximum of 0.03% avoids carbide precipitation
- Post-weld solution annealing.

3. Solution annealing puts carbides back into solution restores corrosion resistance. Austenitic S.S. with stabilization using Nb + Ti or Tantalum and welded with stabilised filler metal gives good strength and corrosion resistance properties.

4. SMAW process is widely used. A large number of electrodes available make the process widely acceptable. Some examples are given below:

- E308-16 electrode–metal transfer is spray type–smooth bead (AC or DCRP)
- Lime covered basic electrodes (only DCRP)–E308-15-globular transfer rough bead
- For heavy flat pieces SAW is used
- For thin sections TIG is excellent
- For sheets spot welding can be used.

Cracking

Interdendritic cracking in the weld area that occurs before the weld cools to room temperature is known as hot cracking or microfissuring. Weld metal with 100% austenite is **more** susceptible to microfissuring than weld metals with duplex structure of delta ferrite in austenite. Susceptibility can be **reduced** by a small increase in carbon or nitrogen content or by a substantial increase in manganese content.

To avoid solidification, cracking, weld metal should have a ferrite content of at least 3-5 ferrite number (FN) and hence filler metal of suitable composition is to be selected. For this purpose Schaeffler diagram is made use of; A modified version of it is *h* shown in Fig. 7.3 which takes care of nitrogen in the metal.

Nitrogen strengthened austenitic stainless steels offer the advantages of:

- Increased strength at all temperatures (cryogenic to elevated)
- Improved resistance to pitting corrsion

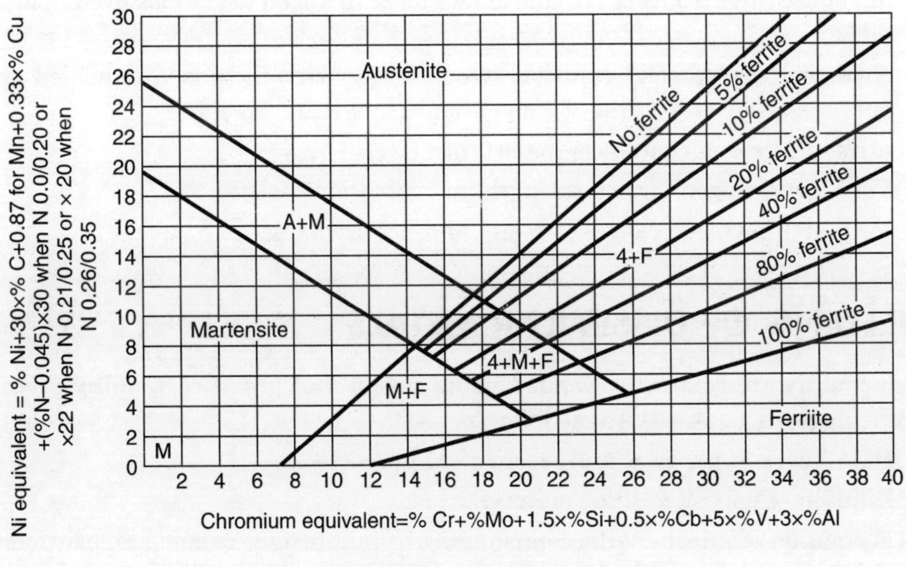

Fig. 7.3 Schaeffler diagram

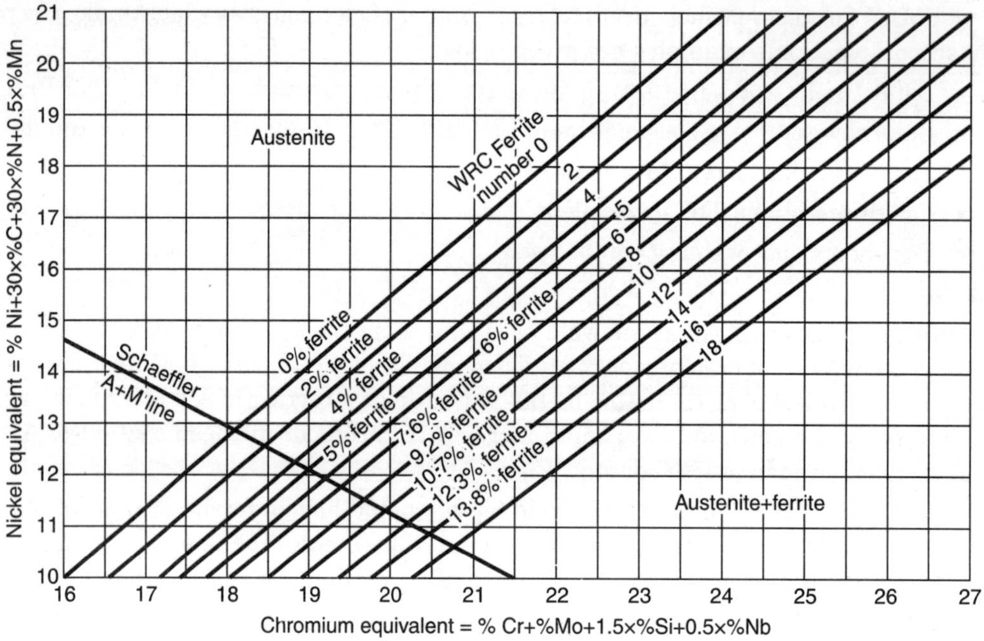

Fig. 7.4 De Long diagram

They differ from conventional austenitic steels in that

- Mn substitutes a part of Ni, this allows more nitrogen to get dissolved in matrix of the alloy.
- Nitrogen acts as solid solution strengthener with increased annealed strength to approximately twice that of conventional austenitic steels.

Control of nitrogen content is important.

- Very low nitrogen lowers strength and corrosion resistance.
- Very high nitrogen causes porosity and hot cracking.

7.5 WELDING OF DISSIMILAR METALS

Dissimilar metals are commonly welded using fusion and pressure welding processes. The major Difficulties encountered are as follows :

1. Differences in physical and mechanical properties.
2. Dilution of deposited filler material.
3. Formation of intermetallic compounds at the interface causing embrittlement of the joint. To eliminate this or reduce, heat input to the weld is reduced.

7.5.1 Guidelines for Welding Dissimilar Metals

In the welding of dissimilar metals the following guidelines are helpful:

1. Minimise heat input to minimise dilution and restrict diffusion.

2. Choose proper filler material compatible with both materials being welded.

3. Reduce dilution by controlling welding process variables related to penetration. Thus minimise penetration. In GMA welding reduce current density so that dip. transfer of metal occurs.

4. Dilution and formation of intermetallic phases can be minimized by applying a layer of compatible material on both the joint faces.

5. In case of the welding of heat treated steels appropriate heat treatment should be used. If one plate is hardenable low-alloy steel, appropriate pre and post weld heat treatment should be used.

If for some reasons heat-treatment is not possible, ductile austenitic filler material must be used (for hardenable materials). This will compensate for lack of ductility in the HAZ.

7.5.2 Tips for Joining Certain Combinations

1. Joining alloy Steels

Joining 2.25 Cr–1 Mo. steel with 1 Cr–0.5 Mo steel or 0.5 Mo steel with plain carbon steel can be best done by using a filler that matches with the lower alloy for good weldability.

2. Joining 'Ferritic steel' with Austenitic steel

This is best done by using austenitic filler rod. Filler metal should have a composition that will stabilize austenite even after dilution, otherwise the carbon will migrate from ferrite and alloy elements from the other plate to form a crack susceplible zone.

3. Joining highly Austenitic Materials

This is successfully done by using a filler material which is highly ferritic such as electrode type 312 (29 Cr—9 Ni). This will leave sufficient ferrite in the weld metal to avoid hot cracking. If one base metal is highly ferritic then a highly austenitic electrode (310) can be used to avoid weld which will contain large quanties of ferrite.

4. Joining stainless steel to plain carbon steel

Plain carbon steel is first coated with a layer of austenitic steel like 309 (25 Cr–12 Ni) using TIG or MMA processes with usual precautions. In service, problems arise, due to different thermal expansion coefficients of plain carbon and stainless steels. Large thermal stresses are built-up due to unequal expansions and contractions. Because of high solubility of carbon in austenitic stainless steels, carbon from low alloy steel will have a tendency to migrate during welding to austenite regions. This will result into decarburized zone in ferritic steel just adjascent to the interface. This may lead to service failures.

5. Welding of aluminium to steel

This is a very common situation in industrial applications. The steel part is first coated with aluminium and the joint is completed using TIG welding using aluminium based filler wires. The arc is directed towards the aluminium member during welding. The molten weld pool flows over the aluminium coating on steel without melting too much of the steel. Thus the formation of intermetallic compounds can be eliminated. The aluminium coating on steel should be thick enough to avoid burning near the edges.

6. Applications of explosive and friction welding

Explosive and **friction** welding can avoid the formation of intermetallic compounds and are used for dissimilar metals welding. Similarly **flash** butt welding has the advantage that the intermetallic phases are squeezed out of the joint while in the molten state.

7.6 HARD SURFACING AND CLADDING

A. Hard Surfacing

1. Hard surfacing is the application of a durable surface layer to a base metal to impart properties like resistance to impact wear and erosion or pitting and corrosion or any combination of these factors. Hard surfacing can be applied by arc welding.

2. Hard facing materials for wear resistance tend to suit specific types of wear like abrasive or sliding wear or build desired dimensions.

3. Electrodes used for such applications are called hardfacing electrodes, covered by AWS A 5.13–1970 used as surface filler metal for gas and TIG welding, and coated electrodes for arc welding.

4. The hardfacing electrodes are designated on the basis of hardness of weld deposit *e.g.*,

Type	*Hardness range BHN*	*Applications*
A	250—280 (Hard)	Moderate hardness: used in gears/machine parts.
B	350 — 380 (Harder)	Brake shoes, cams, rollers, large wheels.
C	280 — 320 (Harder)	Brake shoes, cams, rollers, large wheels.
D	600–625 (Hardest)	Metal cutting/forming tools, punches, dies, crushers, hammers crane wheels.

The above electrodes A, B, C and D give martensitic deposit and impart hardness in as-weld condition at normal cooling rates in air.

5. To obtain desired results for a specific application it is necessary to understand the effect of base metal dilution and cooling rate on the hardfacing deposit. Base metals having high carbon and hardenable elements like Cr and Mo are likely to develop underbead cracks, due to hydrogen from the rc. Low hydrogen, hardfacing electrodes are to be used in such cases.

6. Hardfacing deposits respond to mechanical and thermal treatments. The operation introduces distortion which can be countered by proper fixturing, bead sequencing and pre-heating the base metal.

7. Hardfacing materials may be classified as follows.

(*a*) Alloy steels (Cr, Ni, W and Mn) : Austenitic or martensitic are available in the form of electrodes. Martensitic deposits may be heat treated to get desired properties.

(*b*) Complex alloys (stellite) are used as cast rods or flux coated electrodes. Mainly used in wear resistance applications.

(*c*) Tungsten carbide (one of the hardest materials) used for cutting tools.

8. Semi-austenitic alloys provide balanced composition of good **wear** and impact resistance and is most widely used of all hardfacing materials. These are iron based alloys containing upto 20% alloying elements C = 0.1–0.2% and Cr = 5–12%). The deposit, if cools slowly gets time for austenite to transform to martensite and is less ductile, if cools fast by using short beads, gives soft and tough austenite.

9. Austenitic Mn-steels are used to built-up worn Mn-steel parts. They are used where resistance to severe impact and abrasion are required.

10. Austenitic stainless steel deposits provide resistance to corrosion and chipping from repeated impact forces. Protect turbine blades from corrosion and cavitation erosion. Also used as buffer layer for other hardfacing materials to avoid brittle bond.

11. Tungsten carbide deposits are suitable for cutting tools, tools for earth and rock cutting, chromium carbides used for hard surfacing when corrosion resistance is also required.

12. Hardfacing processes and applications. (Slow cooling rates prevent underbead cracking).

Processess	*Applications* / *Precautions if any*
1. Oxy-acetylene	Hardfacing, Cracking is minimised by flame pre-heating used for small delicate parts requiring thin layers.
2. Manual Metal Arc	Common for repair hard facing. Gives deep penetration deposits.
3. TIG	Requires little pre-heating, used for high alloy steels, Cr and stainless steels, Ni-base alloys, Copper and Co-base alloys. Aust-Mn. steels.
4. MIG	Often used for cladding and build-up. Not very common for hardfacing. Specially suited for aluminium bronze overlays.
5. SAW	Good wear resistance with single layer. DCRP low deposition rate and thin beads. DCSP gives high deposition rate and thick deposits.

3. The major problem in hardfacing is the peeling-off of the deposited layer, particularly when the base metal contains less than 0.15 per cent carbon. Preheating the base metal and slow cooling will reduce peeling tendency and underbead cracking. Spalling can be avoided by : (*a*) cleaning base metal surface (*b*) preheating base plate and slow cooling (*c*) depositing thin layers and peening each layer to relieve stresses.

B. Cladding

1. Cladding, is similar to hardfacing, but is normally a corrosion resistant overlay. In high pressure applications such as nuclear reactor vessels, cladding provides a combination of

mechanical properties and corrosion resistance. Cladding of low alloy steels with austenitic stainless steels is quite common in nuclear reactor vessels.

2. Cladding Processes and applications

Cladding Processes	*Applications*
1. SAW	Most of cladding is carried out. Alloy addition is through flux, high deposition rate ; Slow welding decreases dilution (1.2–5 mm/s)
2. Plasma Cladding	Well controlled heat input, independently controlled deposit thickness and penetration, high weld purity, clads difficult to weld metals where SAW Fluxes developed, and increased productivity.

Surfaces which are deposited by cladding technique include:

1. Austenitic stainless steels
2. Inconel
3. Nickel and cupro-nickel

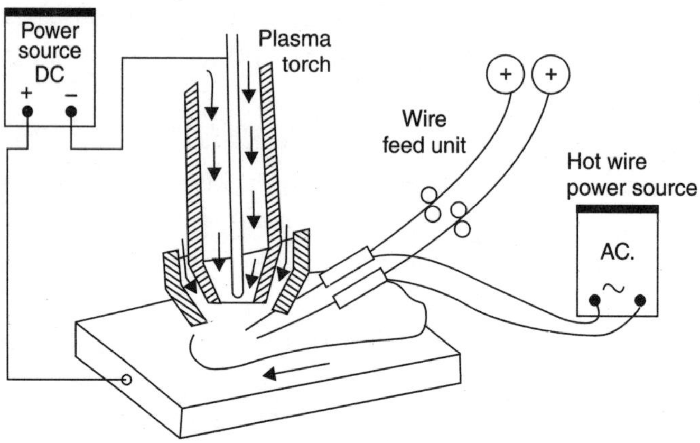

Fig. 7.5 Gas metal plasma hot wire process

3. Cladding integrity

While cladding with austenitic steel on reactor vessels to protect the underlying steels from corrosive environments, ensure that the deposit microstructure contains austenite plus only 3–10% ferrite to avoid solidification cracking. Dilution of deposit may take place when using SAW. SMAW electrode E 309 (23 Cr–12 Ni) to avoid dilution.

Cracking in cladding may expose base metal to corrosive environment. Sometimes the cracks may penetrate the base metal. Causes of cladding degradation are :

– microstructural/phase changes, sensitization, embrittlement, **sigma** phase formation, loss of corrosion resistance.

- low cycle fatigue cracking due to thermal loading.
- carburization and subsequent sensitization.
- loss of adherence (fusion).
- hydrogen embrittlement of weld overlay during shut down and restart.
- stress corrosion cracking due to chlorides and polythionic acids, principally during nuclear vessel shut down periods.

Sigma phase formation can be minimised by keeping the ferrite content of the cladded stainless steel in the range of 3–10 percent. Ferrite phase serves to nucleate sigma phase during post weld heat treatment which increases chances of steel to hydrogen embrittlement.

Embrittlement of austenitic stainless steel cladding material during post welding heat treament is due to both the sigma phase formation and carbide precipitation and is minimised by using low carbon material and by keeping ferrite content at the lower end of the safe ferrite content range.

CHAPTER 8

Welding Procedure and Process Planning

An Engineer entering the field of welded design, usually has the background of mechanical or materials engineering, and has very little understanding of the factors that contribute to efficient welded design as welding technology and weld design are not regular subjects in engineering colleges. A successful welded structure design will:

1. perform its intended functions.
2. have adequate safety and reliability.
3. be capable of being fabricated, inspected, transported and placed in service at a minimum cost.
4. cost includes cost of design, materials, fabrication, erection, inspection operation repair and maintenance.

Efficient and economical designs are possible because of:

1. mechanised flame–cutting equipment (smooth cut edges).
2. press brakes are available to make use of formed plates.
3. a wide range of welding processes and consumables.
4. welding positioners are available that permit low cost welds to be deposited in down hand welding position.

One should avoid over designing or higher safety factors and still safe and reliable design.

In developing a design the following factors are of help:

1. Specify steels that do not require pre or post heat treatment.
2. Use standard rolled sections where possible.
3. Use minimum number of joints and ensure minimum scrap.
4. Use stiffeners properly to provide rigidity at minimum weight of material, use bends or corrugated sheets for extra stiffness.
5. Use closed tubular section or diagonal bracing for torsional resistance.
6. Ensure that the tolerance you are specifying are attainable in practice.
7. Use procedures to minimise welding distortion.
8. To eliminate design problems and reduce manufacturing cost consider the use of steel casting or forging in a complicated weldment.

9. Consider cost-saving ideas.

10. Consider the use of hard facing at the point of wear rather than using expensive bulk material.

11. Save unnecessary weld metal use intermittent welds where necessary. Stiffeners and diaphragms may not need full welding.

12. Divide structure into subassemblies to enable more men to work simultaneously.

13. Use mathematical formulae in design don't use guess work or rule-of-thumb methods.

14. Define the problem clearly and analyse it carefully in regard to the type of loading (steady, impact, repeated-cyclic, tension, compression, shear, fatigue), modulus of elasticity to be considered (tension or shear).

15. Properties of steel sections to consider include, area, length, moment of inertia (stiffness factor in bending), section modulus (strength factor in bending), torsional resistance (stiffness factor in twisting and radius of gyration. Stress is expressed as tensile compressive or shear, strain is expressed as resultant deformation, elongation or contraction, vertical deflection or angular twist.

In the present context we are not discussing the design formulae as it is beyond the escope. For this purpose references on design of welds could be consulted.

8.1 WELDING SYMBOLS

As a production engineer and executive, a knowledge of "location of elements of a welding symbols" is necessary for indicating or interpreting. This will now be discussed in more details in the following paragraphs. Any of the following standards could be used depending upon the situation and case of use.

1. AWS–A24: Symbols for welding and non-destructive testing.
2. BS : 499 (Part II): Symbols for welding.
3. ISO : 2553: Symbolic representation on drawings.
4. IS : 813 (1961): Scheme of symbols for welding.

Basic symbols used in ISO and AWS are identical.

In the AWS system a complete welding symbol consists of the following elements:

1. Reference line (always shown horizontally)
2. Arrow
3. Basic weld symbol
4. Dimentions and other data
5. Supplemental symbols
6. Finish symbols
7. Tail
8. Specification process or other references.

These elements have specified locations with respect to each other on or around the reference line as shown in Fig. 8.1.

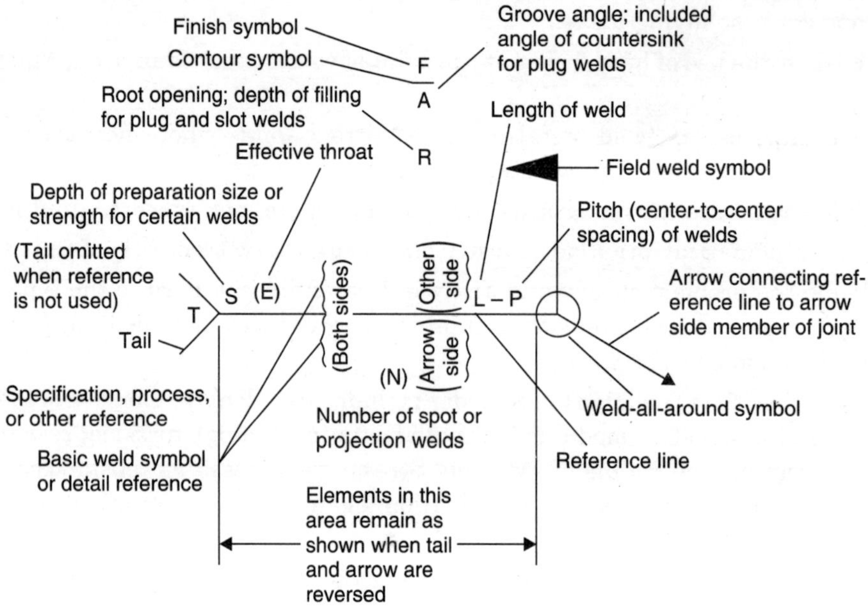

Fig. 8.1 Standard location of elements on the welding symbol

There are two prevailing systems of placing the symbol with respect to the reference line. In USA and UK, the symbol is placed below the reference line for welds on the arrow side. ISO has accomodated both and designate them as A and E (for European system). The designer must be aware of these two systems and take care that his drawing is not misinterpreted.

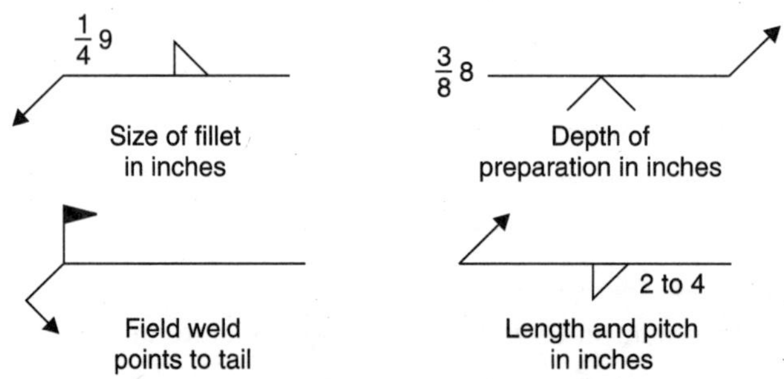

Fig. 8.2 Size location, field weld length, and pitch

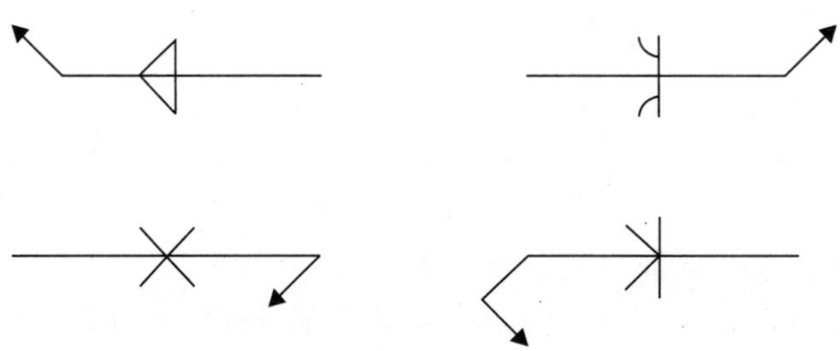

Fig. 8.3 Arrow side, other side reflection

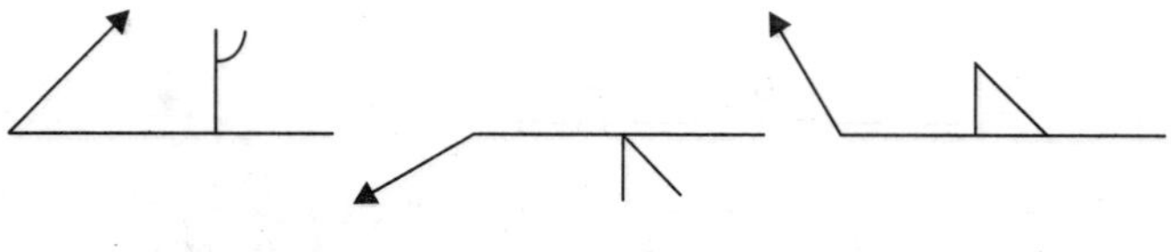

Fig. 8.4 Straight line always on left

8.2 WELDING PROCEDURE SHEETS

In many organisations, the design engineer expected to provide welding procedure sheets alongwith his designs. For this purpose he takes help from the welding engineering and the shop supervisor. To be a good designer he should have the knowledge of welding technology (welding processes, procedures and weldability of metals. He is advised to study this entire book.

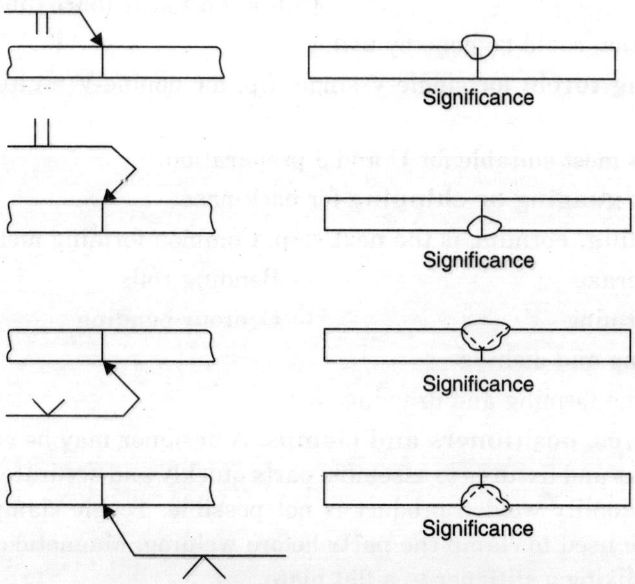

Fig. 8.5 Welding symbols-significance

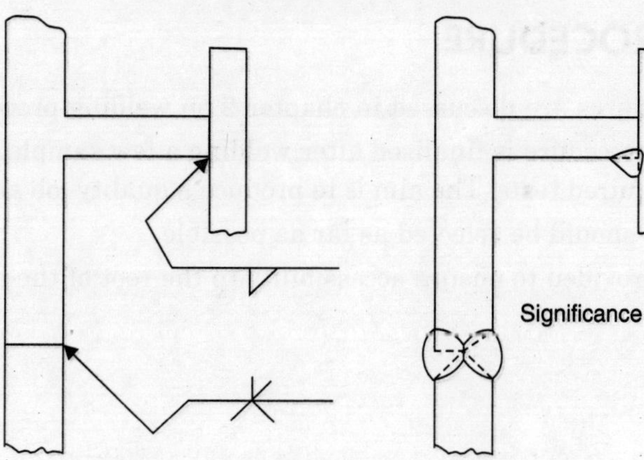

Fig. 8.6 Arrow/side-other/side significance

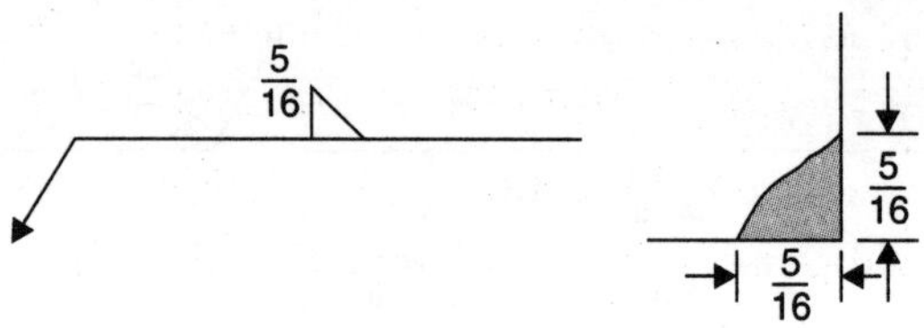

Fig. 8.7 Size of fillet welds

8.2.1 Steps in Preparing Welding Procedure Sheets

1. **Plate preparation.** This includes plate cutting and edge preparation. Plate cutting could be done by using:

Flame-cutting
Shearing
Sawing
Punch press blanking
Nibbling
Cut-off on Lathe (bars/tubes)

Edge preparation could be done by using:

Flame cutting torch; for single-V single tip, for double-V multiple tip torch is preferred.

Edge planer is most suitable for U and J preparation.

Flame or arc guaging or chipping for back-pass.

2. **Plate forming.** Forming is the next step. Common forming methods include:

— Press brake
— Roll forming
— Bending rolls
— Contour-bending
— Flanging and dishing
— Press die forming and drawing.

3. **Jigs, fixtures, positioners and clamps.** A designer may be called upon to design jigs, clamping systems and fixtures to assemble parts quickly and accurately for welding. Without a good fit-up a quality welded product is not possible. Toggle clamps, cam clamps and hydraulic clamps are used to clamp the parts before welding. Magnetic clamps could also be used for instance in fixing a stiffener to a flat plate.

8.3 WELDING PROCEDURE

- Welding procedures are discussed in chapter 2 on welding processes.
- For new jobs, procedure is finalised after welding a few sample joints and subjecting them to the required tests. The aim is to produce a quality job at lowest possible cost.
- Weldable steel should be selected as far as possible.
- A root gap is provided to ensure accessibility to the root of the joint.

- A root face prevents burn through.
- Bevel is usually 30° to 35°.
- J and U preparations save weld metal.
- On butt welds a weld reinforcement of 1.5 mm is adequate.
- Depending upon the application of the joint considerations are given to the following.
  Impact loading
  Fatigue loading
  Problem of brittle fracture
  Torrsional loading
  Vibrational control.

8.4 JOINT PREPARATIONS FOR FUSION WELDING

The objective of edge preparation is to ensure the degree of penetration and ease of welding necessary to obtain sound welds. Type of preparation depends upon:

(*a*) type and thickness of material
(*b*) welding process
(*c*) degree of penetration required for the situation
(*d*) economy of edge preparation and weld metal
(*e*) accessibility and welding position
(*f*) distortion control
(*g*) type of joint.

These factors are considered in many combinations. Demands of the task must be met at economical cost.

8.4.1 Type of Welds

The major type of welds include "Fillet" and "Butt" welds. Fillet welds do not require edge preparation and are almost triangular in transverse cross-section. In butt welds the weld metal lies substantially within the planes of the surfaces of the parts joined. These terms should not be confused with the joint form. Examples of butt and fillet welds are shown in Fig. 8.8.

8.4.2 Joint Preparations for Different Types of Welds

Joint preparations for different plate thickness are shown in Figs. 8.9 to 8.19.

8.4.3 Fatigue as a Joint Preparation Factor

Factors that affect joint preparation are given in Fig. 8.10. Special consideration has been given to fatigue, its causes and precautions taken to eliminate, reduce or minimise it.

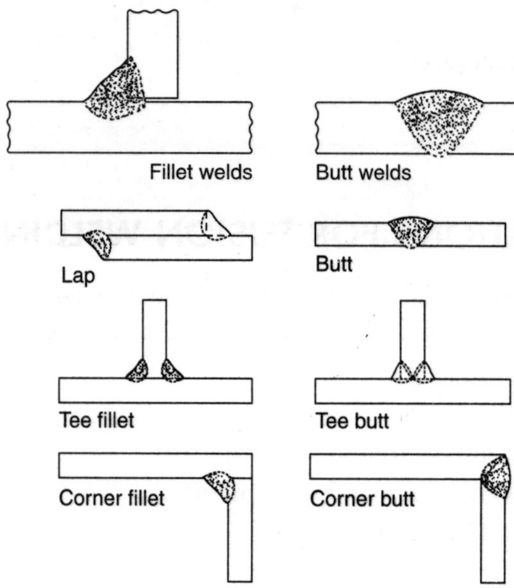

Fig. 8.8 Fillet and butt welds

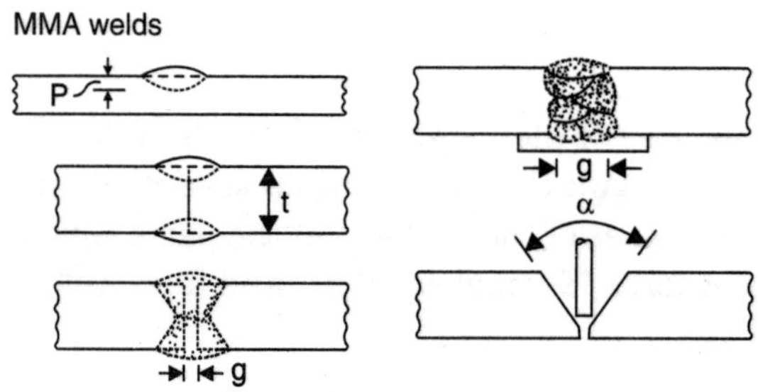

Fig. 8.9 Manual metal arc welds

JOINT PREPARATIONS

1. SQUARE BUTT PREPARATIONS

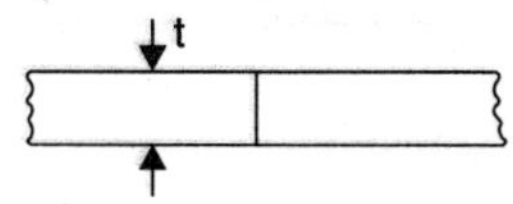

1.1. Close Square Butt

– Thickness 1.25 to 3 mm
– Welded from one side only
– Normal electrodes

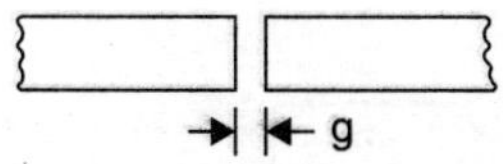

1.2 Open Square Butt

– Thickness $t \leq 6$ mm
– Welded from one side only
– Normal electrodes
– g = 1.5 to 3 mm

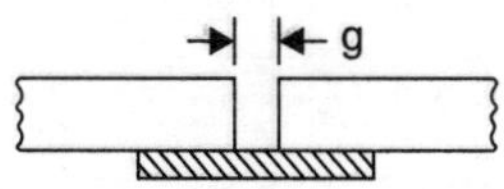

1.3 Square Butt with Integral Backing

– Thickness t = 3 to 12.5 mm
– Normal penetration electrodes
– g = 3 to 8 mm

FATIGUE

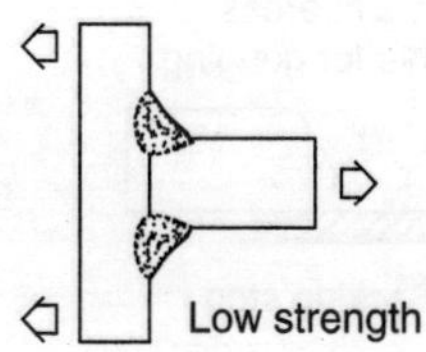

- Lack of penetration and lack of fusion are difficult to detect and they cause fatigue failure of material under fluctuating loads

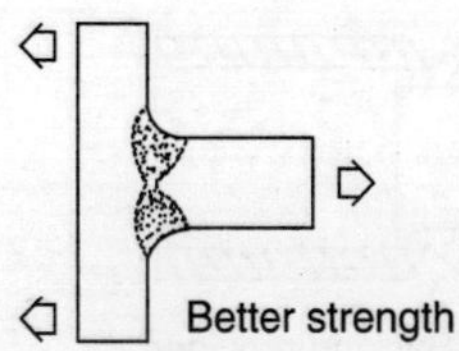

- Susceptibility of a joint to this type of loading depends upon the severity of any notch discontinuity or change in section in the joint

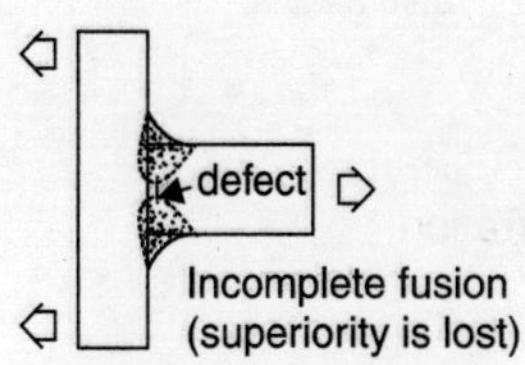

- Unfortunately a weld constitutes a notch. Severity of this notch depends on type of weld and the defect it contains

Fig. 8.10 Factors affecting joint preparation (contd.)

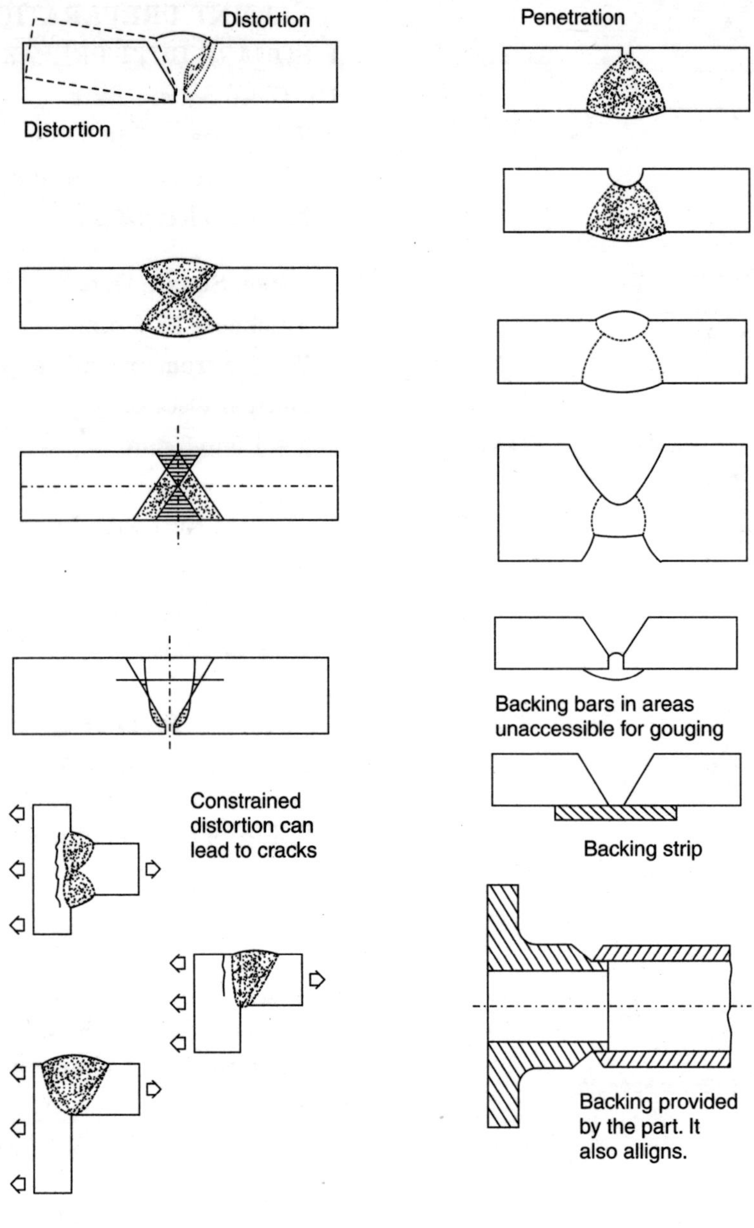

Fig. 8.10 Factors affecting joint preparation

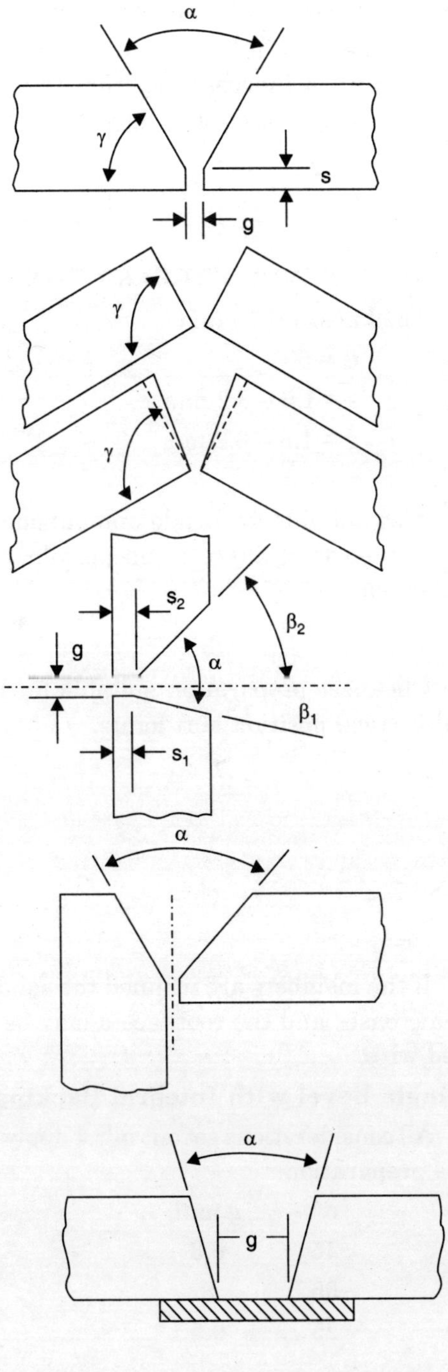

SINGLE V PREPARATION

Thickness $t \leq 19$ mm

Symmetric V

$\alpha = 60°$

$s = 1.5 - 3$ mm

$g = 1.5 - 3$ mm

V FORMED BY INCLINED PLATES

Root face $s = 0$ due to increase in solid angle γ

V-angle could be reduced by reverse bevelling if excessive weld metal is consumed.

Assymmetric V-preparation helps welding in horizontal-vertical position to reduce gravitational effect on the weld pool

$\alpha = 55°$ $\beta_1 = 10 - 15°$

$\beta_2 = 40 - 45°$

$s_1 = 0 - 1.5$ mm

$s_2 = 1.5 - 3$ mm.

Typical values

$\alpha = 45°$ $g = 6$ mm

$\alpha = 30°$ $g = 6$ mm

$\alpha = 20°$ $g = 9.5$ mm.

Fig. 8.11 Single V preparations

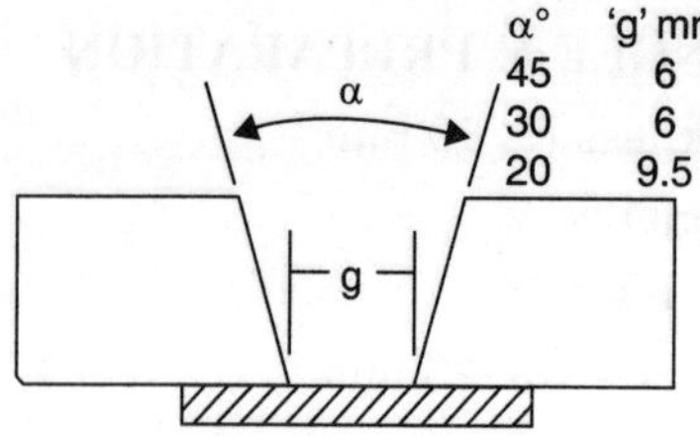

2.1 Single V with Integral Backing

- To ensure full penetration where the joint is inaccessible from the other side, a backing strip may be employed.

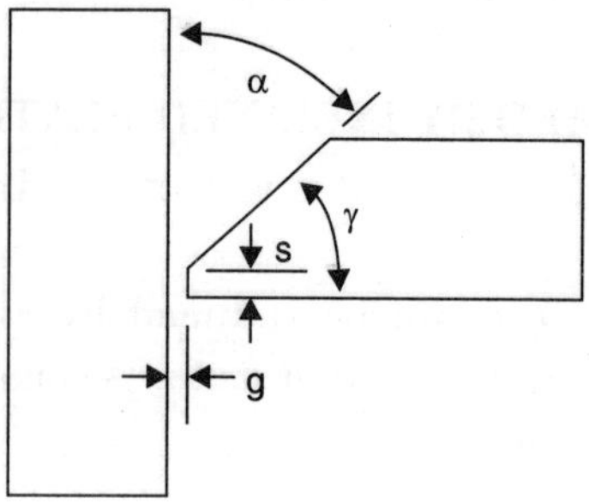

3.0 SINGLE BEVEL PREPARATION

Thickness $t \leq 19$ mm

$\alpha = 50°$

$s = 1.6 - 3.2$ mm

$g = 1.6 - 3.2$ mm

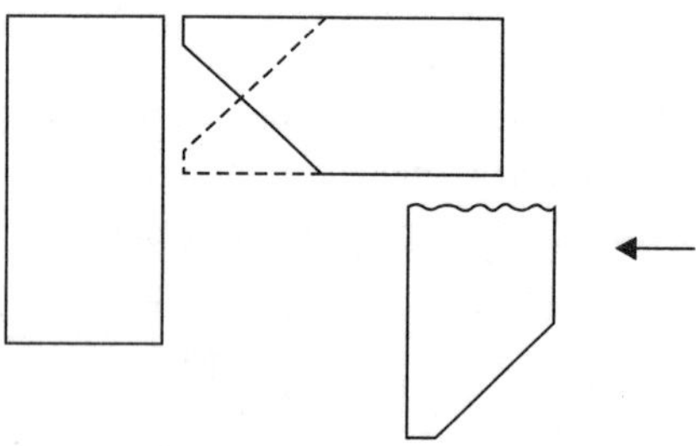

Also suitable for inside and outside corner provided that there is no possibility of lamellar tear.

Cheapest preparation suitable for horizontal-vertical position butt joints.

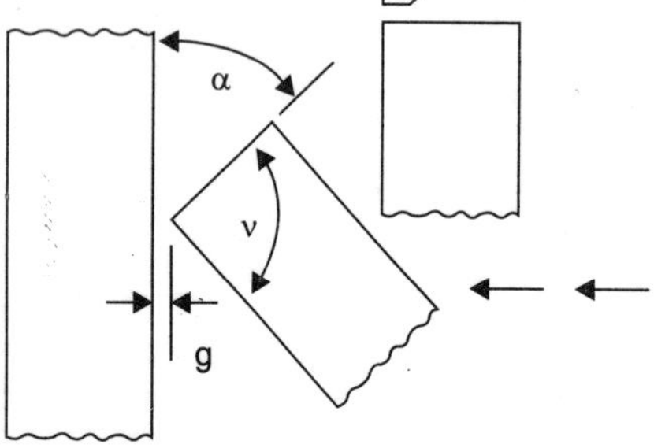

If the members are inclined the solid angle *y* increases and the root-face *s* may be dispensed with.

3.1 Single Bevel with Integral Backing

All considerations set out in 2.1 apply also to this preparation

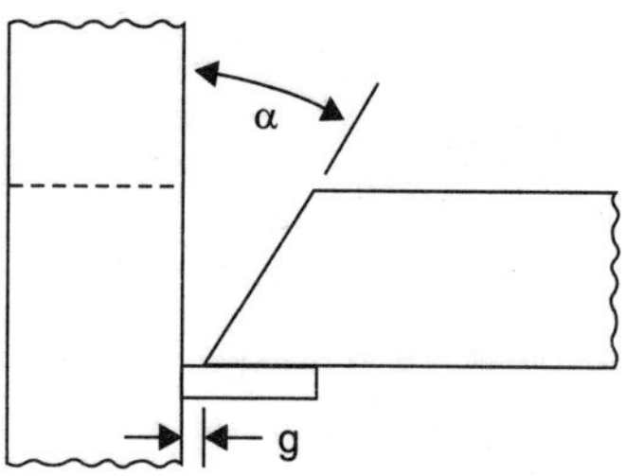

α°	*g* mm
45	6.3
35	8
25	9.5

Fig. 8.12 Single bevel preparation

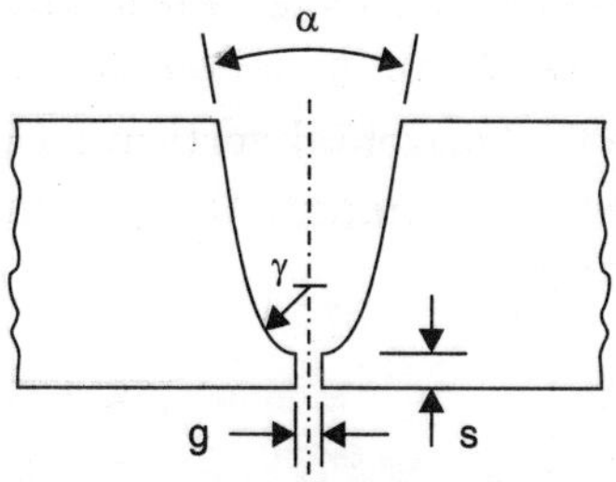

Thickness $t = 19.5 - 38$ mm
$\alpha = 20$, $s = g = 1.6 - 3.2$ mm
$\gamma = 6.3$ to 9.5 mm

4.0 SINGLE U PREPARATION

The objective is to obtain full penetration while welding from one side, lesser volume of weld metal than V prep., distortion is also less. For high efficiency back gouging and welding the other side is necessary. Also needs care during welding due to reduced α.

The shape and dimensions of *u*-basically remain the same relative position of components may change.

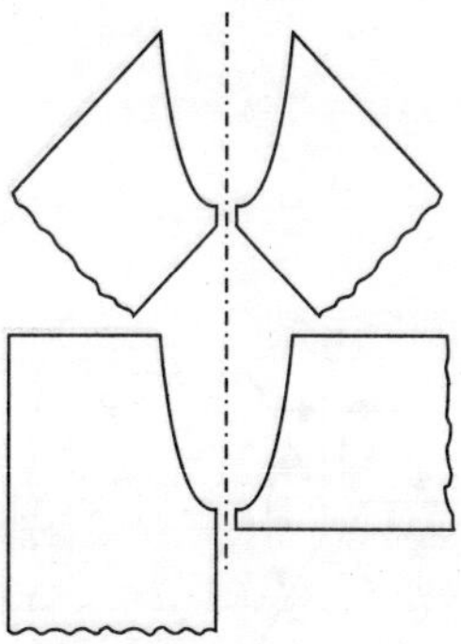

Suitable only for out-side corner

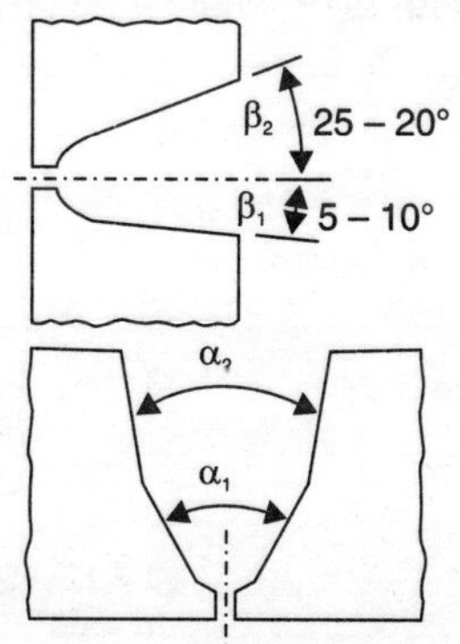

Asymmetric prep. for horizontal-vertical welding

Access and economy in deep grooves
Increase $\alpha_1 = 30 - 40°$
α_2 remains 20°

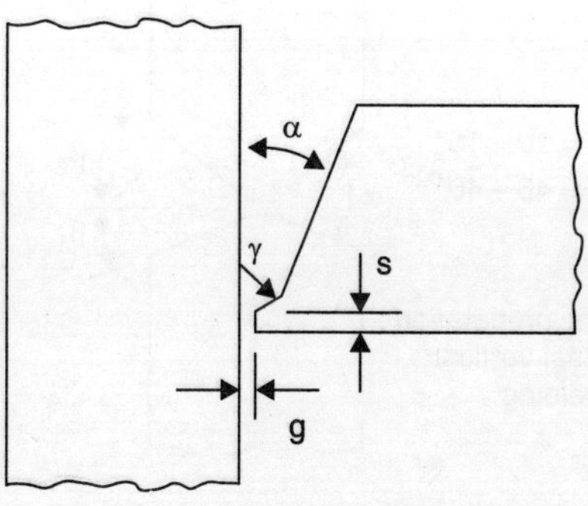

5.0 SINGLE J PREPARATION

This prep. is used for full penet. butt-welds in T and corner joints in plate thicknesses > 19 mm. Lack of fusion may occur, necessitating back gouging for quality joints. As in U prep. a double groove angle $d_1 = 40°$ may be used for very thick plates ($\alpha_z = 20°$). Here thickness $t = 19 - 38$ mm, $\alpha = 20°$

$s = g = 1.6 - 3.2$ mm, $\gamma = 9.5 - 12$ mm.

Fig. 8.13 Single U preparation and single J-preparation

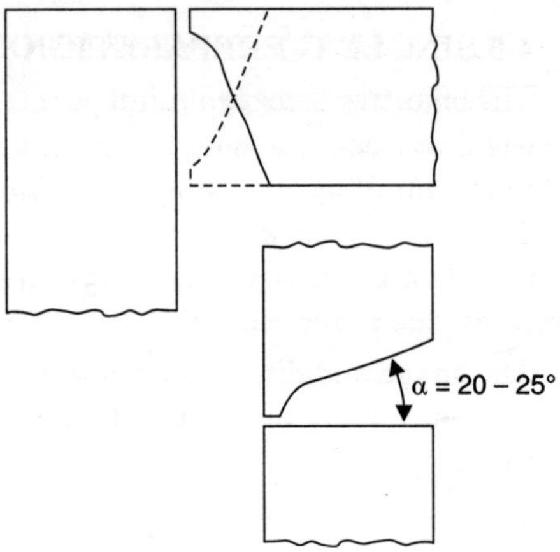

Suitable for inside and outside corner joints provided there is no lamellar tearing.

Also for horizontal-vertical position butt joints. Cheaper to prepare than asymmetric U for this purpose.

Fig. 8.14 Single J preparation

6.0 DOUBLE V PREPARATION

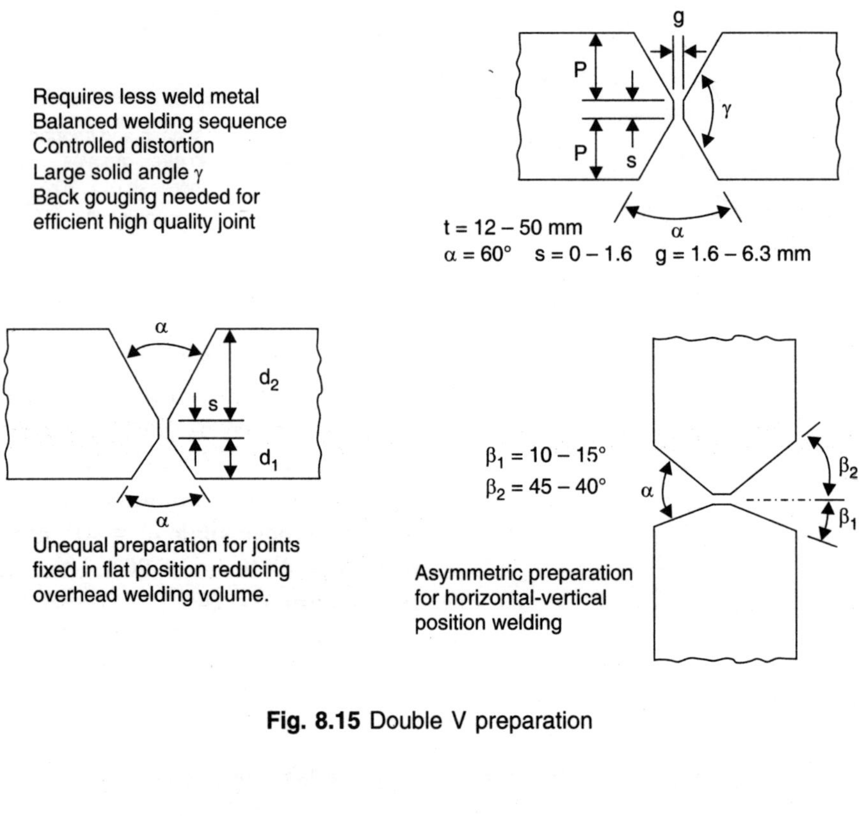

Fig. 8.15 Double V preparation

7.0 DOUBLE BEVEL PREPARATION

Thickness t = 19 to 51 mm

α = 50 – 55°

s = 0 to 1.6 mm

g = 1.6 to 6.3 mm

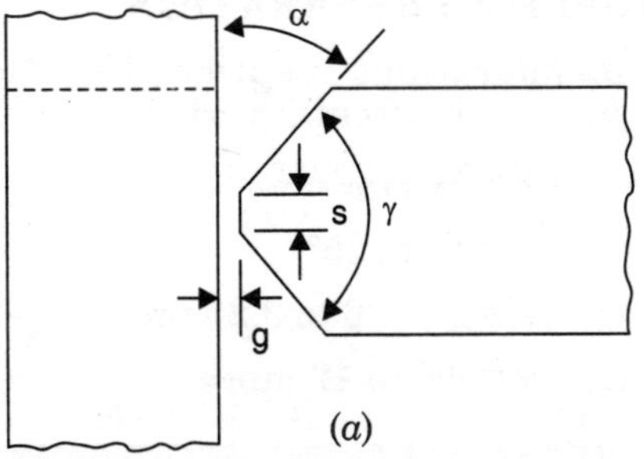

Fig. 8.16 (*a*) Double bevel preparation

Penetration on each side may be different to suit the requirements as in V preparation.

Cheaper to prepare than asymmetric double V for horizontal vertical position butt joints.

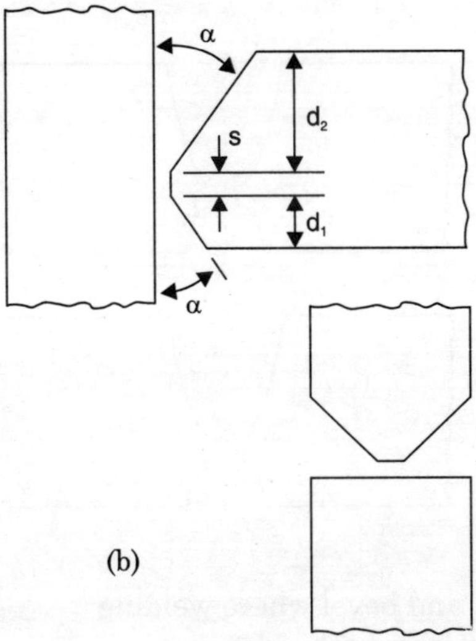

Fig. 8.16 (*b*) Double bevel preparation

8.0 DOUBLE U PREPARATION

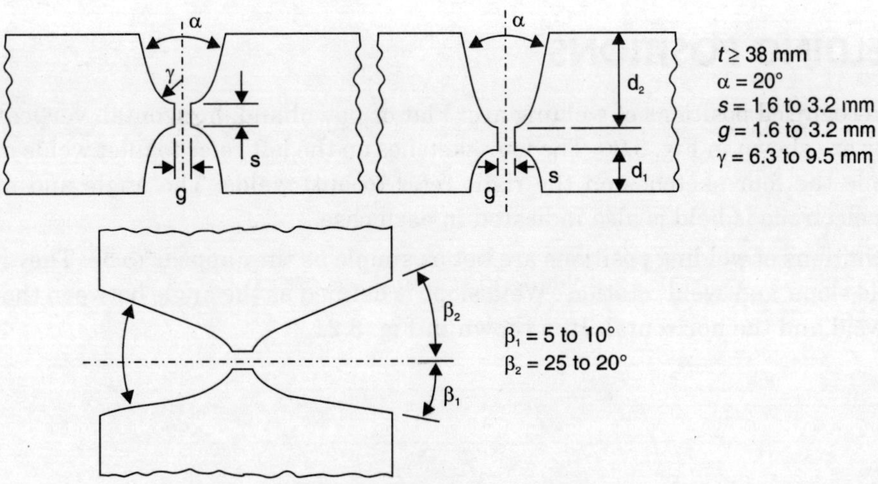

Fig. 8.17 Double U preparation

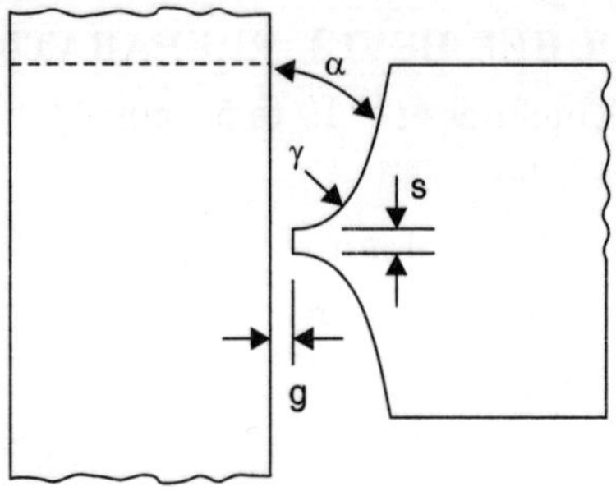

Fig. 8.18 Double J preparation

9.0 DOUBLE J PREPARATION

Considerations mentioned in J-apply here also

$t \geq 38$ mm

$\alpha = 15$ to $25°$

$s = g = 1.6$ to 3.2 mm

$\gamma = 9.5$ to 12 mm

10. MIXED PREPARATIONS

Normal U one side, Flat bottomed U on the other side to facilitate back gouging.

Shallow reverse side allows cheaper V-preparation.

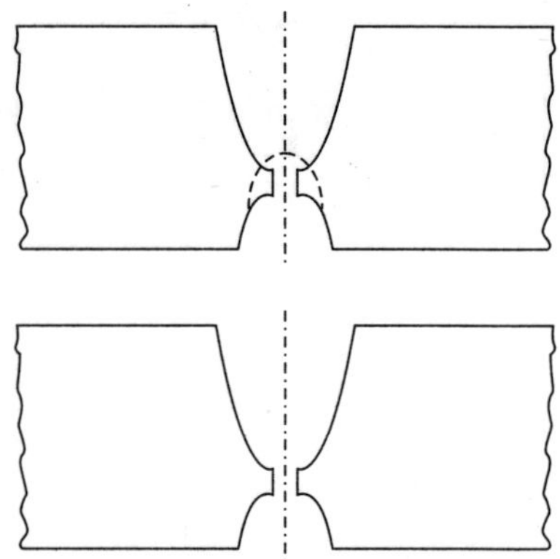

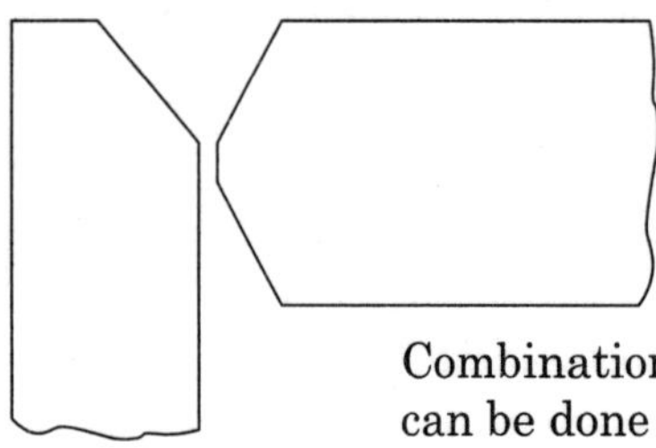

Combination of V and bevel where welding can be done easily from both sides.

Fig. 8.19 Mixed preparations

8.5 WELDING POSITIONS

The four recognised positions of welding are: Flat or downhand, horizontal, vertical and overhead. They are shown in Fig. 8.20. The four sketches on the left refer to fillet welds made in the joints, while the four sketches on the right refer to butt welds. The angle and direction in which the electrode is held is also indicated in each case.

Definitions of welding positions are not as simple as they appear to be. They involve the terms 'weld slope' and 'weld rotation'. Weld slope is defined as the angle between the line of the root of a weld and the horizontal. It is shown in Fig. 8.21.

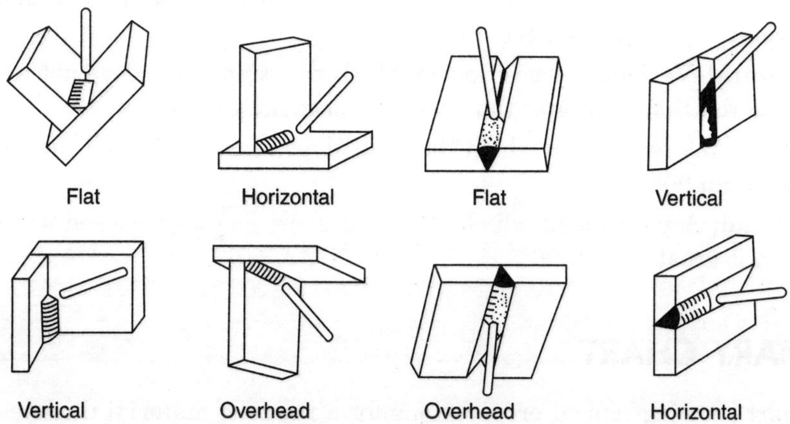

Fig. 8.20 Welding positions for butt and fillet welds

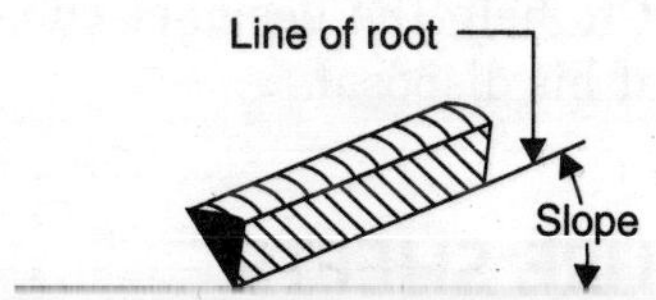

Fig. 8.21 Diagram to illustrate weld slope

Weld rotate is defined as the angle between the upper portion of the vertical reference plane passing through the line of a weld root, and a line drawn through the same root intersecting the weld surface at a point equidistant from either toe of the weld. It is illustrated in Fig. 8.22.

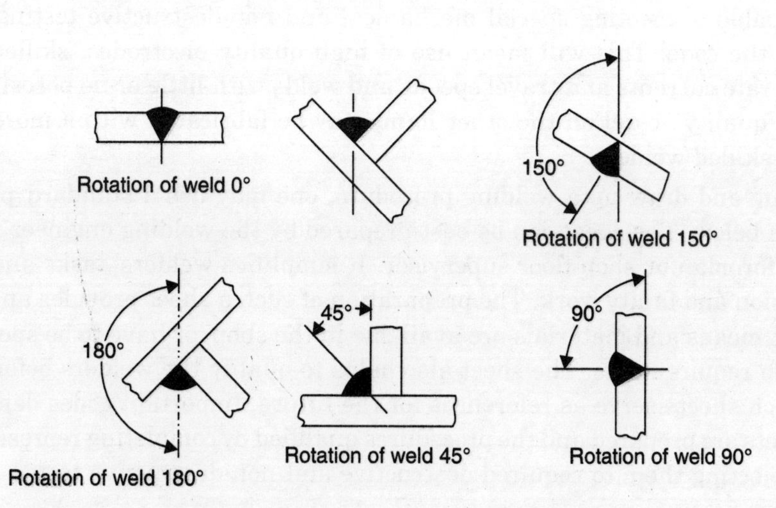

Fig. 8.22 Diagrams to show weld rotation

The welding position are defined as follows:

– **Downhand or flat:** A position in which the slope does not exceed 10° and the weld rotation does not exceed 10°.

- **Inclined:** A position in which the weld slope exceeds 10° but not 45° and in which the weld rotation does not exceed 90°.
- **Horizontal–Vertical:** A position in which the weld slope does not exceed 10°, and the weld rotation is greater than 10°, but does not exceed 90°.
- **Vertical:** Any position in which the weld slope exceeds 45° and the weld rotation is greater than 90°.
- **Overhead:** A position in which the weld slope does not exceed 45° and the weld rotation is greater than 90°.

8.6 SUMMARY CHART

A summary chart showing typical preparations for a range of material thicknesses for major arc welding processes has been provided for quick reference on page 165.

The illustrations given do not cover all possible joints which may be used in practice but the principles have been clarified to help the designer choose the best preparations for the constraints of the choices he has at his disposal.

8.7 WELDING PROCEDURE SHEETS

AWS defines welding procedure, as the detailed methods and practices including all joint welding procedures involved in the production of a weldment. It is very important that before starting to weld, a welding procedure is drawn up, which will ensure acceptable quality welds at the lowest overall cost. Procedures become more stringent and costly as criticality of the job increases. For example, fabrication of a pressure vessel conforming ASME code requires defect-free welds capable of meeting special mechanical and non-destructive testing requirements demanded by the code. This will mean use of high quality electrodes, skilled and certified welders, moderate currents and travel speeds and welds with little or no porosity or undercut. A commercial quality vessel on the other hand may be fabricated with a more liberal procedure and less skilled welders.

To define and draw up a welding procedure, one may use a standard procedure sheet such as shown below. The sheet can be best prepared by the welding engineer in consultation with welding foreman or shop-floor supervisor. It simplifies welders' tasks and prevents last minute confusion and faulty work. The preparation of such a sheet provides an opportunity to check on what means and materials are available in the shop, or have to be specially provided to meet the job requirements. The sheet also helps to qualify the welders before they are put on the job. Such sheets serve as references for the future. Important codes demand that such procedure sheets are prepared and the procedures qualified by completing representative welded joints and subjecting them to required destructive and non-destructive tests.

SUMMARY CHART:

Typical preparations for a range of material thickness.

Material thickness	Process				
	Manual metal arc	Manual CO_2 DIP transfer	Manual CO_2 spray transfer	Mechanised CO2	Submerged arc
20 S.W.G.					
16 S.W.G.				1/32 in.	
1/8 in.		1/16 in.		1/16 in.	
3/16 in.	60° 1/16 in. 1/16 in.	60°			
1/4 in.	60° 1/16 in. 1/16 in.	60°	1/32 in.		
3/8 in.	60°-70° 1/16 in. 1/16 in.	60° 1/16 in.	40°-50° 1/16 in.	40° 1/16 in.	40° 1/16 in.
1/2 in.	60°-70° 3/32 in. 3/32 in.	60° 1/16 in.	40°-50° 1/16 in.	40° 1/16 in.	40° 1/16 in.
3/4 in.	60°-70° 1/8 in. 1/8 in.	60° 1/16 in.	50° 1/8 in. 50°	40° 1/4 in. 40°	40° 1/4 in. 40°
1 in.	60°-70° 1/8 in. 60°-70°	60° 1/16 in. 60°	50° 1/8 in. 50°	40° 1/4 in. 40°	40° 1/4 in. 40°
1½ in.	60°-70° 1/8 in. 60°-70°	60° 1/16 in. 60°	50° 1/8 in. 50°	60° 1/2 in. 60°	40° 1/4 in. 40°
3 in.	20° 1/4 in. r 1/8 in. 20°	60° 1/16 in. 60°	50° 1/8 in. 50°	30° 1/4 in. r 1/2 in. 30°	30° 1/4 in. r 1/4 in. 30°

Typical Procedure Sheet for Smaw

(*a*) Welding procedure number

(*b*) Related specification and/or drawing number

(*c*) Material to be welded; specification number or composition

(*d*) Metallurgical condition of material

(*e*) Type of weld

(*f*) Preparation of parts:

  (*i*) Angle of bevel

  (*ii*) Root face (*iii*) Root radius

(*g*) Cleaning before welding

(*h*) Set-up of joint (gap, included angle, tolerance on alignment etc.)

(*i*) Particulars of backing strip or bar

(*j*) Welding position and direction

(*k*) Make, type and classification of electrode

(*l*) Electrical supply and electrode polarity

(*m*) Size of electrode for each run

(*n*) Length of run per electrode

(*o*) Current for each run

(*p*) Open circuit voltage

(*q*) Arc voltage

(*r*) Preheating procedure

(*s*) Time between runs

(*t*) Number and arrangement of runs

(*u*) Welding sequence

(*v*) Technique for depositing each run

(*w*) Method of inter-run cleaning

(*x*) Mechanical working of runs

(*y*) Preparation of root before welding reverse side

(*z*) Postweld heat treatment.

8.7.1 Type of Joints

There are six common types of joints, namely, butt, tee, cruciform, lap, corner and edge. These are illustrated in Fig. 8.23, which also illustrates three main types of weld, namely, butt, fillet, and edge. A typical butt weld is shown in the butt joint. A fillet weld is approximately triangular in transverse cross-section, and is used in tee, cruciform, lap and corner joints. An edge weld is a weld in an edge joint, and it covers a part or the whole of the edge widths.

Design of welded joints is based on several considerations, some of which are:

(*a*) Manner of stress tension, shear, bend, torsion.

(*b*) Whether loading is static or dynamic; whether fatigue is involved.

(*c*) Whether subjected to corrosion or erosion.

(*d*) Joint efficiency, which is defined as the ratio of the strength of the joint to that of the base metal, expressed as a percentage.

(*e*) Economy; amount of weld metal required to complete the joint and whether high deposition processes and procedures can be used.

(*f*) Constriction factors: accessibility, control of distortion and shrinkage cracking, production of sound welds.

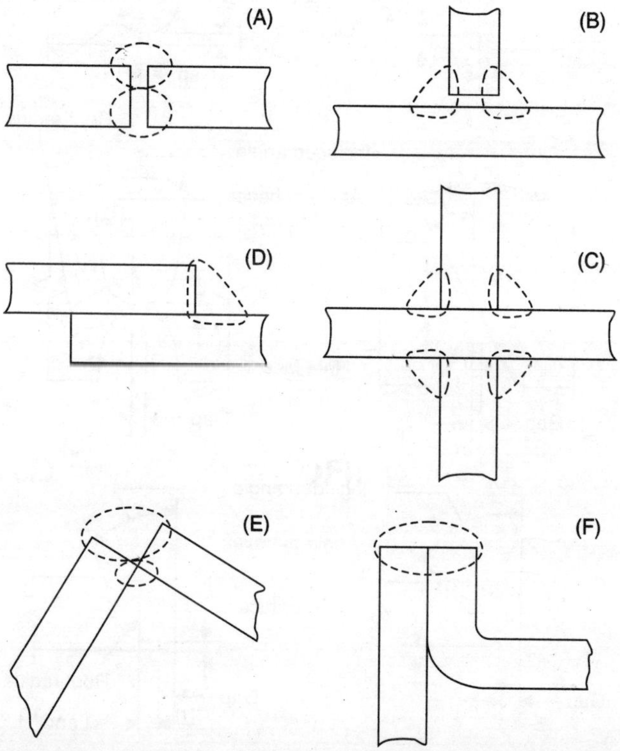

Fig. 8.23 Major types of joints: (A) Square butt weld (B) Square tee-joint and fillet welds (C) Cruciform joint with four fillet welds (D) Lap joint with single fillet weld (E) Full open corner joint with fillet welds (F) Edge joint with edge weld.

Various types of joints and welds used in welded strictures are given in Figs. 7.9–7.19 (Chapter 7).

8.7.2 Welding Parameters

To devise a welding procedure, one must choose correct welding parameters, *i.e.,* electrode size, current characteristics and value, welding speed, arc length, angle of electrode, welding position and welding technique. The following notes are meant to help one to arrive at an acceptable procedure.

(*a*) *Electrode size.* Each size has a specific current capacity range, which is indicated on the package by the electrode producer. Use of currents above the range will cause the covering to overheat and breakdown, resulting in increased spatter and low weld quality. Lower currents will give insufficient penetration.

Electrode size depends on joint thickness, edge preparation and welding position. Largest size that gives quality welds at high production rate should be preferred.

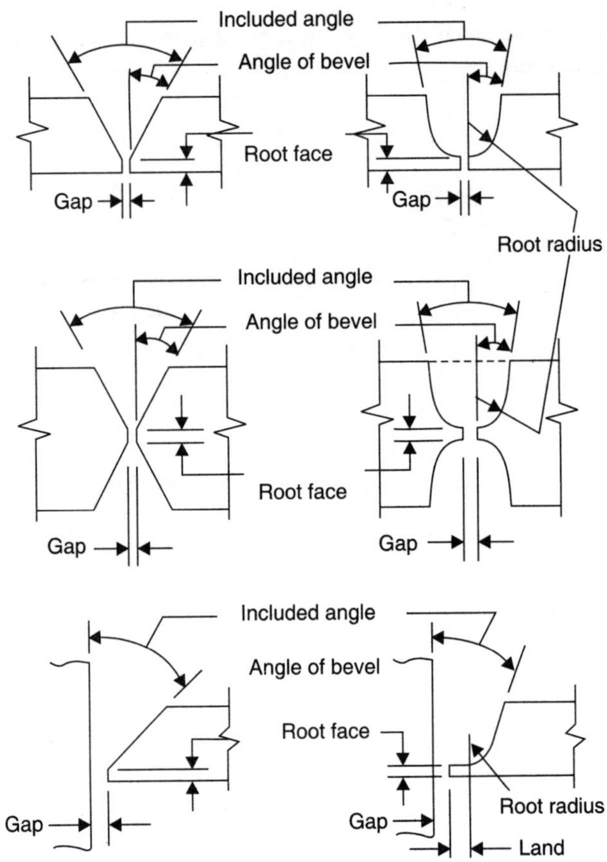

Fig. 8.24 Terms pertaining to typical weld preparations

For vertical and overhead welding, smaller diameter electrodes have to be used to restrict the size of the weld puddle, since there is a tendency for the molten metal to flow out of it due to the force of gravity. The largest size which an average welder can manage in these positions is 4 mm diameter in the case of non-iron powder type electrode (say E6013), and 3.15 mm diameter in the case of an iron-powder type (E7018). A skilled welder can weld satisfactorily in vertical and overhead positions with 5 mm diameter electrodes of E6013 as well as E7018 class.

The electrode size is also dictated by the consideration of accessibility to the root of the joint. In a V-grove, for example, electrodes small enough to give correct arc length and to reach the root have to be used for the initial passes, followed by larger size to complete the weld. In a T-joint, on the other hand, a larger diameter electrode (6 mm or 8 mm) can be used for the initial pass, since the access to the root it easy.

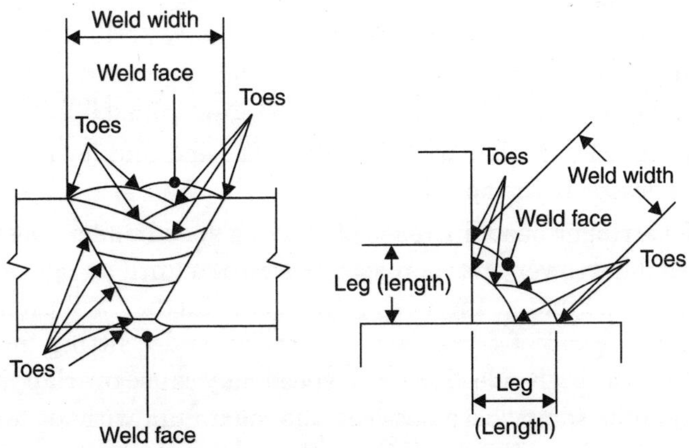

Fig. 8.25 Term pertaining to welds

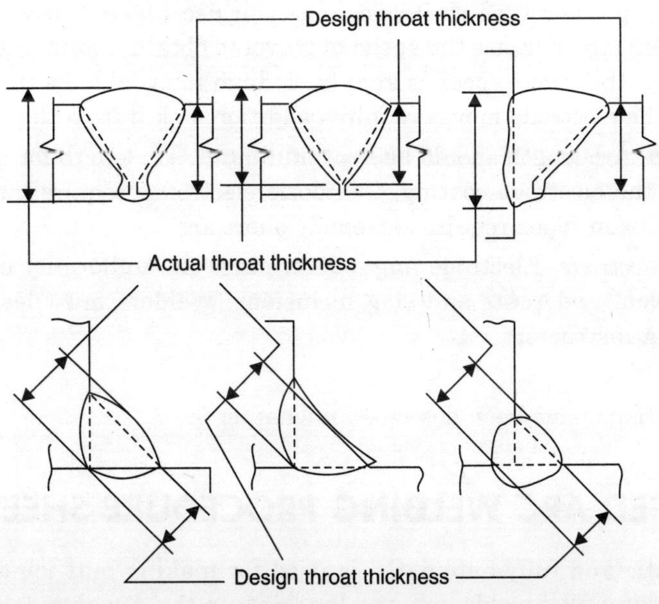

Fig. 8.26 Actual and design throat thicknesses of welds

In some cases, the electrode size has to be restricted to avoid the possibility of burn-through, caused either by bad fit-up (large gap at the root) or thinness of the material. In some metals and alloys, the weldability considerations require that the heat input is restricted by using electrodes of smaller sizes than normally used.

(*b*) *Current-type and amount.* The various factors which must be considered in choosing AC or DC, and the polarity in DC, are explained in chapter 4 article 4.2. Current values to be used are indicated under Welding Currents (Table 4.3 p. 77)

Where previous experience is not available, the safest course is to follow the manufacturer's recommendation regarding the type of current, polarity in the case of DC and the amount of current to be used.

(*c*) *Welding speed.* By welding speed is meant the arc travel speed. For a given electrode size and current, the speed is higher with the stringer bead and lower with the weave bead. The wider the weave, lesser is the speed.

In the case of a stringer bead, increase of welding speed under constant arc voltage and current makes the bead narrower and increase penetration until an optimum speed is reached at which penetration is maximum. Increasing the speed further will cause a reduction in the penetration. Too high a speed of travel also results in undercutting, more so when this is coupled with current on the high side. Too low a speed may cause overlapping and overwelding. The travel speed should be somewhere between the maximum without underwelding and the minimum without overwelding. Fillet welding affords a wider latitude with regard to travel speed, but it should be suitably adjusted to obtain the required size of fillet weld.

Electrode melt-off rate is one of the most important factors influencing arc speed. With high-deposition iron powder type electrodes, one can use higher currents to obtain higher melt-off, and considerably increase the speed of travel to obtain a weld bead of a given size. In sheet metal working, the travel speed is kept fairly high to avoid burn through but filling the crater properly as the electrode moves requires additional skill from the welder.

(*d*) *Arc length.* Arc length should be kept minimum. Arc length for quality weld deposit also depends upon the electrode coating. Cellulosic electrodes require larger arc than rutile and basic. Low hydrogen types require extremely short arc.

(*e*) *Angle of electrode.* Electrode angle determines the uniformity of fusion, weld bead contour, freedom from undercuts and slag inclusions. Welders must learn this skill under experienced welding instructors.

Welding Positions

Welding positions have been described in chapter 7.

8.8 SUBMERGED ARC WELDING PROCEDURE SHEETS

SAW, semi-automatic and fully-automatic, is used for making butt joints in the downhand position and for making fillet welds in T and lap joints in the downhand and horizontal-vertical positions as shown in Fig. 8.27.

Normally this process cannot be used in vertical and overhead position, because of the difficulty of preplacing the flux.

It is important to bear in mind that the SAW process demands accurate edge preparation and fit-up. In MMAW, irregularities in this regard are taken care of by the manual welder, though they do result in increased welding time and a large consumption of electrodes. In SAW, on the other hand, the operation is automatic, welding currents are high and the arc is deeply penetrating. Moreover, since the joint is submerged under the flux, the operator is unable to adjust the procedure to accommodate joint irregularities. A poor fit-up in a butt joint can cause the granular flux to spill through the root gap. It can also give rise to burn-through and slag inclusions.

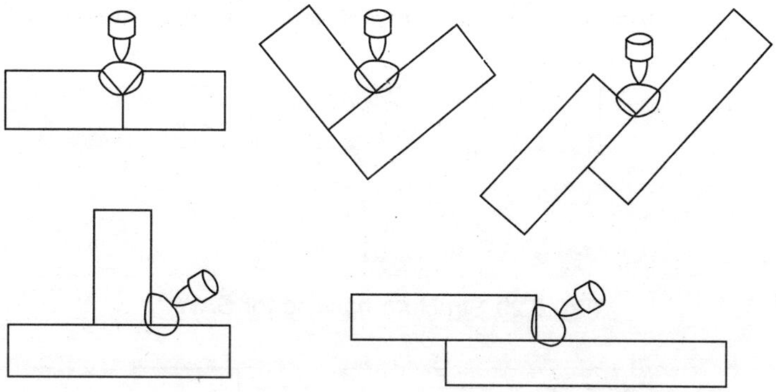

Fig. 8.27 Joint and positions suitable for SAW

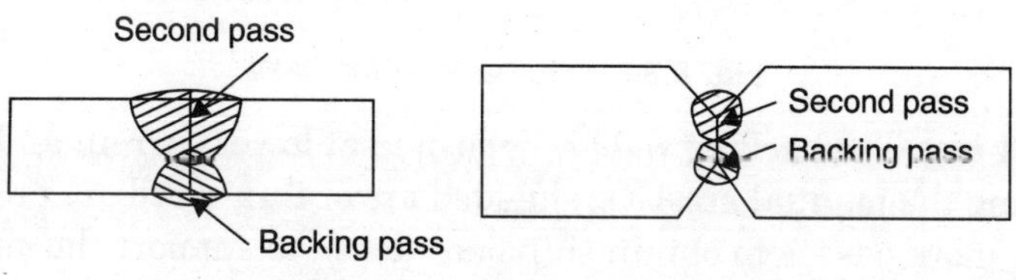

Fig. 8.28 Base metal backing for SAW

Shops using SAW are advised to make edge preparations with automatic thermal cutting equipment (oxy-acetylene or plasma-arc), or by machining. In the absence of such facilities, SAW becomes a slow and unproductive operation with frequent interruptions and increased proportion of weld rectification.

In SAW, the weld puddle is of large size and remains in a molten condition for a long time. The welding procedure must ensure that this molten puddle is supported and contained until it has solidified at the root of the weld. This precaution is a must when full joint penetration has to be achieved in a butt joint. The technique used for this purpose is termed weld backing.

8.8.1 Weld Backing Techniques

The various commonly used techniques involve use of the following: (1)Base metal backing; (2) Structural backing; (3) Weld backing; (4) Backing strip; (5) Copper backing; (6) Flux backing; (7) Backing tapes.

1. Base metal backing. The root face is kept sufficiently thick as shown in Fig. 8.28, to support the weld pool without burn-through. This technique is used for square or partially bevelled butt joints, for fillet welds and for plug or slot welds. Care has to be taken that the root faces of grove welds are in close contact. The first pass, deposited sometimes with lower current, acts as a backing for the second pass deposited with higher current to get through penetration.

2. Structure backing. In certain cases where design permits, another structural member can serve as a backing for the weld, as shown in Fig. 8.29. It is very important that the contact surfaces of the joint are clean and the contact is intimate in order to avoid porosity and slag inclusions. The weld must also provide sufficient depth of fission in the backing member.

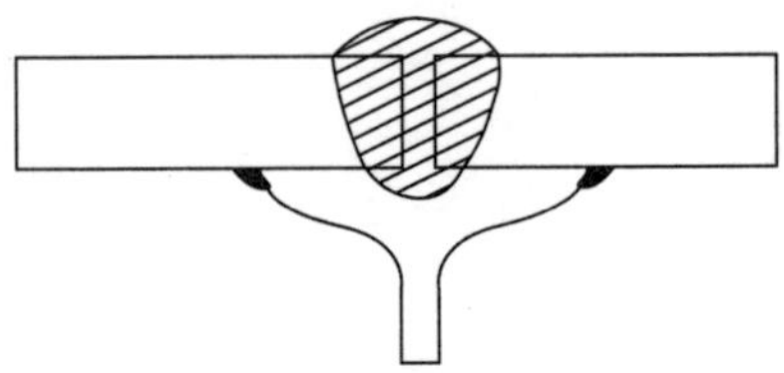

Fig. 8.29 Structure backing for SAW

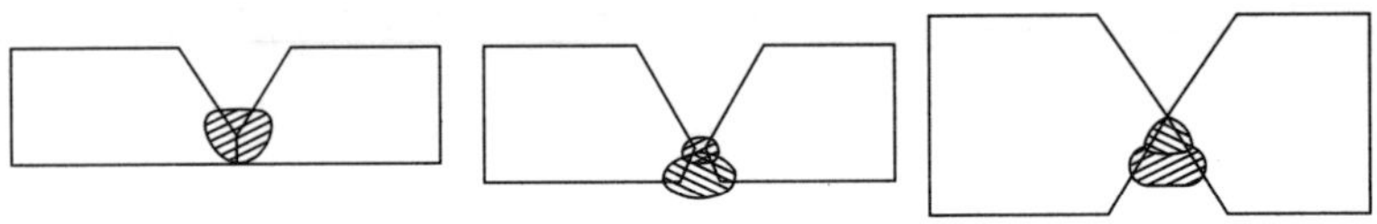

Fig. 8.30 Weld backing for SAW

3. Weld backing. The backing weld is deposited at lower current and with a moderately penetrating arc using the manual arc, CO_2 shielded arc or flux-cored arc process (see Fig 8.30). It may be in one or more passes to obtain sufficient depth to support the submerged-arc weld. The backing weld may be retained in the joint if it is of suitable quality. If otherwise, it may be removed by oxygen on arc gouging, by chipping or by machining after the submerged-arc welds have been deposited. The resulting groove is filled up with a submerged-arc weld.

4. Backing strip. The backing strip is of metal that is compatible with the one being welded. The weld metal fuses into the backing strip, so that it becomes an integral part of the joint as shown in Fig. 8.31. In this case, it is termed a permanent backing. In case it is intended to be a temporary backing, it may be removed finally by machining. Suitable root opening must be kept to ensure full penetration. It varies between 1.6 and 4.8 mm, depending on joint thickness. It is important that the contact surfaces between the plates and the strip are clean and the contact is intimate; otherwise porosity and leakage of molten weld metal may occur.

5. Copper backing. Copper backing shown in Fig. 8.32 has several advantages. Its high thermal conductivity enables it to extract the heat rapidly from the molten weld pool. Also the molten steel weld metal does not fuse with the copper material. Hence it only serves as a temporary backing.

The copper backing bar is either as long as the joint; or it is of short length and designed to slide underneath the travelling arc. In still other applications, it may be in the form of a rotating wheel.

For high production applications, the copper bar is provided with internal water circulation to maintain it relatively cool. The bar is usually grooved as shown in the figure to obtain weld reinforcement on the underside of the joint. It is important to ensure that the copper bar has sufficient mass to prevent melting of the copper material, which can result in contamination of the weld with copper. It must be borne in mind that mechanical properties of steel weld metal deteriorate when the Cu content exceeds a certain limit.

6. Flux backing. As shown in Fig. 8.33, dry granular SA flux is placed in a trough of flexible sheet material. This sheet material rests on a rubberised canvas hose, which can be inflated to hold the flux tightly against the back of the joint. This technique will be discussed in detail while describing the one-side SAW used in Japanese shipyards.

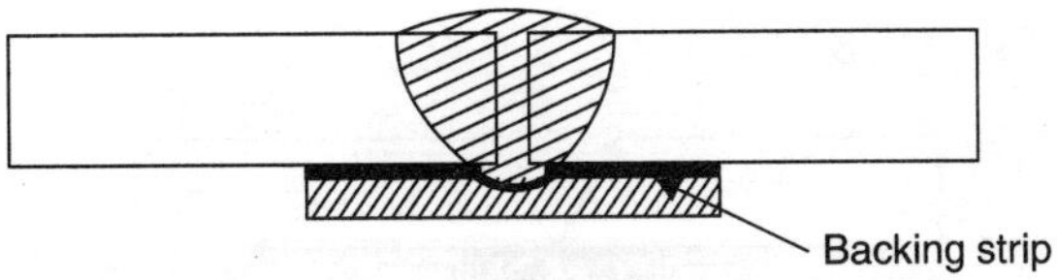

Fig. 8.31 Backing strip for SWA

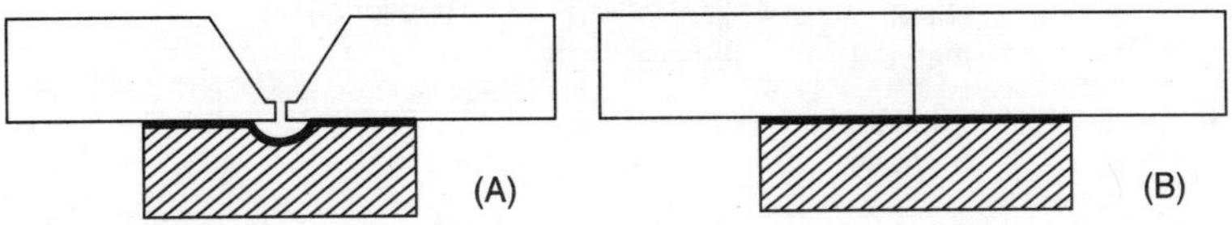

Fig. 8.32 Copper backing for SAW: (A) V-groove butt; (B) Square butt

7. Backing tapes. Ceramic back-up tapes consisting of a ceramic material on an aluminium foil backing are available in the U.S.A. The exposed aluminium foil edges are covered with pressures sensitive adhesive covered with a removable liner. Lengths of strips are 0.5 to 1.0 metre. These can be easily applied to joints or seams to provide shielding or back-up for oneside welding and root pass back-up for two-side welds to be deposited by TIG, MIG and other arc processes. By using these tapes, arc gouging and further backside joint operations such as griding are eliminated or minimised. They avoid the use of expensive and clumsy fixtures, back-up bars and gas purging of weld.

8.8.2 Butt Welds

To make a full penetration butt weld in sheet metal without burn-through, steel or copper backing bar must be used. The joint is then completed with a single weld pass deposited from one side.

With copper backing, a square butt joint without root gap is used. The procedure data are given in Table 8.1.

Table 8.1. Data for SA butt welds with copper backing

Plate thickness t, mm	*Electrode dia. mm*	*Current amps. Electrode + ve*	*Voltage V*	*Speed mm/sec.*
1.6	2.4	350	23	50
2.0	2.4	400	24	42
2.4	3.2	500	30	40
3.6	3.2	650	31	30

Joint fit-up with steel backing is shown in Fig. 8.34 which shows that a small root opening is helpful. The procedure data are given in Table 8.2.

Plates up to 12.7 mm thickness and with square edges can be butt welded with a single pass using a steel backing strip. It is advisable to keep a root opening, because when the edges are butted together tightly, the resultant weld has a high build-up. Alternatively, a grove can be provided. Procedure data are given in Table 8.2.

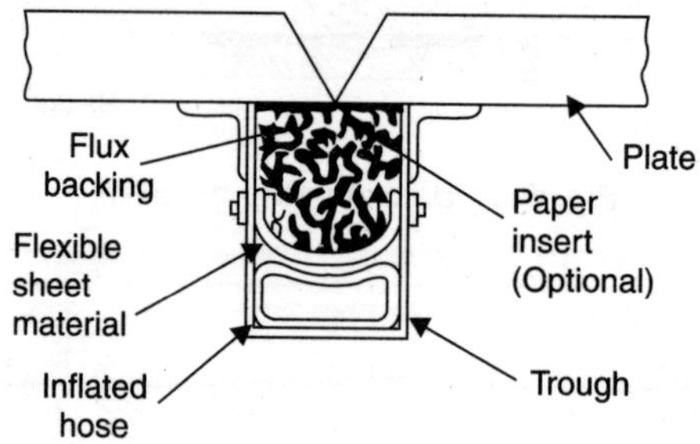

Fig. 8.33 A method of producing flux backing for SAW

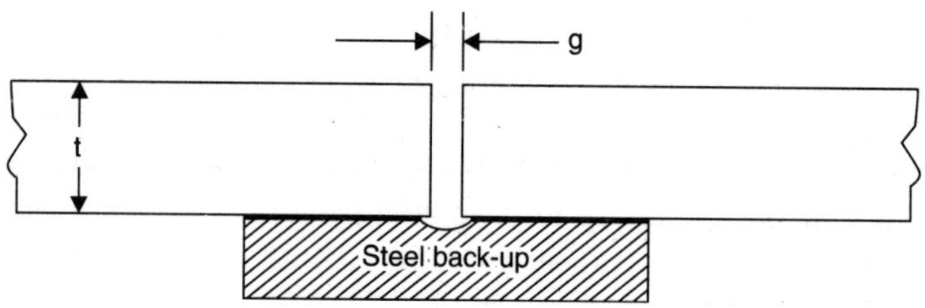

Fig. 8.34 Joint fit-up for butt welds in sheet metal

Table 8.2. Data for SA butt welds with steel backing

Plate thickness mm/sec. t, mm	*Root opening g, mm*	*Electrode dia. mm*	*Current amps. Electrode + ve*	*Voltage V*	*Speed mm/sec*
1.6	0–0.8	3.2	450	25	45
2.0	0–0.8	3.2	500	27	33
2.4	0–1.6	3.2	550	27	25
3.6	0–1.6	3.2	650	28	20
4.8	1.6	5.0	850	32	15
6.4	3.2	5.0	900	33	11
9.5	3.2	5.6	950	33	10
12.7	4.8	5.6	1,000	34	8

Plates in the thickness range of 6.4–15.9 mm and with square edge butted together tightly, can be conveniently butt welded with two passes, one from each side as shown in Fig. 8.35. The first pass deposited at a lower current serves as a backing for the second pass. It is important that the two passes penetrate into each other sufficiently to prevent lack of fusion and slag inclusion in the central region. Procedure data are provided in Table 8.3.

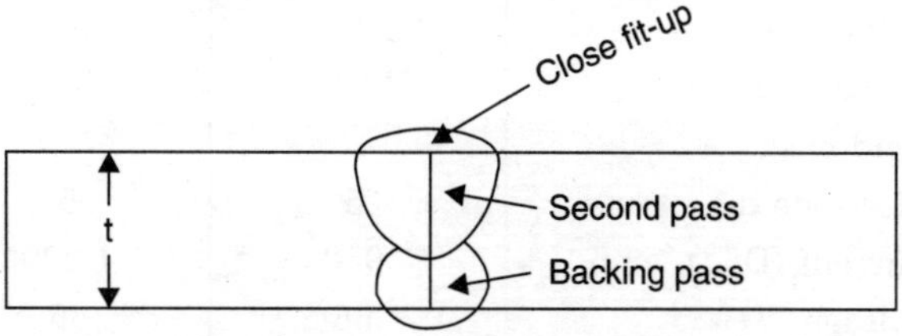

Fig. 8.35 Square butt weld in two passes, one from each side

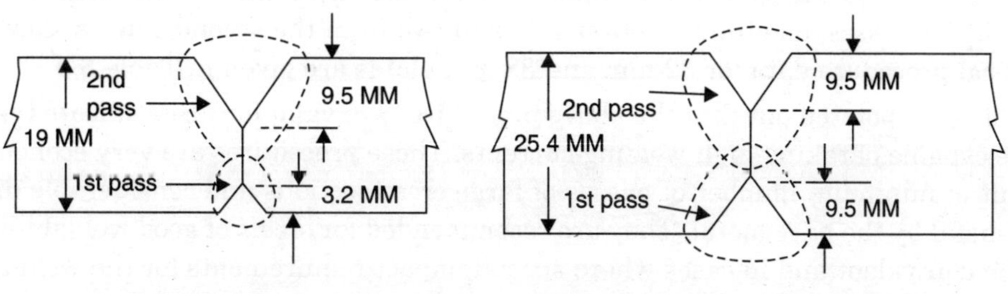

Fig. 8.36 Parameters for two-pass 19 mm and 25.4 mm t butt welds

Table 8.3. Data for two-pass square butt weld, one from each side

	Baking pass				*Second pass*			
Plate thickness t, mm	*Electrode dia. mm*	*Current amps.*	*Voltage V*	*Speed mm/sec.*	*Electrode dia.*	*Current amps. mm*	*Voltage V*	*Speed mm/sec.*
6.4	4.0	475	29	20	4.0	575	32	20
9.5	4.0	500	33	14	4.0	850	35	14
12.7	5.0	700	35	11	5.0	950	36	11
15.9	5.0	900	36	9	5.0	950	36	9

The above-described procedure can be extended to plates of 19 mm and 25.4 mm thickness by providing 60° V-groves on both the sides and sufficiently large root face as shown in Fig. 8.36. Procedure data are given in Table 8.4.

Table 8.4. Data for 19 mm and 25.4 mm t butt welds

	18 mm t	*25.4 mm t*
First pass		
Electrode dia., mm	5	5
Current (DC^+), amp	700	850
Voltage, V	35	35
Speed, mm/sec	12	5.5
Second pass		
Electrode dia., mm	5	5
Current (DC^+), amp	950	1,000
Voltage, V	36	36
Speed, mm/sec	6	7

When plate thickness increases further, it becomes necessary to increase the V-groove and deposit the passes, one from the first side and two from the second side as shown in Fig. 8.37. Typical procedure data for 32 mm and 38 mm plates are given in Table 8.5.

It must be pointed out that the above procedures are valid for fused silicate type fluxes, which are capable of taking high welding currents. These procedures are very economical and they result in minimum number of passes of large cross-sections and considerable dilution of the weld metal by the base metal. They are recommended for steels of good weldability having low carbon equivalent and in cases where special impact requirements for the weld metal are not specified.

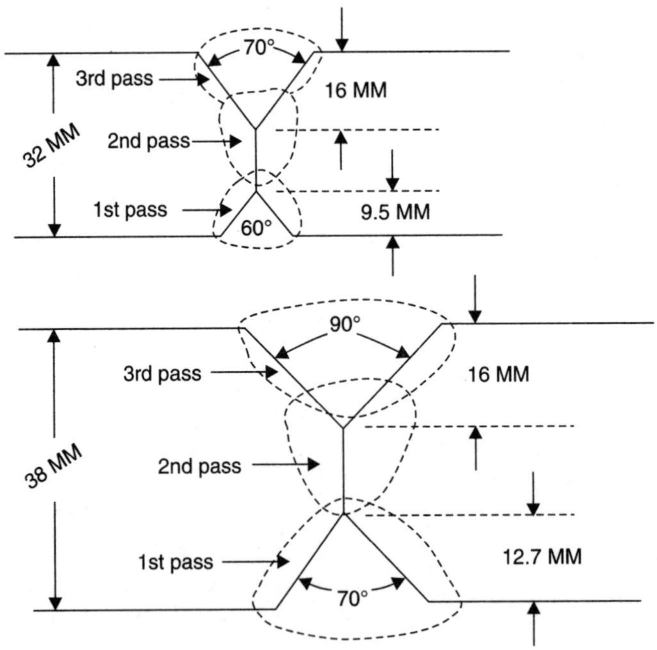

Fig. 8.37 Parameters for three-pass 32 mm and 38 mm t butt welds

Table 8.5. Data for 32 mm and 38 mm t butt welds

Plate	*First pass*				*Second pass*			
thickness t, mm	*Electrode dia. mm*	*Current amps.*	*Voltage V*	*Speed mm/sec.*	*Electrode dia.*	*Current amps. mm*	*Voltage V*	*Speed mm/sec.*
32	5	850	35	5.5	5	1,000	36	5
83	5	1,000	36	4	5	1,000	36	4
					Third pass			
					Electrode dia. mm	*Current amps.*	*Voltage V*	*Speed mm/sec.*
					5	850	35	4
					5	950	34	3

For welding steels of difficult weldability, or where stringent weld metal impact requirements are specified, procedures involving basic type of flux, multiple passes of limited cross-sections deposited with low currents, and minimum dilution by the base metal are recommended. For plates of 16, 25.4 and 38 mm thickness, for example, the joint fit-up is made as shown in Fig. 8.38. First two passes are deposited manually with a 4 mm basic low-hydrogen type electrode. With these passes serving as a backing SA weld passes are deposited at a speed of 7 mm/sec using 4 mm diameter electrode, 550 amps, 28 V. The number of SA passes for 16, 25.4 and 38 mm thick joints are 5, 12 and 26 respectively. After the vee is filled up, the manual weld at the root is completely gouged out and the groove is filled up with a SA pass.

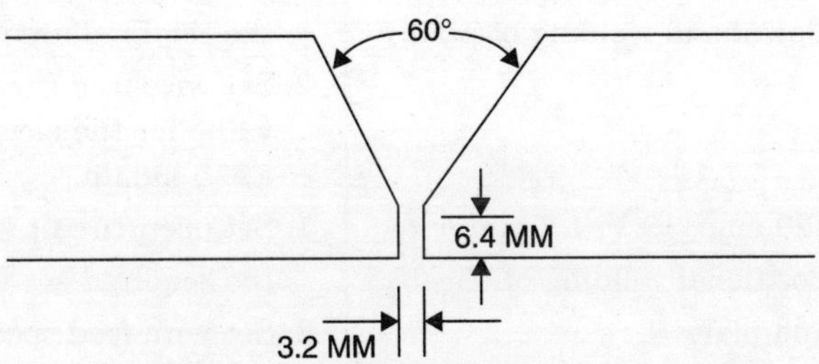

Fig. 8.38 Joint fit-up for multi-pass butt weld

8.9 WELDING PROCEDURE FOR MIG/CO_2 WELDING

As with other arc welding procedures, a good MIG/CO_2 welding procedure starts with correct edge preparation and joint fit-up. The joint surfaces must be free from rust, scale grease, oil, paint and other foreign materials. For making full penetration joints by welding with spray transfer technique from both sides, it is necessary to gouge out the root from the second side before starting to weld that side. When welding is done only from one side, suitable weld backing must be provided. Sometimes weld backing can be avoided by making the root pass with the short-circuiting technique to obtain uniform penetration and depositing the fill-up passes by high current spray transfer technique.

The welding equipment must be assembled and the welding parameters set according to the manufacturer's instructions. All gas and water connections must be absolutely leak-proof. If the shielding gas gets contaminated with air or water, the arc becomes erratic and pores appear on the weld.

The gun nozzle size and the shielding gas flow rate must be correctly set according to the material being, welded and its joints design. Some joint designs demand longer nozzle-to-work distance than normal; in such cases one must use higher gas flow rates than those recommended by the equipment manufacturer or as specified in standard procedures, and a gas nozzle of adequate size to cover the welding area. On the other hand, smaller nozzle sizes may be used for welding in confined areas or in the root of a thick joint. The electrode-feed rolls and the contact tube must be compatible with the size and composition of the electrode, as recommended by the manufacturer. If the contact tube is worn in usage, it must be replaced before the gun starts getting heated due to bad electrical contact between it and the electrode.

Electrode extension is the distance between the end of the contact tube and the gas nozzle opening, which is between 6.4 and 9.5 mm for normal spray-type welding. In special applications, the contact tube may be flush with or protruding from the gas nozzle. For example, when using the short-circuiting arc, the contact tube may extend 3 mm beyond the end of nozzle. Further guidance on procedures using contant-voltage power source is given in Table 8.6.

Table 8.6. Guidance on MIG/CO_2 welding procedure

Arc type	*Typical conditions and applications*	*Procedure*
Spray-type arc	360 amp, 34 V, 1.6 mm wire. Downhand welding of plate	1. Set open-circuit voltage to a little above the required arc voltage; *e.g.,* 38 V. 2. Set wire-feed speed* to the recommended value for the electrode size and material, *e.g.* 5 m/min.
Short-circuiting arc	120 amp, 19 V, 1.2 mm wire. Positional welding of sheet and plate	1. Set open-circuit voltage to a little above the required arc voltage, *e.g.* 20 V. 2. Set wire-feed speed* to the recommended value for the electrode size and material, *e.g.* 2.5 m/min. 3. Set choke (tune the circuit) to get required crispness and heat of arc.

The wire-feed-speed determines the welding current.

Following the setting of Table 8.6, trial bead welds should be deposited to arrive at correct arc voltage and the electrode-feed rate (current). In the short-circuiting procedure, the choke should be finally adjusted to obtain good arc start and a stable arc with minimum spatter.

■ QUESTIONS

8.1 What features a successful weld design must possess. List the factors that are of help in developing a weld design.

8.2 With a neat sketch state the elements that a complete welding symbol contains according to ISO and AWS system.

8.3 What is welding procedure sheet? Discuss the steps taken in preparing a welding procedure sheet. Discuss joint preparations for fusion welds.

8.4 What is meant by welding position? With neat sketches explain the different types of welding positions. Define the terms "weld slope" and "weld rotation" in this regard.

8.5 How do you define welding procedure? Why is it important to draw-up welding procedure before the welding is carried out.

8.6 What are the main elements of an "standard procedure sheet"? What are the benefits of using a standard procedure sheet?

8.7 Discuss the types of joints used in welds. State the factors which are considered in the design of welded joints.

8.8 How do you select welding parameters? Such as :

(*a*) Electrode size (*b*) Current type and amount

(*c*) Welding speed (*d*) Arc length

(*e*) Electrode angle (*f*) Welding positions.

8.9 Briefly discuss the special considerations in welding procedure development for SAW. What type of weld backings are in common use for SAW.

8.10 Explain the difference between the various types of backings used in SAW.

(*a*) Backing strip and copper backing

(*b*) Flux backing and backing tapes.

8.11 Briefly explain the TIG and MIG welding procedure.

CHAPTER 9

Weld Quality

As the welded joints are finding applications in critical components where the failure results into a catastrophy, the inspection methods and acceptance standards are increasing. Acceptance standards represent the minimum weld quality and are based upon test of welded specimens containing some discontinuities, usually a safety factor is added to yield the final acceptance standard. A good research effort is being directed to correlate the discontinuities with the performance.

In the present discussion we shall study the weld discontinuities commonly observed in the welds, their causes, remedies and their significance. Small imperfections, which cause some variation in the normal average properties of the weld-metal are called discontinuities. When the discontinuity is large enough to effect the function of the joint it is termed a defect.

Standard codes do permit limited level of defects based on fracture mechanics principles, taking consideration the service conditions of the fabrication. Inspite of all this, the fabricator

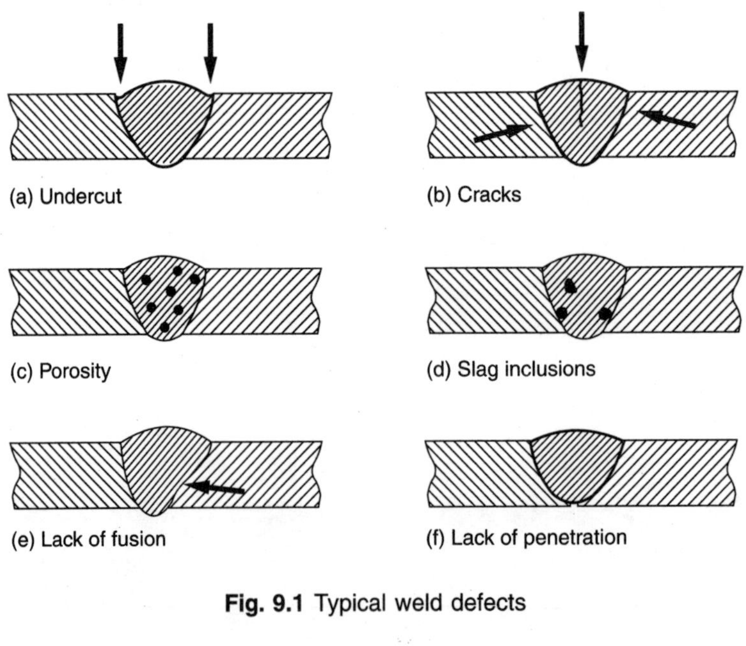

Fig. 9.1 Typical weld defects

must strive to prevent the occurrence of weld defects in the first instance and to rectify them if they do occur. There are many types of defects which have been classified in various documents (*e.g.*, BS499 part I, 1965). For our purpose we shall be discussing the most important ones shown in Fig. 9.1. These are undercuts, cracks, porosity, slag inclusions, lack of fusion and lack of penetration.

9.1 UNDERCUTS

The term is used to describe a groove melted into the base metal adjacent to the toe of a weld and left unfilled by the weld metal. It also describes the melting away of the sidewall of a welding groove at the edge of a layer or bead. This melting away of the groove forms a sharp recess in the sidewall in the area in which the next layer or bead must fuse. (Slag may be "keyed" into this undercut which, if not removed prior to subsequent passes, may become trapped in the weld.) An undercut, therefore, is a groove that may vary in depth, with, and sharpness at its root.

9.2 CRACKS

Cracks are linear ruptures of metal-under stress. Although sometimes wide, they are often very narrow separations in the weld or adjascent base metal. Usually little deformation is apparent. Three major classes of cracks are generally recognised: hot cracks, cold cracks, and macrofissures. All types can occur in the weld or base metal.

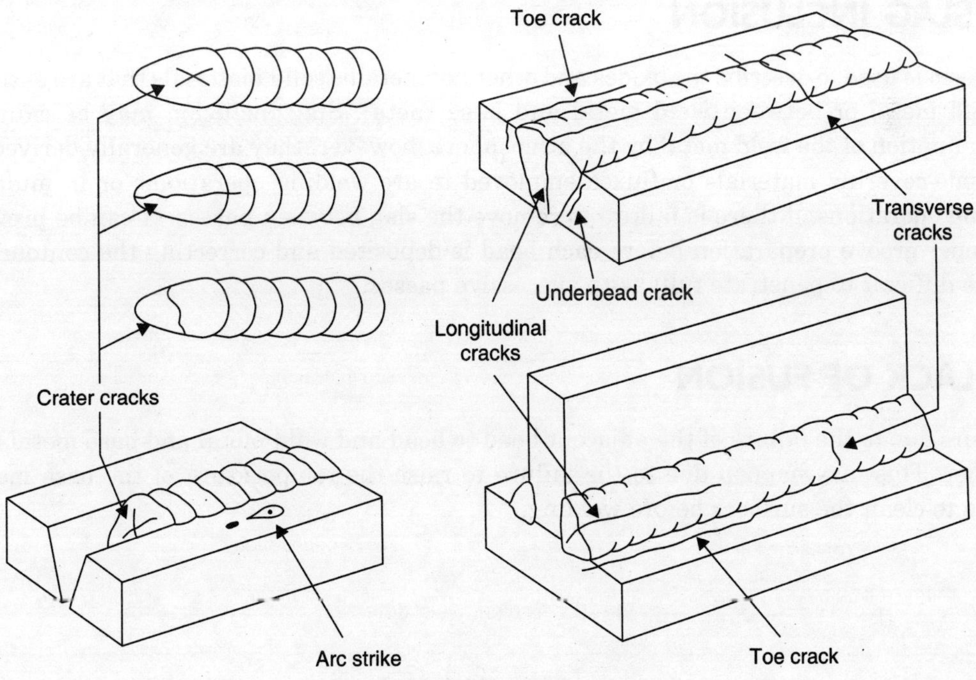

Fig. 9.2 Types of cracks in welded joints

Fig. 9.2 illustrates a variety of cracks including underbead cracks, toe cracks, crater cracks, longitudinal cracks, and transverse cracks. The underbead crack, limited mainly to steel, is base metal crack usually associated with hydrogen. Toe cracks in steel can be of similar origin. In other metals (including stainless steel), cracks at the toe are often termed edge of weld cracks, attributable to hot cracking in near the fusion line. Crater cracks are shrinkage cracks which result from stopping the arc suddenly.

9.3 POROSITY

Porosity is the presence of a group of gas pores in a weld caused by the entrapment of gas during solidification (when solidification is too rapid). They are small spherical cavities, scattered or clustered locally. Sometimes, the entrapped gas may form a single large cavity which is termed as a blow hole.

Causes:

1. Lack of deoxidisers
2. Base metal sulphur content being high
3. Presence of oil, grease, moisture or mill scale on the joint surface
4. Excessive moisture in flux
5. Inadequate gas shielding
6. Low current or long arc
7. Rapid solidification of weld deposit

9.4 SLAG INCLUSION

This term is used to describe the oxides and other nonmetallic solid materials that are entrapped in weld metal or between weld metal and base metal. Slag inclusion may be caused by contamination of the weld metal by the atmosphere, however, they are generally derived from electrode-covering materials or fluxes employed in arc welding operations; or in multilayer welding operations, if there is failure to remove the slag between passes. It can be prevented by proper groove preparation before each bead is deposited and correcting the contours that will be difficult to penetrate fully with successive passes.

9.5 LACK OF FUSION

It occurs due to the failure of the adjacent bead to bead and weld metal and base metal fusing together. This may happen due to the failure to raise the temperature of the base metal or failure to clean the surfaces before welding.

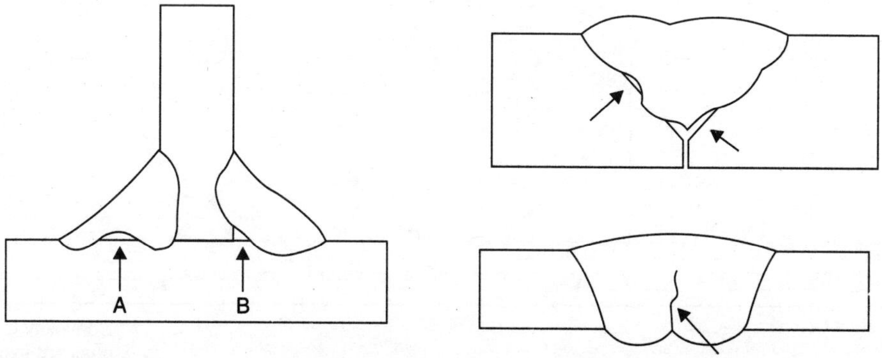

Fig. 9.3 Types of lack of fusion

9.6. LACK OF PENETRATION

This defect, occurs when the weld metal fails to reach the root of the joint and fuse the root faces completely. It is caused by using incorrect electrode size with respect to the form of the joint, low welding current, inadequate joint design and fit-up. It occurs more often in vertical and overhead welding positions.

9.7 FAULTY WELD SIZE AND PROFILE

A weld, otherwise deposited correctly without a defect may not be acceptable due to the shape of its profile. Excessive or lack of reinforcement are both defective. Defective profiles on butt welds are shown in Fig. 9.4 while Fig. 9.5 describes desirable, acceptable and defective profiles on fillet welds. These faults arise from the use of an incorrect welding procedure and could be eliminated if the following factors are considered:

(*a*) correct joint preparation and fit-up

(*b*) proper electrode size and welding current

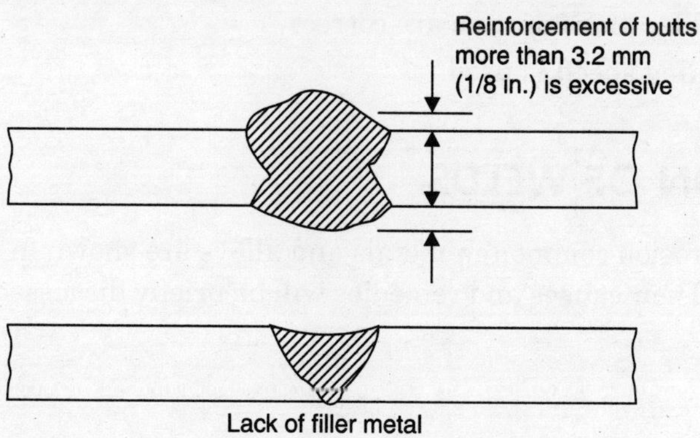

Fig. 9.4 Excessive reinforcement, Lack of filler metal

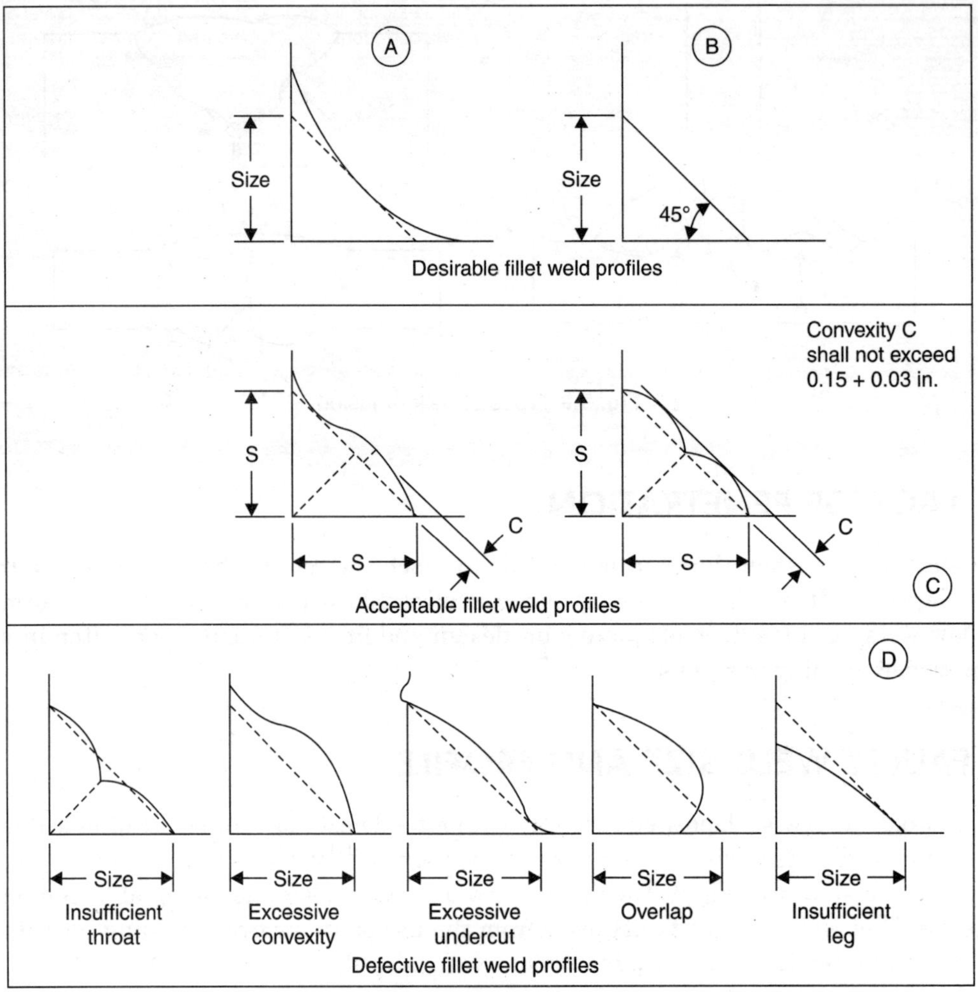

Fig. 9.5 Desirable, acceptable and defective fillet weld profiles

(*c*) number and locations of runs are correct

(*d*) correct welding speed is used.

9.8 CORROSION OF WELDS

Different types of corrosion common in metals and alloys are shown in Fig. 9.6. Some of these are related to welds. Their causes and remedies will be briefly discussed in the following paragraphs.

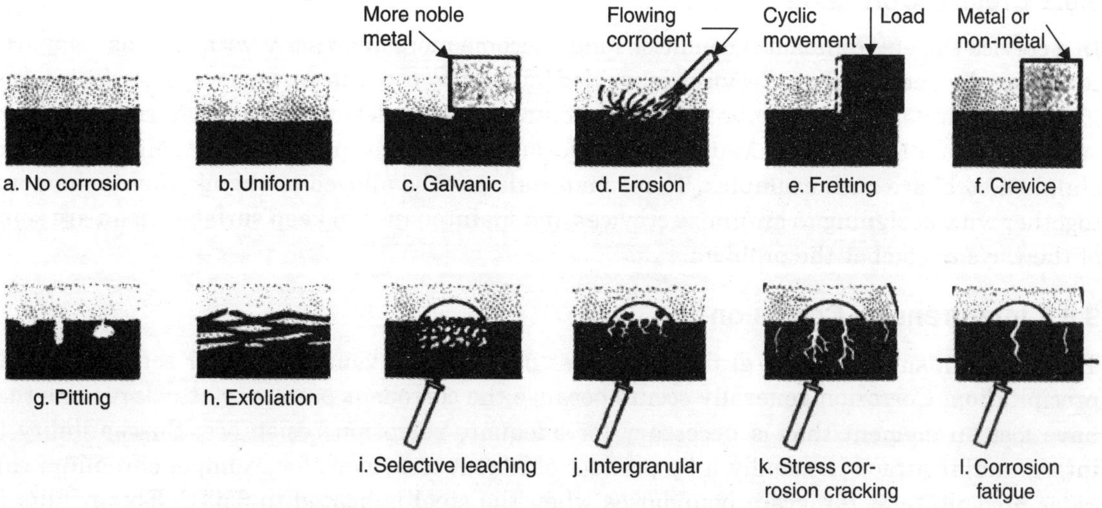

Fig. 9.6 Types of corrosion commonly found in metals and alloys

9.8.1 Galvanic Corrosion

This corrosion occurs when two metals in contact are exposed to a conductive medium. The electrical potential difference acts as a driving force to corrode one of the metals in the couple as electric current flows. **Active** metals corrode more than the **noble** metals.

Galvanic corrosion can occur in welds when the filler metal is of different composition than the base metal. It may occasionally occur because of cast weld metal and wrought base metal. Comparatively larger area of the noble compared to active metal will accelerate the attack. This situation is shown in Fig. 9.7.

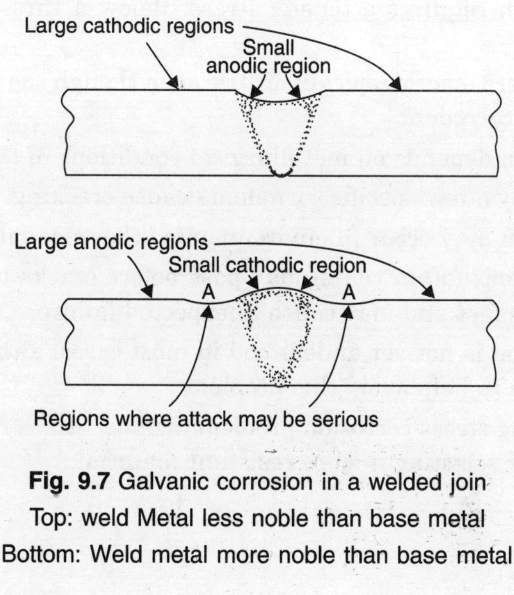

Fig. 9.7 Galvanic corrosion in a welded join
Top: weld Metal less noble than base metal
Bottom: Weld metal more noble than base metal

9.8.2 Crevice Corrosion

In a crevice the environmental conditions may become more aggressive with time as compared to the nearby open surface. Crevices in welded joints may occur in various ways: surface porosity, cracks, undercuts, inadequate penetration and design defects. Some materials are more susceptible to it than others. Materials that form oxide film for protection *e.g.*, aluminium and stainless steel are such examples. These materials may be alloyed to change their behaviour, together with designing to minimize crevices and maintenance to keep surfaces clean are some of the ways to combat the problem.

9.8.3 Intergranular Corrosion

The atomic mismatch at the grain boundaries makes it a favoured place for segregation and precipitation. Corrosion generally occurs because the corrodent prefers to attack regions that have lost an element that is necessary for adequate corrosion resistance. Susceptibility to intergranular attack is usually a by product of a heat treatment for example chromium carbides precipitate at the grain boundaries when the steel is heated to 650°C. This results in intergranular corrosion in a band array from weld where the temperature reached is 650°C. This problem can be avoided by post weld annealing.

9.8.4 Stress Corrosion

A combination of tensile stress and corrosive medium gives rise to cracking of a metal. Many alloys are susceptible to this attack, but fortunately the number of alloy-corrodent combinations that cause it are relatively few. Stresses that cause this arise from residuals stresses due to cold work, welding, thermal treatment and may be due to externally applied forces during assembly and service. Cracks may follow intergranular or transgranular path. There is a tendency of crack branching. The following list gives some characteristics of stress corrosion cracking:

(*a*) Stress corrosion requires a tensile stress. Below a threshold stress cracks do not occur.

(*b*) Cracking appears macroscopically brittle even though the material may be ductile in the absence of corrodent.

(*c*) Stress corrosion depends on metallurgical conditions of the alloy.

(*d*) In a given alloy a few specific corrodents cause cracking.

(*e*) Stress corrosion may occur in environments otherwise mild for uniform corrosion.

(*f*) Long time periods (often years) may pass before cracks become visible. The cracks then propagate fast and may cause unexpected failure.

(*g*) Stress corrosion is not yet understood in most cases, although there is now a large amount of data to help avoid this problem.

Methods of fighting stress corrosion problem include: stress relieving, removing critical environmental species or selecting a more resistant material.

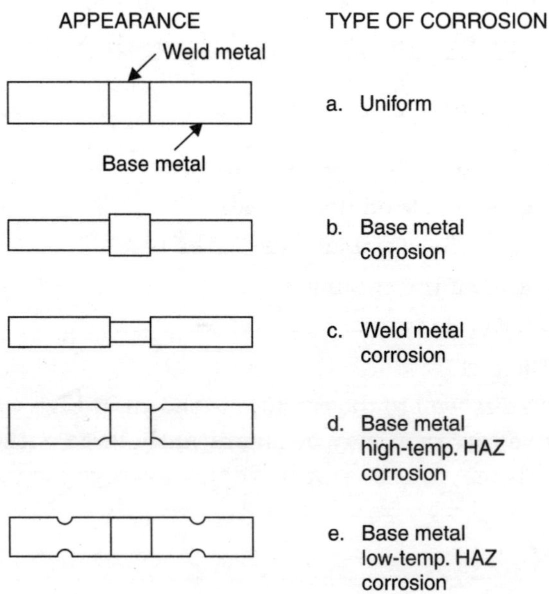

Fig. 9.8 Types of corrosion in a welded joint

9.9 CORROSION TESTING OF WELDED JOINTS

A welded specimen may corrode uniformly over its entire surface (Fig. 9.8*a*). The weld metal may corrode less than the base metal (Fig. 9.8*b*) or more than the base metal (Fig. 9.8*c*) depending upon the composition of weld metal during solidification. In addition the base metal may corrode adjacent to weld metal in the HAZ. During high-temperature welding stresses will develop just adjacent to weld metal and corrosion occurs in HAZ just touching the weld-metal (Fig. 9.8*d*). At low temperature welding the corrosion may be intergranular away from weld-metal in HAZ touching the base metal (Fig. 9.8*e*).

9.9.1 Factors Affecting Corrosion Resistance of Welded Joints

1. Metallurgical structure composition of base-metal and weld-metal.
2. Thermal and mechanical treatment history before welding.
3. Welding process.
4. Welding procedure (manual, automatic, number of passes, welding speed, current and voltage.
5. Shielding gas composition and flow rate.
6. Size and geometry of weld deposit.

While reporting corrosion data for a welded joint, the items in the above list should also be reported.

The most common corrosion resistance evaluation method is to measure the weight lost during exposure to corrodent and convert it to an average corrosion rate using the formula

$$R = \frac{KW}{ADT}$$

where R = corrosion rate in depth of attack per unit time

K = constant (value depends on units used)

W = the weight lost by the specimen during the test

A = total surface area of the specimen

D = specimen material density

T = duration of the test.

The above formula suits well to the conditions shown in Figs. 9.8*a*, 9.8*b*, 9.8*c*. For Figs. 9.8*d* and 9.8*e*, the selective corrosion may be significantly large without resulting in a large amount of weight loss. This may cause error in finding average corrosion rate.

■ QUESTIONS

9.1 Briefly explain the meaning of weld quality. Discuss the factors that determine weld quality.

9.2 With neat sketches discuss the defects in welds their causes and remedies.

9.3 With neat sketches discuss the faulty weld profiles in butt and fillet welds.

9.4 Discuss the various types of corrosions common in metals and alloys related to welds. Discuss their causes and remedies.

9.5 What is stress corrosion? State some characteristics of stress corrosion cracking. List the methods of fighting stress corrosion problems.

Chapter 10

Testing and Inspection of Welds

All types of welded structures from jet engines to metal trash cans are expected to perform some function. The joints comprising these structures must possess some service related capabilities. To test that the required function will be met some tests are conducted. The **ideal** test is the observance of the structure in actual practice. This is usually not possible. Therefore some tests are made on standard specimens to assess the behaviour of the structure in service. Laboratory tests should be used with caution because the size, configuration, environment, type of loading may not be identical to the actual situation. When selecting a test, its function, time and cost factors should be considered.

10.1 TENSILE PROPERTIES

Tension and bend tests are used to evaluate the breaking strength and ductility of a material and to determine that the material meets the specification requirements. Welding causes changes in the metallurgical structure and mechanical properties of a given material. Tension and bend tests are made to assess the suitability of the welded joint for service and are also used to qualify welding procedures for welders according to specific code requirements. In the following paragraphs tension and bend tests according to AWS specifications will be dicussed.

10.1.1 Tension Tests for base metal

Longitudinal or transverse Test. Specimens oriented parallel to the direction of rolling are designated longitudinal, those oriented at right angles to the rolling direction are called transverse. These tests are conducted on the base material.

10.1.2 Weld Tension Test

The tension test for welds is not like that for the base metal because the weld test section is heterogeneous in nature containing base metal, heat affected zone and weld metal. To obtain correct assessment of the strength and ductility several different tests have to be carried out, using different specimens shown in Fig. 10.1. The following tests are commonly carried out.

All Weld-metal tension test. Specimen locations are shown in Fig. 10.1. The details of the specimen dimensions are shown in Fig. 10.2.

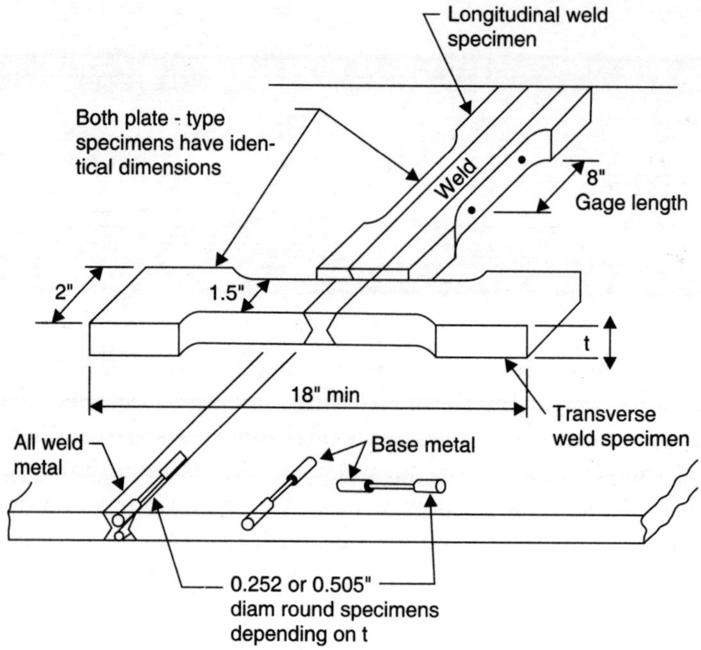

Fig. 10.1 Typical test specimens for evaluation of welded joints (dimensions in inch units)

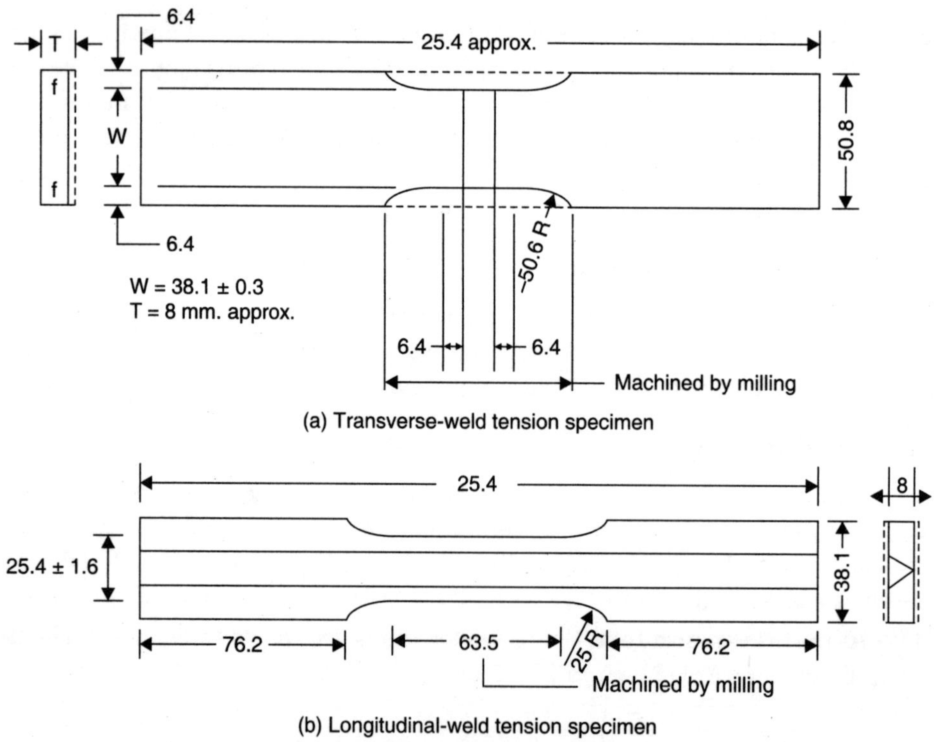

Fig. 10.2 Tension test specimens with dimensions in mm

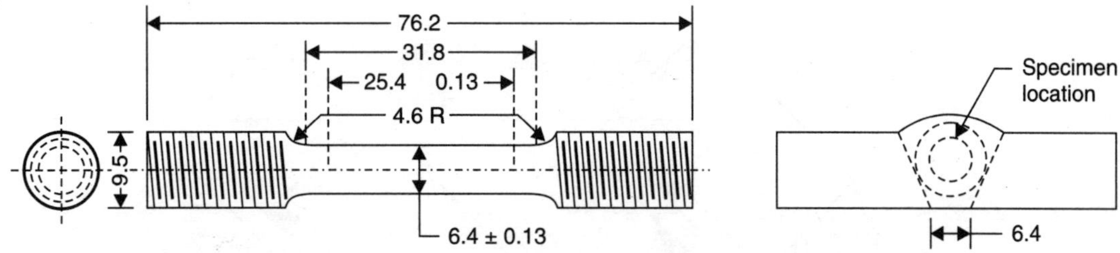

(c) All weld metal tension specimen

Fig. 10.2 Tension test specimens with dimension in mm

Transverse butt-weld test. This test shows that the weld metal is stronger than base metal if the failure occurs in the base metal. It fails to give comparative idea about different types of electrodes. When the weld strength is lower than the base metal, the plastic strain occurs in the weld joint. Ultimate strength is thus obtained but no idea about the joint ductility is obtained from this test. Ideally there is no uniform straining within the specified gauge length and therefore, it is not possible to obtain a reliable measure of yield strength across a welded joint.

Longitudinal-butt-weld test. Here the loading is parallel to the weld axis. It differs from all-weld-metal test in that it contains weld, HAZ and base metal along the gauge length. All these zones must strain equally and simultaneously. Weld metal elongates with the base metal until failure occurs. This test thus provides more information about the composite joint than the transverse test specially when base metal and weld-metal strengths differ significantly.

10.1.3 Tension-shear Test

Fillet weld shear test. Tension-shear tests may be used to evaluate the shear properties of fillet welds. Such tests are usually intended to represent completed joints in weldments and so are prepared using similar procedures. Two basic specimen types, transverse and longitudinal, are employed (see Fig. 10.3).

Of the transverse-shear specimens, double lap specimens are preferred because they are more symmetrical and therefore the stress state under load better approaches pure shear. In the single lap joint, pure shear loading requires special test fixtures to align the specimen or prevent bending, particularly for thick plates where eccentric loading becomes significant. Consequently, single lap specimens are generally not used for plates over 6 mm thick. The data obtained from transverse fillet weld tests are the weld shearing strengths, reported as either load per lineal millimetre of weld or megapascals based on the weld throat.

The longitudinal fillet weld shear test measures the strength of the filled weld when the specimen is loaded parallel to the axis of the weld. The weld shearing strength is reported as load per lineal millimetre of weld for welds which fail.

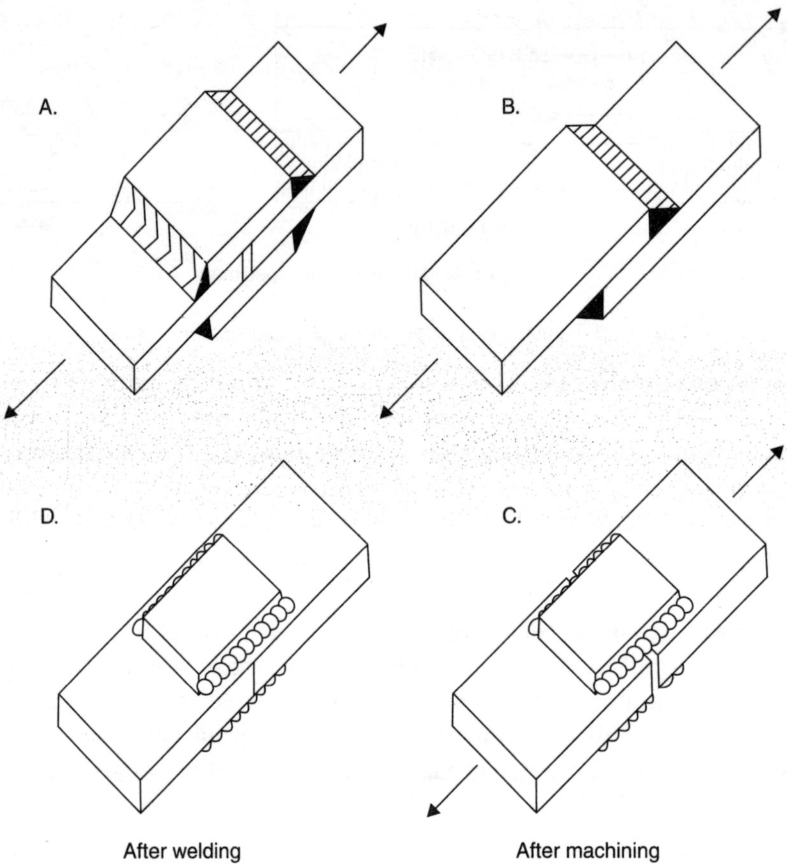

Fig. 10.3 Various types of tension-shear specimens

10.1.4 Tension Tests for Resistance Welds

Tension-Shear Test. The tension-shear test is the most widely used method for determining the strength of resistance spot welds. It is also used for evaluation of weld schedules for ferrous and nonferrous alloys. The test specimen in Fig. 10.4 is made by overlapping suitable size coupons and making a spot weld in the center of the overlapped area. A tensile test machine is used to make the test.

The test is used mainly to establish ultimate shear strength when the specimen is tested in tension. When this test is used in combination with the cross-tension test (Fig. 10.5), the cross-tension strength/tension-shear strength ratio is referred to as a measure of ductility.

When gages less than about 1 mm (0.04 in.) are tested, a plug will usually be pulled from one sheet. This condition is typical of the fracture due to the eccentric loading caused by the overlapped sheets. As the thickness of the sheets or strength increases, the weld will fracture by shearing across the nugget (weld metal) at the interface.

When the thickness becomes large such as 4.8 mm (0.19 in.) and greater, the wedge grips of the test machine should be offset to reduce the eccentric loading which is accentuated by the thickness of the specimen. A more precise shear load will be imposed on the spot weld, thus minimizing a tension or peeling component.

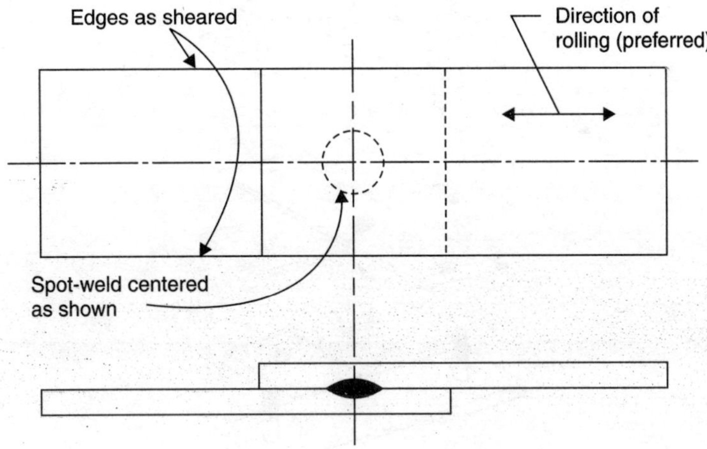

Fig. 10.4 Test specimen for tension shear

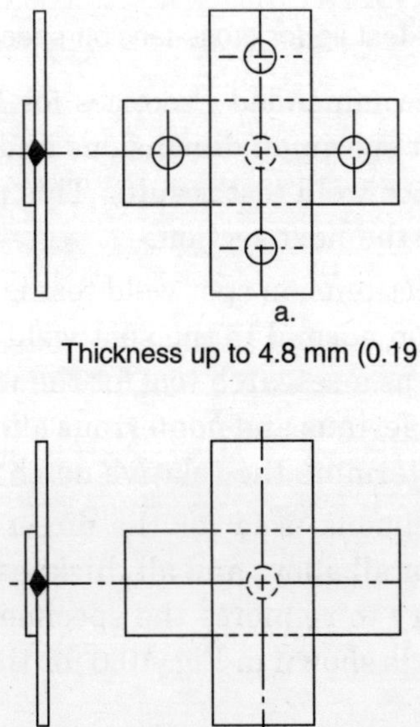

Fig. 10.5 Cross-tension test

The tension-shear test is commonly used in production assurance testing because it is an easy and inexpensive test to perform. Coupons welded at regular intervals are tested to a prior established standard of test results.

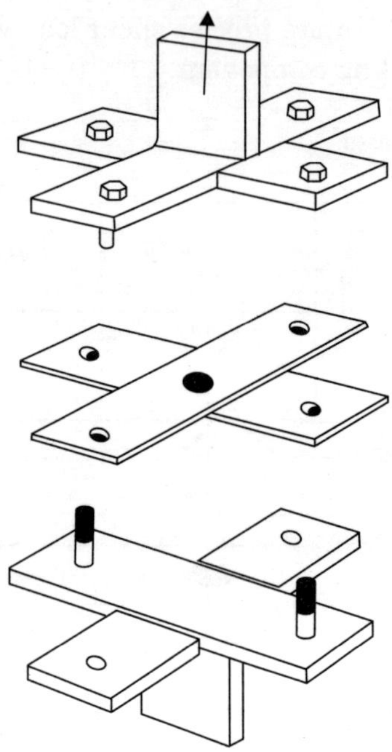

Fig. 10.6 Test jig for cross-tension specimens

The reader is directed to Recommended Practices for Resistance Welding. AWS C1.1, for more details with respect to test specimen dimensions and test fixtures as well as statistical methods for evaluating resistance weld test results. This publication is also applicable for the direct-tension test described in the next section.

Direct-Tension Test. The direct-tension spot weld test is used to measure the strength of welds for loads applied in a direction normal to the spot weld interface. This test used mostly for weld schedule development and as a research tool for the weldability of new materials. The direct-tension test can be applied to ferrous and nonferrous alloys of all thicknesses. The direct-tension test specimen is used to determine the relative notch sensitivity of spot welds.

There are two types of specimens used for the direct-tension test. The cross-tension specimens of Fig 10.5 can be used for all alloys and all thicknesses. When the metal gage is less than 1 mm (0.04 in.), it is necessary to reinforce the specimen to prevent excessive bending. Test jig for cross-tension specimens is shown in Fig. 10.6 for thicknesses up to 4.9 mm and Fig. 10.7 for greater thicknesses.

Peel Test. A variation of the direct-tension test is the peel test which is commonly used as a production control test. The test is shown in Fig 10.7(*b*). The size of the plug or button is measured or correlated with weld sizes having known strengths that are produced by satisfactory production weld schedules. This weld test is fast and inexpensive to perform. Howerver, high strength or thicker specimens may fracture at the interface without producing a plug.

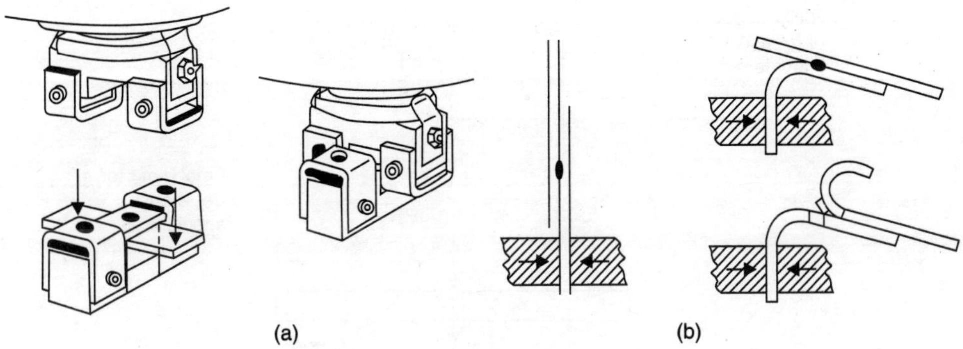

Fig. 10.7 Jig for cross-tension test (t > 4.8 mm)

10.2 BEND TESTS

Bend tests on corner, but, lap and tee welds are shown in Fig. 10.8(*a*).

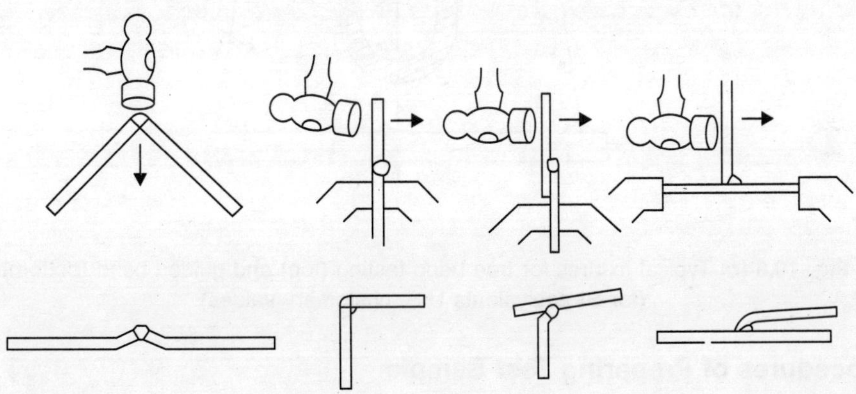

Fig. 10.8 (*a*) Bend tests

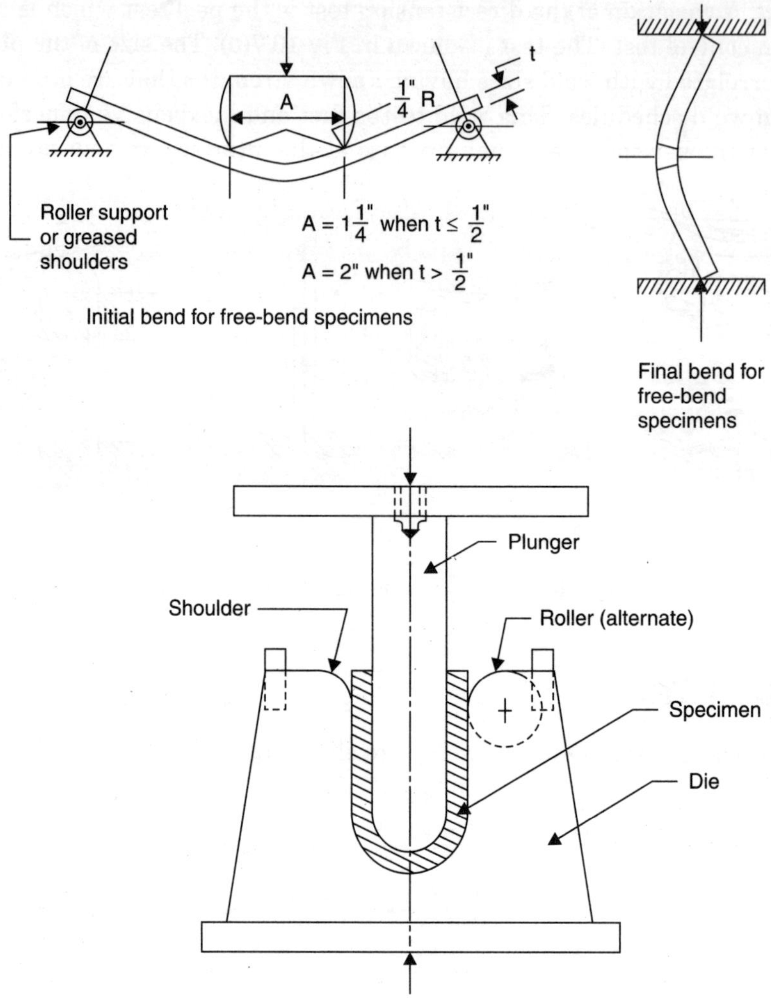

Fig. 10.8 (*b*) Typical fixtures for free bend testing (top) and guided bend (bottom). (for SI equivalents U.S. customary values)

10.2.1 Procedures of Preparing Test Sample

Procedure for butt welds specimen preparation is given step-wise as follows:

1. Cut the coupon from the center of the plate approximately 5.08 cm wide along the length of the weld (Fig. 10.9). Use a shear or cutting torch depending on the thickness of the material. Steel plates of 4.76 mm should be cut with a cutting touch.
2. Save the material from each side for use on the next joint.
3. Cut the weld into sections 7.62 cm (3 in.) long (Fig. 10.10). Use a cutting torch if the material is thicker than the capacity of the shear available. For most SMAW, a cutting torch will be required.
4. Grind the cut sections and finish with a fine file.

5. Check the sectioned surfaces for defects.
   (*a*) Undercut
   (*b*) Lack of fusion
   (*c*) Slag inclusions
   (*d*) Prosity
6. Show test pieces to the instructor for evaluation and recording.

Remember that the final test will be by bending. Bend test requires much more material and will be done under the guidance of the instructor.

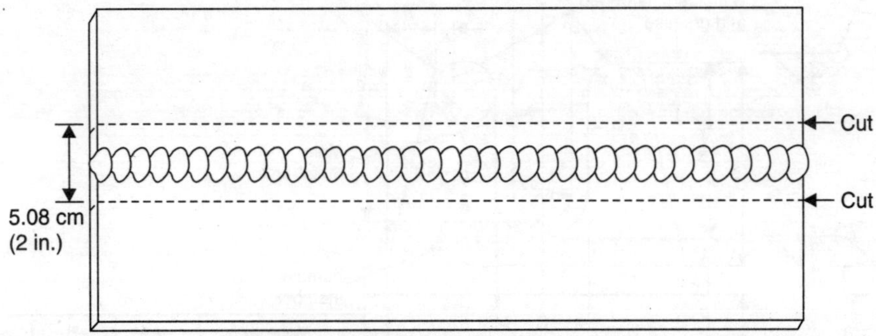

Fig. 10.9 Cutting test samples

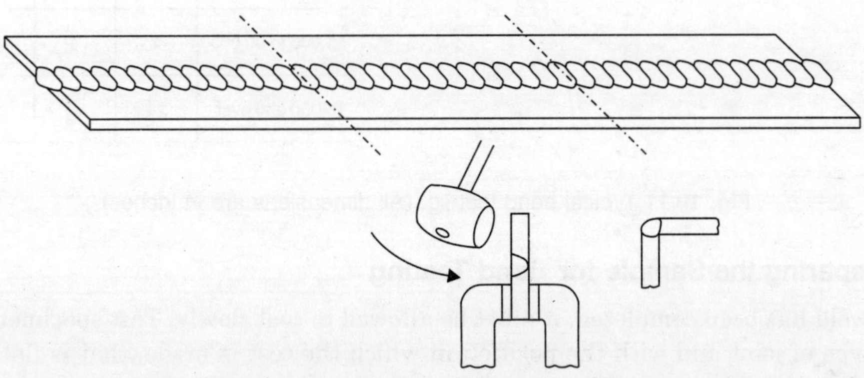

Fig. 10.10 Sample cut into equal pieces

10.2.2 Guided Bend Tests

The guided bend test for plate and pipe requires a special test jig to hold the specimen in place while the bending takes place. Specifications for the test jig design and the bending procedure for specific materials must be followed. Various organizations have designed bending jigs and prescribed procedures for testing different materials. Some of these organizations are:

AWS	American Welding Society Standard for Qualification of Welding Procedures and Welders for Piping and Tubing. D10.9 - 69.
ASME	American Society of Mechanical Engineers Code for Boilers and Pressure Vessels.

API American Petroleum Institute Standard for Welding Pipe Lines and Related Facilities.

A typical guided bend jig and test samples are shown in Fig. 10.11. This device can be used with a hydraulic jack or manual jack that has a force of about 703 kg/cm^2 (10,00 psi).

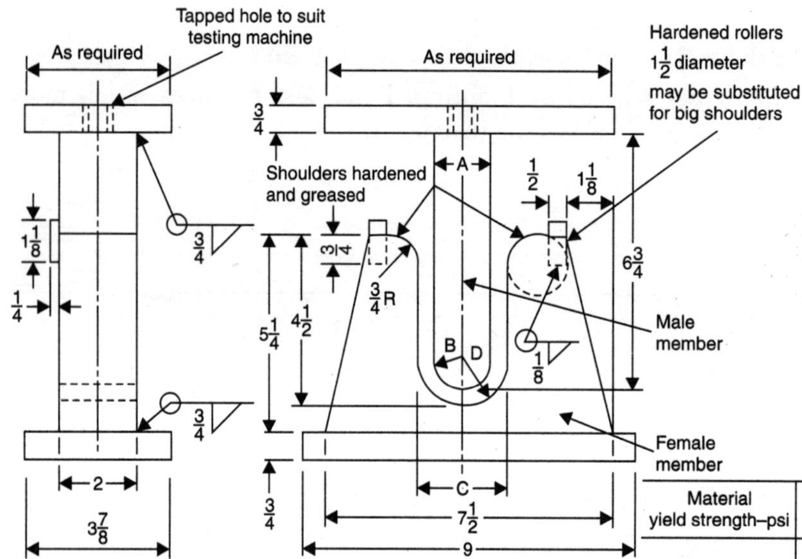

Material yield strength–psi	–A– (inches)	–B– (inches)	–C– (inches)	–D– (inches)
50,000 and under	$1\frac{1}{2}$	$\frac{3}{4}$	$2\frac{3}{8}$	$1\frac{3}{16}$
55,000 to 90,000	2	1	$2\frac{7}{8}$	$1\frac{7}{16}$
90,000 and over	$2\frac{1}{2}$	$1\frac{1}{4}$	$3\frac{3}{8}$	$1\frac{11}{16}$

Fig. 10.11 Typical bend test jig. (All dimensions are in inches)

10.2.3 Preparing the Sample for Bend Testing

Once the weld has been completed, it must be allowed to cool slowly. Test specimens will vary with the type of joint and with the position in which the test is made, that is flat plate (Fig. 10.12) or all position box pipe (Fig. 10.13). For all test coupons, the reinforcement of the weld must be removed completely and the edges rounded slightly (Fig. 10.14).

The grind or file marks from the reinforcement removal should travel lengthwise on the bend test specimen. The sides of the specimen should be smooth and the corners rounded to a maximum of 3.17 mm radius (Fig. 10.15). This smoothness and roundness will allow the specimen to slide freely in the bending jig. Any deep scratches or grooves running lengthwise in the specimen in the weld area are potential breaking points (stress riser).

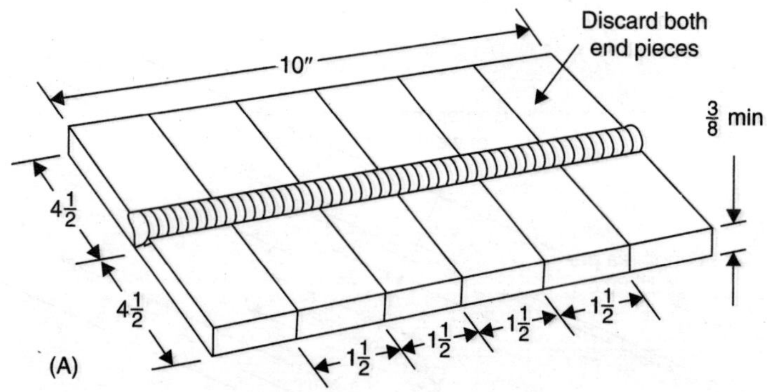

Fig. 10.12 Flat plate test. (All dimensions in inches)

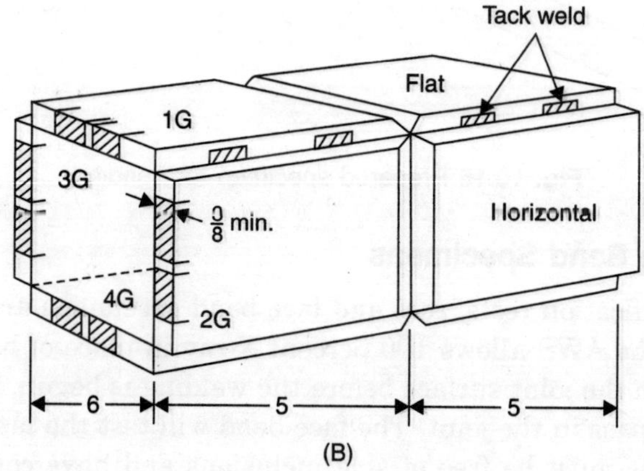

Fig. 10.13 Fixed box pipe all position test. 1G-1 Flat position root bend 1G-2 Flat position face bend 2G-3 Horizontal position root bend 2G-4 Horizontal position face bend 3G-5 Vertical position root bend 3G-6 Vertical position face bend 4G-7 Overhead position root bend 4G-8 Overhead position face bend.

Fig. 10.14 Reinforcement removal

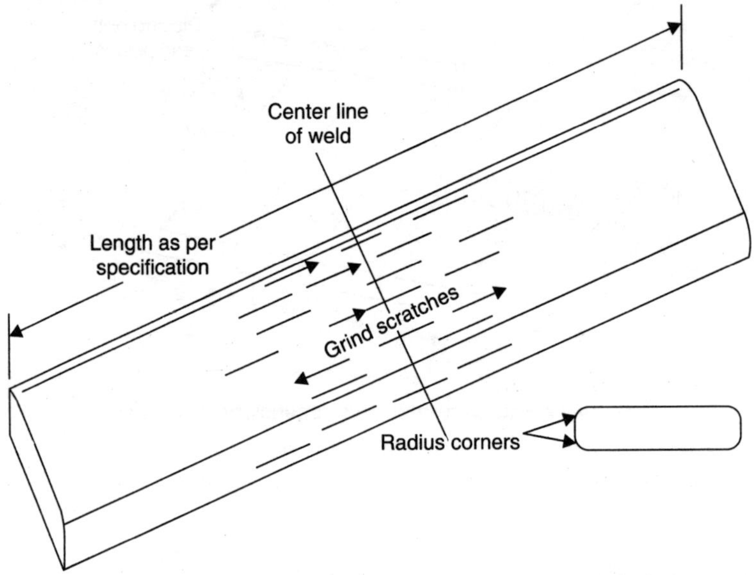

Fig. 10.15 Prepared specimen for bending

10.2.4 Root and Face Bend Specimens

For most welding qualification tests, root and face bend specimens are required (Figs. 10.16 and 10.17). However, the AWS allows 100 percent X-ray in place of bend tests. These specimens may be located on the joint surface before the welding is begun. The root bend will test the quality of the first pass in the joint. The face bend will test the last pass or passes in the joint. Satisfactory welds must be free of slag inclusions and have complete fusion. In most tests, a total distance of 3.2 mm discontinuity (crack, inclusion, or lack of fusion) is acceptable.

If the defect is longer than 3.2 mm in any direction, the test piece is considered to be a failure. For example, the 6G position pipe test requires the removal of four test pieces. If the number of defects in one test sample adds up to more than 3.2 mm in length, the test is a failure.

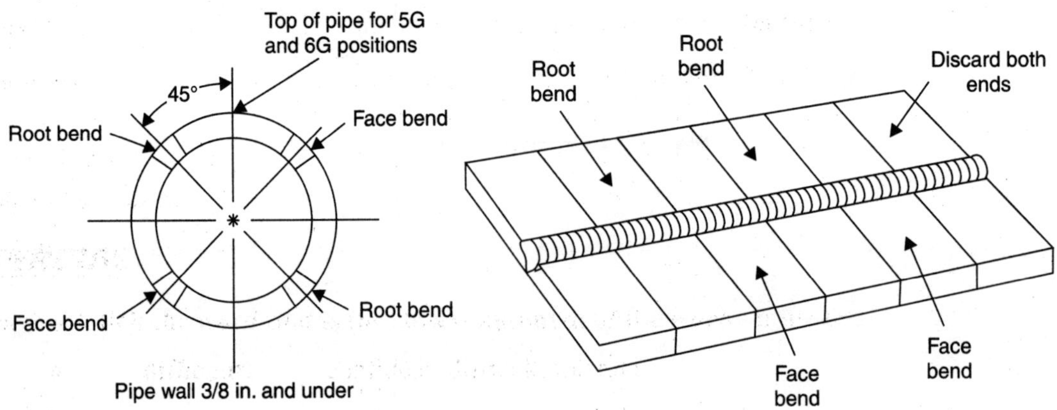

Fig. 10.16 (*a*) Pipe root and face. Plate root and face

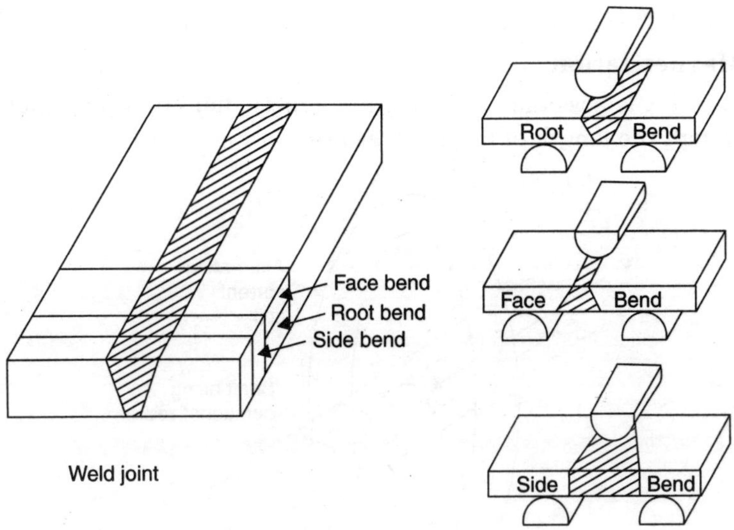

Fig. 10.16 (*b*) Relative orientations of face, root, and side-bend tests from a welded plate

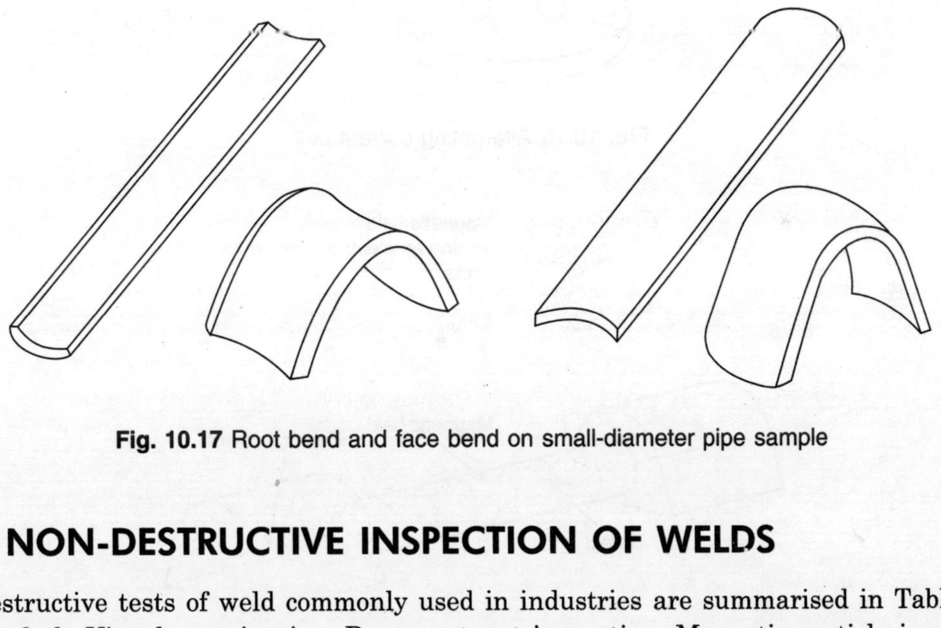

Fig. 10.17 Root bend and face bend on small-diameter pipe sample

10.3 NON-DESTRUCTIVE INSPECTION OF WELDS

Non-destructive tests of weld commonly used in industries are summarised in Table 10.1. They include Visual examination, Dye-penetrant inspection, Magnetic-particle inspection. Radiography and ultrasonics. The last three tests are more common and will be described in the following paragraphs.

10.3.1 Magnetic Particle Inspection

Magnetic particle inspection, as the name implies, requires the use of a magnetic field. The work to be checked must be able to accept magnetism. This process is therefore limited to magnetic metals. It is also limited to surface or near-surface faults. Steel castings, forgings, and sections that have been welded are the most common parts to be inspected by the magnetic particle process. There are several variations of this process.

Longitudinal Magnetization

By using a coil it is possible to include a magnetic field in a part that has the lines of force running through the length of the shaft as seen in Fig. 10.20.

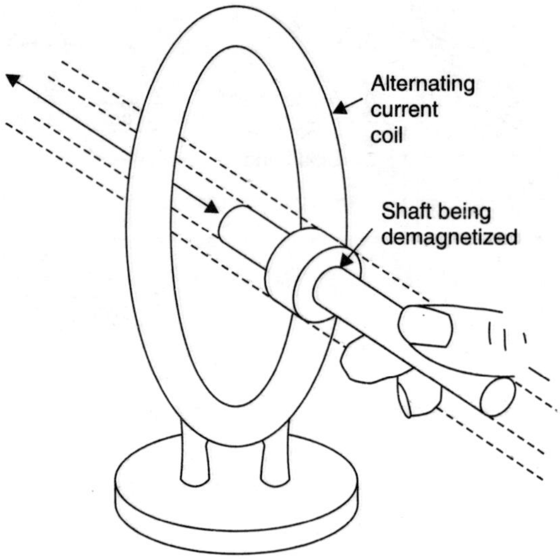

Fig. 10.18 Alternating current coil

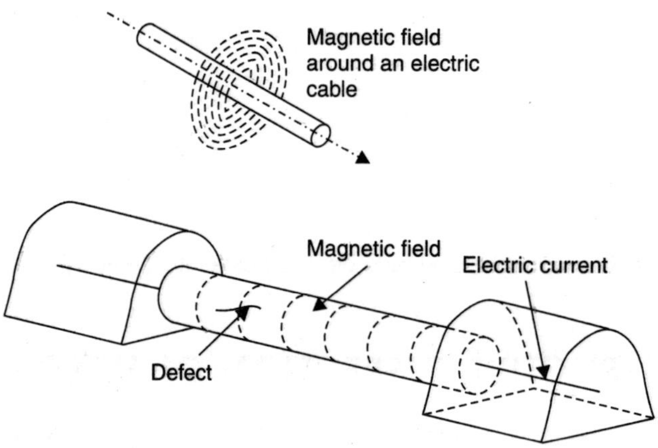

Fig. 10.19 Circular magnetization of a shaft

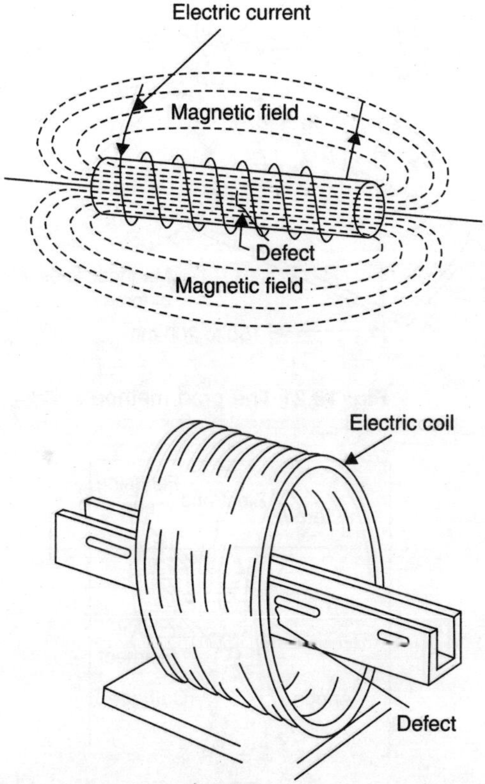

Fig. 10.20 Longitudinal magnetic inspection

10.3.2 Radiographic Inspection

Radiography uses X-rays or gamma rays, which have the ability to penetrate materials that absorb or reflect ordinary light. X-rays are created under controlled conditions by bombarding a specific area with a flow of electrons. Gamma rays are produced by radioactive isotopes. These isotopes never stop giving off radiation; therefore, they must be stored in special shielded containers.

The ability of a material to absorb radiation is dependent upon its density and the wavelength of radiation being used. Lead absorbs more radiation than iron and iron absorbs more than aluminium. This absorption of radiation also varies with the thickness of a piece of material. A thinner piece of material will absorb less radiation as the rays pass through the object; therefore, more radiation will escape through the object. A film placed behind the object to be inspected will be affected more in thin sections than thick sections. Defects in the part being examined will allow more radiation to pass through it and the defect will then be visible on the film.

A radiograph is the recorded image produced on a photographic plate by X-ray. A simplified version of the process is shown in Fig. 10.22. The flaw in the specimen will not absorb as much radiation as does the rest of the part. Therefore, a darker image is present on the film where the flaw exists.

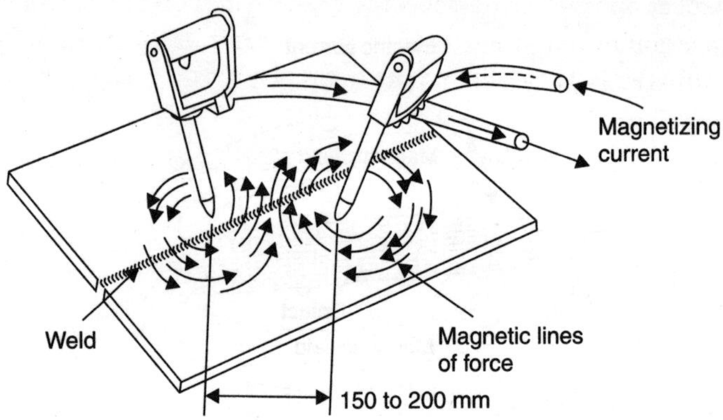

Fig. 10.21 The prod method

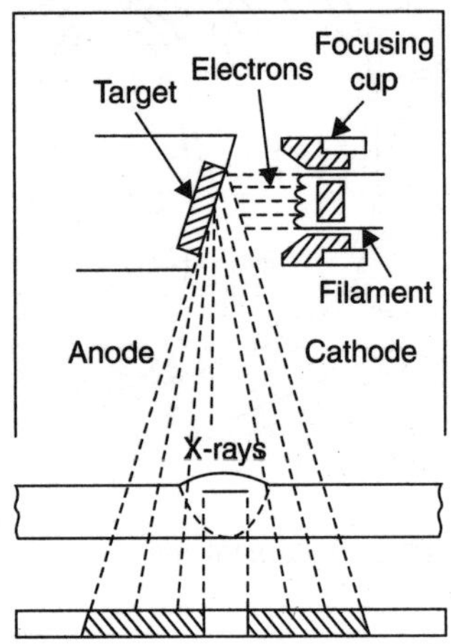

Fig. 10.22 Operation of an X-ray device

One of the most important facts to remember when working in the area where X-ray or gamma ray equipment is being used is that this process is very dangerous. If excessive radiation is absorbed by the body, sickness and even death can be the result.

Fig. 10.23 shows a simplified version of an X-ray tube. X-ray tubes used in industry consist of two electrodes located in a vacuumed glass tube.

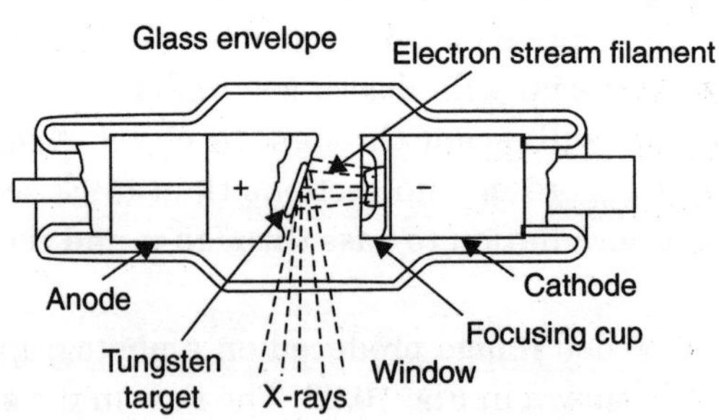

Fig. 10.23 Construction of an X-ray tube

The X-ray inspection process has become a very common method of inspection in industry today. Aircraft inspection of major sections of the aircraft are successfully accomplished by X-ray. The pipeline industry is very dependent upon the X-ray process to ensure that each weld on the pipe is sound.

The pipeline industry uses X-ray units that will swing completely around the circumference of a weldment on the pipe. On completion of the travel around the pipe, complete picture of that entire weld is presented on the radiogram (X-ray film). The films are maintained as a permanent record of the inspection. They are numbered to identify each weld on an entire pipeline and may be referred to at a later date if a breakdown of the pipe occurs.

10.3.3 Ultrasonic Inspection

Ultrasonic Inspection makes use of the science of acoustics in frequencies above the upper audible limit of approximately 15,000 cycles per second.

The basic operation of ultrasonic inspection is the conversion of pulsating electronic waves into ultrasonic sound. These sound waves are introduced into the material to be tested through a quartz crystal. The crystal is set into a special search unit that not only sends out the sound but also acts as a receiver to accept reflections of that sound on its return. If the signal sent out runs into a defect in the material, a return signal comes back to the receiver in less time than it would have had it travelled the full distance to the other side of the part and back.

A cathode ray tube (CRT) is incorporated in the ultrasonic equipment to provide a visual indication on the screen of the initial signal and reflected signals. Fig 10.24 shows a diagram of the CRT screen with pips of the initial pulse, discontinuity, and back surface reflection. Fig. 10.25 shows the basic cathode ray tube construction.

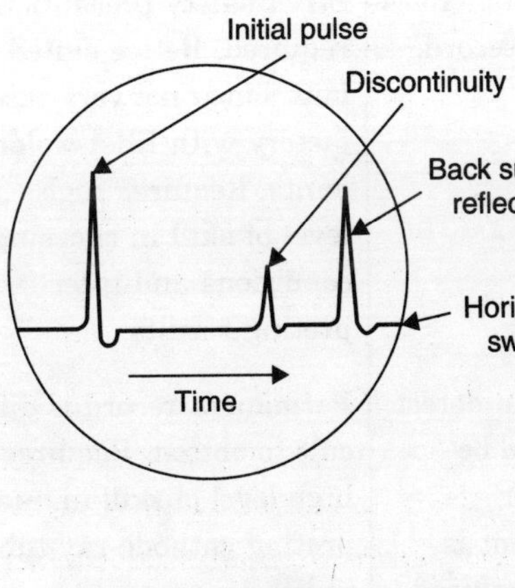

Fig. 10.24 Cathode ray tube

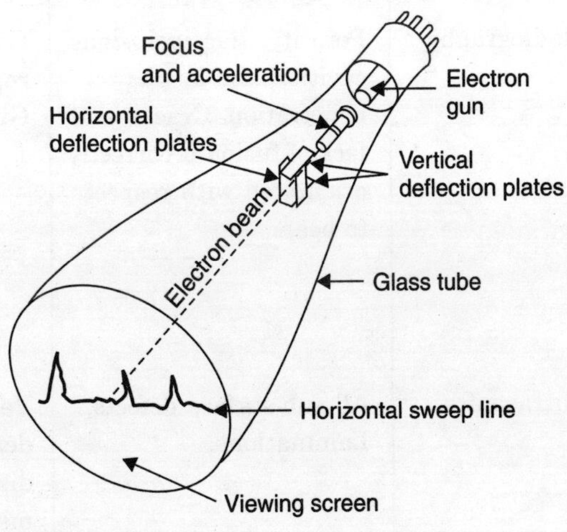

Fig. 10.25 Cathode tube construction

The pulses that are sent out by the quartz crystal may span a time of two millionths of a second or less and may vary in cycles of transmission from 60 to 1000 times per second. The return signals, shown as pips on the CRT, will be spaced in proportion to the distance between the points in the material they represent. For example, a pip representing a defect close to the back surface reflection indicates a defect that is close to the far edge of the part being inspected.

As with all electronic non-destructive testing methods, a considerable amount of skill is required to operate the ultrasonic inspection unit. As is the case with many skilled tasks, technique, practice, and experience determine the efficiency with which the inspection is completed. This inspection method is becoming more useful in the welding industry as new techniques for scanning welds are being perfected.

Table 10.1 Summary of the methods of non-destructively testing welds

Method	*Defects detected*	*Advantages*	*Limitations*
Visual	Inaccuracies in size and shape. Surface cracks and porosity, undercut, overlap, crater faults.	Easy to apply at any stage of fabrication and welding. Low cost both in capital and labour.	Does not provide a permanent record. Provides positive information only for surface defects.
Dye-penetrant	Surface cracks which may be missed by naked eye.	Easy to use. No equipment required. Low cost both in materials and labour.	Only surface cracks detected with certainty. No permanent record.
Magnetic-particle	Surface cracks which may be missed by naked eye. May give indication of subsurface flaws.	Relatively low cost. Portable. Gives clear indication.	Only surface cracks detected with certainty. Can be used only on ferromagnetic metals. Can give spurious indications. No permanent record.
Radiography	Porosity, slag inclusions, cavities, and lack of penetration. Cracks and lack of fusion if correctly orientated with respect to beam.	Can be controlled to give reproducible results. Gives Gives permanent record.	Expensive equipment. Strict safety precautions required. Better suited to butt joins - not very satisfactory with fillet-welded joints. Requires high level of skill in choosing conditions and interpreting results.
Ultrasonics	All sub-surface defects, Laminations.	Very sensitive - can detect defects too small to be discovered by other methods. Equipment is portable. Access required to only one side.	Permanent record is difficult to obtain. Requires high level of skill in interpreting cathode-ray-tube indications.

■ QUESTIONS

10.1 Briefly discuss the necessity of conducting destructive testing of welds. Why standard specimen are used for testing? State the basic considerations in choosing a test of mechanical properties.

10.2 What tests do you suggest to determine the strength and ductility of a welded joint? Why several different tests are carried out to determine correct strength and ductility of a welded joint?

10.3 With neat sketches explain the weld-tension tests all weld-metal tension test, transverse butt-weld test, longitudinal butt-weld-test.

10.4 With meat sketches explain the various types of tension shear tests for fillet welds.

10.5 With neat sketches discuss the various tests carried out to assess the strength properties of spot welds. What is cross-tension test? How is it carried out?

10.6 Explain the difference between free bend and guided bend tests. How their specimen are prepared. Differentiate between root-bend and face-bend specimen, pipe root and face bend and plate root and face bend tests. How their specimen are prepared?

10.7 Name the tests commonly used for the inspection of welds. For each test summarise the defect it detects, its advantages and limitations.

10.8 With neat sketches describe briefly the following non-destructive tests.

(*a*) Magnetic particle inspection

(*b*) Radiographic inspection

(*c*) Ultrasonic inspection.

Chapter 11

Welding of Pipelines and Piping

In the industrial world, the term piping is usually understood to cover pipe; tubing; fittings such as tees, elbows, flanges and reducers; valves and hearders used in oil refineries, power stations, nuclear plants, chemical and petrochemical plants and other industrial plants.

The term pipelines usually applies to long transmission pipelines designed to conduct liquids such as water, crude oil and petrol, and gases such as natural gas.

Today, piping systems and pipelines in industry are almost fully welded. Threaded joints are rarely used. Flanged joints are used only where sections have to be opened for internal inspection or replacement.

Piping and pipelines are dealt separately in this section. Penstocks are also considered to be transmission pipelines, but for convenience they are dealt with in the section on power generating plant.

11.1 PIPING

Industrial pipings are critical items in a production plant and they frequently operate under high pressures, high temperatures and in corrosive atmospheres. The efficiency, productivity and safe operation of plants depend to some extent on how effectively, piping systems withstand the rigours of service. Serious consideration has to be given to the selection of grades and sizes of materials, design, fabrication, erection, testing and inspection. Guidance is provided by various codes and standards applicable to weld piping systems prepared by technical societies, trade associations and standardisation bodies. For example, the American National Standards Institute (ANSI) has issued Code for Pressure Piping, which covers Power Piping, Industrial Gas and Air Piping, Pertoleum Refinery Piping, Oil Transportation Piping, Refrigeration Piping, Chemical Industry Process Piping, Nuclear Power Piping, Gas Transmission and Distribution Piping Systems.

Piping connected to boilers are covered in several sections of the ASME Boiler and Pressure Vessel Code. The American Petrol Institute (API) has issued a standard for Field Welding of Pipe-Lines. ASME Guide for Gas Transmission and Distribution Piping Systems is another useful publication. The American Welding Society has published the following recommended welding practices :

(*a*) Welding of Austenitic Chromium-Nickel Steel Piping and Tubing, D10.4 (1966).

(*b*) Welding of Chromium-Molybdenum Steel Piping, D 10.8 (1961).

(*c*) Recommended Practices for Gas Shield-Arc Welding of Aluminium and Aluminium Alloy Pipe, D10.7 (1960).

(*d*) Welding Ferrous Materials for Nuclear Power Piping, D10.5 (1959).

(*e*) Gas Tungsten-Arc Welding of Titanium Piping and Tubing, D10.6 (1959).

To ensure satisfactory welding of piping installation, it is first necessary to establish and qualify the welding procedure covering base metal specifications, filler metals, edge preparation and joint fit-up, pipe position, welding process, process parameters, welding techniques, preheat, interpass and postheat schedules, and final inspection and testing. It is also necessary to qualify the welders for the welding procedure adopted. Standard procedures for the qualification of welders and welding machine operators are given in relevant codes, for example, in section IX of the ASME Boiler and Pressure Vessel Code.

Pipe materials and fittings are available in standardised specifications, sizes and with standard tolerances. Pipes are available in long lengths as seamless or welded pipes. Pipings are longitudinally welded in a tube mill from strips by using the electric resistance butt or high-frequency resistance welding process, while pipes for pipelines are welded along their long seams in a pipe mill by the automatic submerged-arc or MIG/CO_2 process. In the erection of pipings and pipelines, welding is restricted to girth joints or to joints between pipes and their attachments. Hence in the following sections, only girth welding techniques are described.

The metals used for piping are : carbon steel, wrought iron, C-Mo steels, Cr-Mo alloy steels, cryogenic steels, stainless steels, Al and its alloys, Ni and its alloys, Cu and its alloys and Ti and its alloys.

Carbon steel. Carbon steel piping is mostly welded by the manual metal-arc process using E6010 or E7018 class of electrodes. For critical applications which demand full penetration welds, split or solid backing rings are provided on the inside, or the well-penetrated root pass is made with the TIG process as described in Chapter 5. This technique applies to all metals. MIG/CO_2 process using gas mixture of CO_2 and argon is used on less critical piping, where full root penetration and fusion are not essential. In shop fabrication of thick-walled pipe having O.D. of more than 200 mm, automatic submerged arc welding is used for the filling passes, after the root pass has been completed with the manual metal-arc or TIG process. If backing rings are used and the fit-up is good, the entire joint can be made by the SA process. Generally preheating is not necessary if the carbon content of the steel is below 0.30%. If the wall thickness exceeds 19 mm, postweld heat treatment is usually recommended. It consists of heating to 600– 650°C and holding for one hour per 25 mm of wall thickness, with a minimum holding time of 30 min, and then cooling in still air. For further details, relevant codes must be consulted.

During manufacture of boiler units large number of tube butt welds have to be made with the tubes positioned at any angle from horizontal to vertical, and being often in positions of restricted access. Automated orbital TIG welding machines with automatic cold wire feed have been developed for this purpose. A typical orbital TIG welder has a weldhead, covering tube sizes in the 25–50 mm O.D. range and requires only 44.4 mm clearance between adjacent tubes. It features an integral wire-feed system, *i.e.*, the wire-feed facility is mounted on the head and rotated with the electrode block. Arc-voltage control provides a means of maintaining a constant preset distance between electrode and workpiece. These facilities allow for a number of continuous orbits (*i.e.*, multiple weld pass) to be made around the tube joint. Such a machine can be applied on pipings of all industrial metals. Lately welding heads capable of joining tubes 18.2 mm O.D. with a clearance of only 16.8 mm have been produced.

Wrought iron. Wrought iron piping has low carbon content (0.12% maximum). It is usually welded by the manual metal-arc process. It is advisable to use low welding currents and speeds. Preheating and postheating are generally not required.

C-Mo steel. The welding processes used for these steels are the same as those used for carbon steels. For manual welding, electrodes of E7010-A1, E7016-A1 or E7018-A1 are used. For SA welding, the Mo alloy of the weld-metal is derived either from the wire or the flux. Preheat and postheat data are given in Chapter 10 while discussing the weldability of these steels.

When used in service temperatures exceeding 425°C, C-Mo steels have been known to undergo graphitisation, *i.e.*, the carbon transforms to nodules of graphite, which substantially reduces the toughness of the steel.Though such unfavourable phenomenon can be suppressed by stress-relieving the welded joints at 720°C for four hours, use of C-Mo steel pipings for high temperature applications is being discouraged.

Cr Mo steels. These grades are mostly used for service in the 400–593°C temperature range. They are usually welded by the manual metal-arc process, using low-hydrogen type low-alloy steel electrodes of matching alloy contents.

For submerged-arc welding, it is advisable to use neutral flux and alloyed wire in preference to alloyed flux and neutral wire, because in the latter case, the alloy balance in the weld deposit gets upset during multi-pass welding at high interpass temperatures.

Low-temperature steels. The types of steel used for various low-temperature service pipings are given in Table 11.1. They are usually welded by the MMA process. The suitable AWS classes of electrodes are indicated in the Table. Preheating is a must for Ni steels, because nickel renders the steel to get air-hardened. Preheat and postheat data are given in Chapter 5.

Table 11.1. Steels and electrodes for low-temperature service

Min. temp. °C	*Type of steel*	*AWS class MMA Electrode*
– 46	Fine-grained fully deoxidised steel	E7016–E7018
– 60	2.25% Ni steel	E8015–C1
– 100	3.5% Ni steel	E8015–C2
– 196	9% Ni steel	ENiCrFe–2

Martensitic stainless steels. These are hardenable steels and are susceptible to cracking during welding. Preheat and postheat operations are necessary. The postweld heat treatment must immediately follow the completion of welding without withdrawing the preheat.

Welding data are given in Table 11.2. If for some reasons postheating is not possible, type 310 or 309 stainless steel filler wire must be used.

Table 11.2. Recommendations for wrought martensitic stainless steel pipes

Type of steel	*Chemical composition (%)*		*Recommended electrode or welding rod*	*Preheat and interpass temperature °C*	*Postheat temperature °C*
	C	*Cr*			
12Cr	0.15 max.	11.5 – 13.5	E, ER410	320 – 370	705 – 760
			E, ER310 or E, ER309	200 – 320	705 – 760
12Cr	0.08 max.	11.5 – 13.5	E, ER410	150 – 260	705 – 760
			E, ER310 or E, ER309	150 – 260	705 – 760
13Cr	over 0.15	12.0 – 14.0	E, ER410 or E, ER430	320 – 370	705 – 760
			E, ER310 or E, ER309	200 – 320	705 – 760

Ferritic stainless steels. These steels are less susceptible to cracking during welding than the martensitic types, but they may become embrittled due to the high temperatures attained during welding and consequent grain growth. To remove embrittlement, the steel is annealed for one hour between 705 and 790°C, and then quenched or air-cooled. The welding data are given in Table 11.3.

Table 11.3. Recommendations for welding ferritic stainless steel pipes

Type of steel	*Chemical composition (%)*			*Recommended electrode or welding rod*	*Preheat and interpass temperature °C*	*Postheat temperature °C*
	C	*Cr*	*Other*			
12 Cr, Al	0.08 max.	11.5 – 14.5	0.10 – 1 0.30 Al	E, ER430	Not necessary	Highly recommended
				E, ER310 or E, ER309	Not necessary	Recommended
16 Cr	0.12 max.	14.0 – 18.0	....	E, ER310 or E, ER309	Not necessary	Recommended
27 Cr	0.20 max.	23.0– 27.0	0.25 max. N	446	150–200	Essential
				E, ER310 or E, ER309	Not necessary	Recommended

Al and its Alloys. These alloys are commonly welded by the TIG process and in some cases by the MIG process. Before attempting to weld pipings, welders must undergo training and gain some experience. In welding horizontally positioned fixed piping, the molten metal sinks due to its high fluidity. Aluminium backing rings and consumable insert rings are sometimes used to obtain good root penetration. Preheating is generally not necessary, but may be used with advantage when the diameter exceeds 60 mm. Preheat temperature ranges between 280 and 300°C. Some Al alloys are unfavourably affected when preheated above 200°C. Hence, high preheat temperatures must be used with care.

Ni and its alloys. These alloys are commonly used in piping because of strength properties, good corrosion resistance to many acids, and easy weldability. They can also be readily welded to ferritic and austenitic steels. The welding processes commonly used are : MMA, TIG and MIG. Backing rings should not be used, because they promote crevices, root cracks and corrosion. Consumable insert rings should be preferred. During root pass welding, the inside of piping must be purged with inert gas, which can be helium, argon, hydrogen or their mixtures.

It is important to remember that Ni and its alloys are susceptible to embrittlement by accidental presence of lead, sulphur, phosphorus and some low-melting metals.

Copper and its alloys. They are commonly welded by oxyacetylene, MMA, TIG and MIG processes. It is advisable to use backing rings whenever possible, because of the high fluidity of molten copper. Because of the high heat conductivity of copper, preheating with a gas torch is necessary when large diameter or heavy-walled pipes are being welded. Red brass and yellow brass are preferably welded by the oxyacetylene process to minimise vaporisation of zinc. Cupronickel 30 (*i.e.,* 70:30 alloy) is extensively welded and used for water pipe and condenser tubing on ships, because of its superior resistance to sea water corrosion. The most suitable welding processes for this alloy are MMA and TIG.

Ti and its alloys. Welding of these materials demands special techniques and specialised skill on the part of the welder. Pipes of wall thickness 1.6 mm and below are normally welded by the TIG process without filler wires. For heavier pipes, filler metals are used. Unless the filler wire is thoroughly cleaned and handled with care, it can contaminate the weld. Contamination also occurs if the hot end of the wire is withdrawn from the gas shield and exposed to atmosphere during intermittent deposition. Special care must be taken that there is 100% root penetration all over the joint. A small root defect can develop into a crack during service and lead to serious failure.

Dissimilar metals. Pipings of dissimilar metals often welded in power plants, oil refineries, nuclear plants, etc. The metals commonly involved are carbon steels, low-alloy steels, stainless steels and nickel and its alloys. Normal welding procedures can be used in these cases, because the melting points of these metals are fairly close. The main considerations are filler metal compositions and preheat/postheat temperatures. For dissimilar joints involving non-ferrous alloys, the filler metal and welding procedure must be carefully determined after studying the metallurgical aspects of the joint in question.

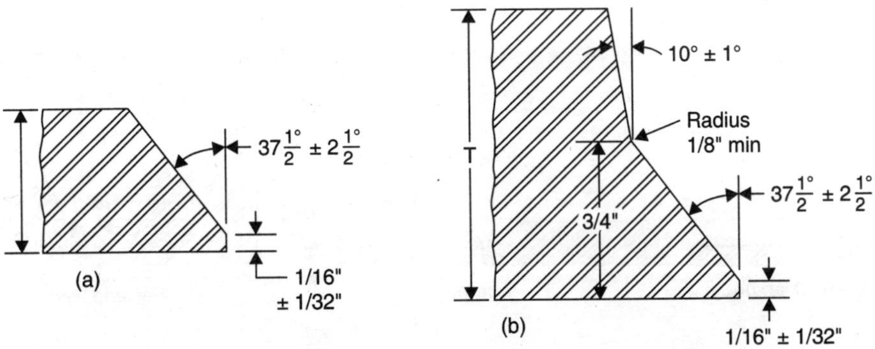

Fig. 11.1 Edge preparations of pipe end for MMA welding

Sometimes, it helps to butter the joint edge metal having the higher melting point before final welding. For example, when carbon steel is to be joined to silicon-bronze, the carbon steel is buttered with silicon-bronze weld deposit. When the metals to be joined have widely different melting points, brazing, braze welding or soldering should be resorted to.

11.2 JOINT DESIGN

As stated earlier, the usual joint to be welded in pipings is the circumferential butt joint. To weld such a joint by the MMA process, the pipe edge can be square or slightly chamfered when the wall thickness is below 5 mm for carbon steel, and below 3.2 mm for stainless steel. Thicknesses greater than these and up to 22 mm should have their edges prepared as at (*a*) in Fig. 11.1, while thicknesses greater than 22 mm should have edge preparation as at (*b*) in the same figure.

In critical applications where carbon and low-alloy steel piping stainless steel piping and most non-ferrous piping is to be TIG welded, joint preparations including consumable insert rings as shown in Fig. 11.2 are used. In all the cases shown, U or flat-land bevel preparations are employed, because they help to minimise excessive shrink.

For butt joints between unequal wall thicknesses (for example, between a pipe and a cast steel fitting or valve body), codes recommend that a smooth taper be provided on the edge of the thicker member.

Fillet-welded joints are often used for pipe sizes 50 mm in diameter and smaller, and for joining pipe to flanges, pipe to valves and pipe to socket joints. Three examples are shown in Fig. 11.3.

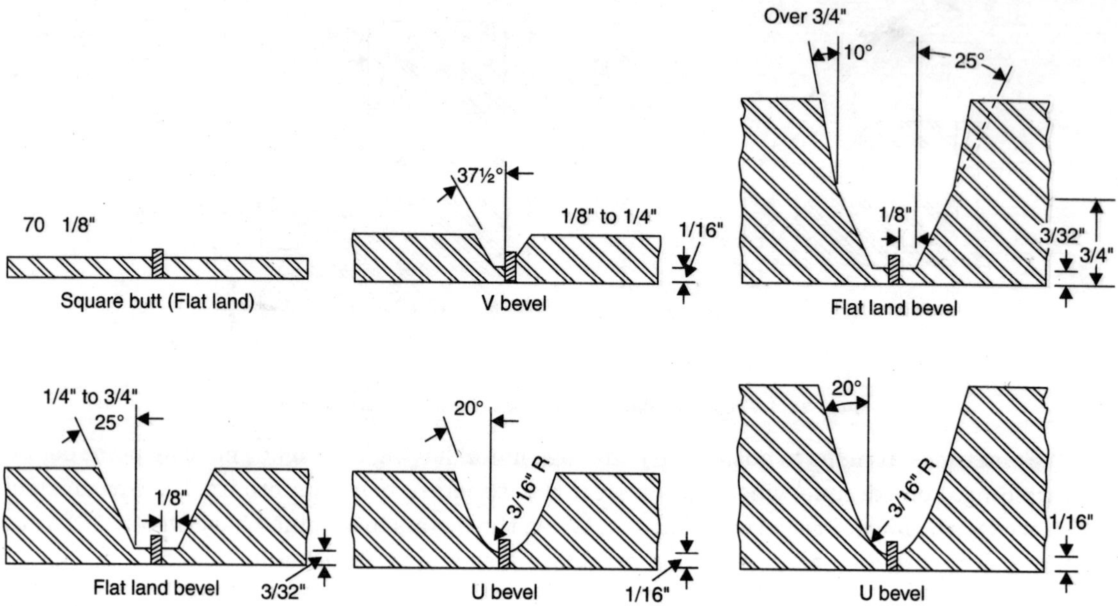

Fig. 11.2 Joint fit-up using consumable insert for TIG welding

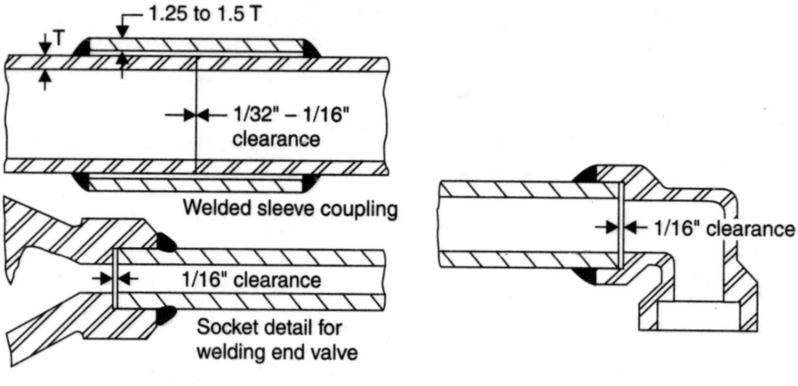

Fig. 11.3 Examples of fillet-welded joints

11.3 BACKING RINGS

Backing rings are commonly employed for welding carbon steel and low-alloy steel piping by the MMA process in steam power plants and other applications. While split rings are sometimes used for non-critical applications, solid flat or taper-machined backing rings are preferred for critical applications. Some designs of backing rings and the manner in which they are fitted are shown in Fig. 11.4. The figure shows that the pipe-end must also be suitably machined on the inside diameter. Chemical composition of the ring is important as also the seat contact between the pipe-end and the ring. Guidance for the correct use of baking rings is available in relevant codes. Backing rings are rarely used for piping in oil refineries and chemical plants.

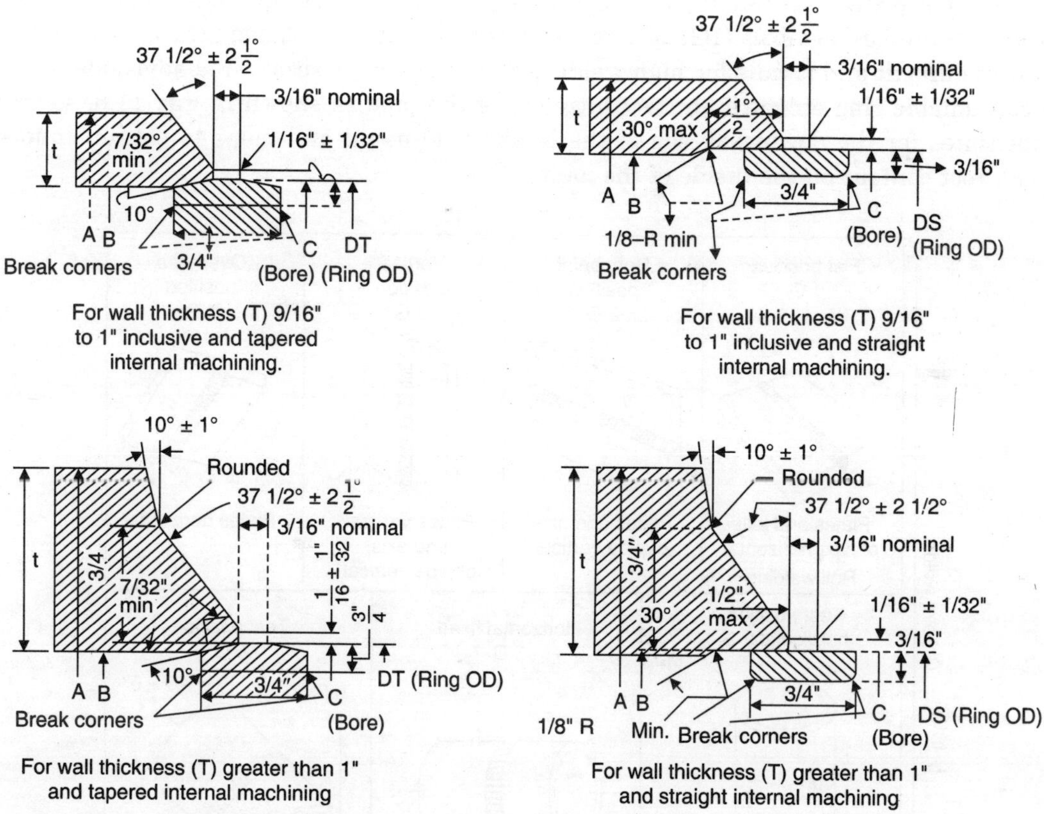

Fig. 11.4 Edge preparation using flat or taper machined solid backing rings

Where the weld joint quality and especially its corrosion resistance are important, consumable insert rings are placed at the root, as mentioned earlier and illustrated in Fig. 11.2 and fused with a TIG torch, so that a sound root weld pass results. This technique dispenses with the addition of filler metal, which could interfere with the welding operation and cause lack of penetration. The subsequent passes, if required, are then deposited by the TIG process using a filler wire or by the MMA process. If instead of using an insert, the pipe-end is suitably machined at the root and autogenously welded, cracking or porosity is likely to occur because of the unfavourable base metal composition. Use of a consumable insert ring of properly balanced composition and dimensions:

(*a*) provides the best welding conditions even in horizontal fixed or 5G position, (*b*) minimises human element and thereby ensures weld uniformity, (*c*) gives the most favourable weld contour which can resist cracking arising from weld metal shrinkage, and (*d*) gives weld-metal composition which can guarantee optimum mechanical properties and corrosion resistance.

At this point, it is pertinent to mention that the various pipe welding positions are defined by standard symbols (1G, 2G, etc.) as shown in Fig. 11.5. Among these, 5G position is the most difficult and it calls for high welding skill. For this position, it is advisable to insert the consumable ring eccentric to the centreline of the pipe as shown in Fig. 11.6, so that it compensates for the downward sag of the liquid weld-metal and helps to obtain uniformly smooth root contour on the inside of the joint.

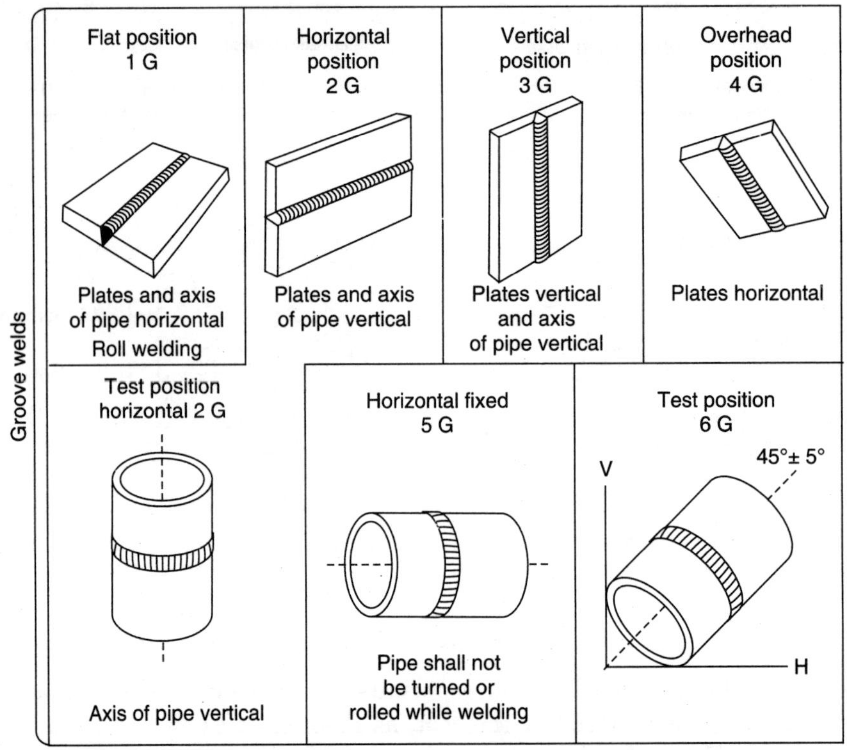

Fig. 11.5 Standard symbols for designating welding position

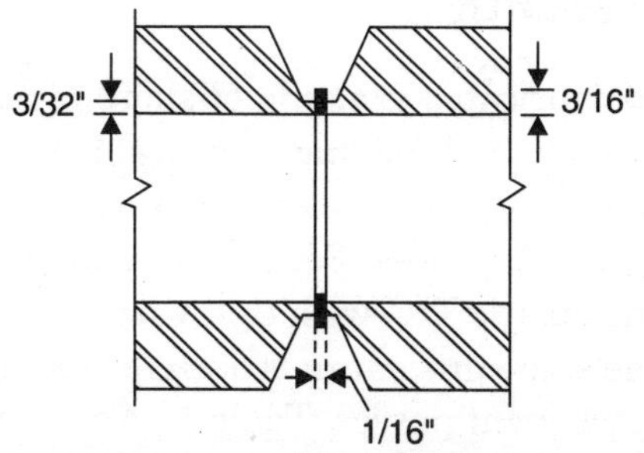

Fig. 11.6 Eccentric insertion of consumable insert ring for 5G position pipe welding

Consumable insert rings of proper shapes, diameters and chemical compositions to suit various metals and applications are provided by manufacturers in advanced countries.

In the installation of piping systems, tees, laterals, wyes and vessel openings have to be welded, and they normally involve intersection joints. Since such joints are difficult to weld, standard welding fittings supplied by manufacturers are used. These fittings possess bursting strengths equivalent to those of pipes of the same weight and they are designed to be connected by simple putt welds. Some examples of such fittings are shown in Fig. 11.7. Manufacturers also provide factory-made nozzles, necks, outlets, tees, etc., specially designed for welding to simplify the fabrication of piping.

11.4 HEAT TREATMENT

Preheating, concurrent heating and postweld heating are important steps in the welding of pipings; and their successful performance in service often depends upon correct heat treatment. The heat treatment procedure includes consideration of the maximum temperature to be attained, time at maximum temperature, rates of heating and cooling, and the width of the heating band. The usual methods of heat treatment are : (*a*) oxyfuel, (*b*) electric resistance heating, (*c*) induction heating and (*d*) heating in furnace.

In the oxyfuel method, a simple gas torch is adequate for small diameter pipes. For larger pipes and connections, ring burners are more effective. For temperature control, temperature indicating crayons are used. In this method, surface thermometers or electrically operated pyrometers are used to control automatically the current flow to the heating units. Thermocouples are usually attached to the metal to be heated by induction heating. The thermocouple wires are then connected to control equipment, which may automatically control the time-temperature cycle and even program the heating and cooling rates of the metal.

During postweld heat treatment, it sometimes becomes necessary to support the welded pipe sections suitably, to prevent deformation and distortion. This is accomplished in the shop by placing adjustable roller-type supports under the parts being welded as near to the joint as possible, allowing sufficient space for the placement of the heating apparatus over the joint. In field work, where the welds are made in position, chain falls or other suitable rigging secured to the building or other supporting structures are used to accomplish the same objectives.

11.5 OFFSHORE PIPEWORK

A company in the Netherlands fabricates exacting offshore pipework using several automatic TIG and MIG welding installations, each having a turntable with two sets of adjustable roller beds. For 100 mm diameter pipes, manual TIG is used for the root pass and automatic TIG with wire feed for filling and capping passes. For 300 mm diameter pipes, the same procedure is used for the root pass and automatic TIG with cold wire feed for filling and gapping passes: For 300 mm diameter pipes, the same procedure is used for root pass, and automatic MIG with a flux-cored wire is used for subsequent passes. Table 11.4 shows a procedure for C: Mn pipe in which two types of flux-cored wires can be used for the MIG passes, one for temperatures down to 25°C and the other for temperatures below –25°C. The latter deposits a 2.5% Ni steel weld-metal with Charpy-V notch value of 47 J minimum at –60°C which also meets the COD test requirement.

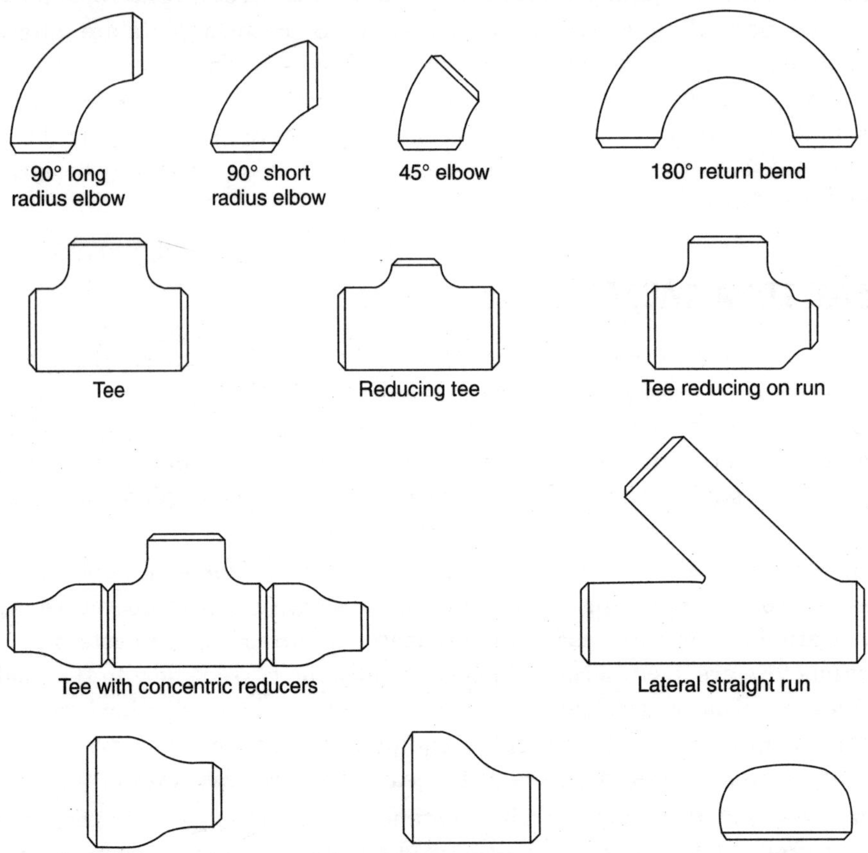

Fig. 11.7 Examples of standard manufactured commercial welding fittings

Table 11.4. Procedure for offshore pipework welding

Material	A333 GR6			
Root pass	TIG hand			
Filling/capping	MIG auto			
Welding position	IG			
Preheat temp. (°C)	100			
Interpass temp. (°C)	300			
Pipe dia. (in.)	4			
Wall thickness (mm)	10.7			
Joint preparation	$V_2 \times 30°$			
Root pass				
TIG wire type	PZ 6500			
Wire dia. (mm)	2			
Welding current (amp)	100			
Filling/capping				
MIG wire type	Flux-cored			
Wire dia (mm)	1.2			
Gas type	Mixed gas, 80/20 (Argon/CO_2)			
Gas flow (1/min)	10			
Welding current (amp)	205	230	225	225
Welding voltage	28	28	27.5	27.5
Wire-feed speed (cm/min)	788	788	788	788
Welding speed (cm/min)	24.5	19.6	23.5	23.5
Total welding time (min)	2.0	2.5	2.1	2.1
Joint preparation and runs	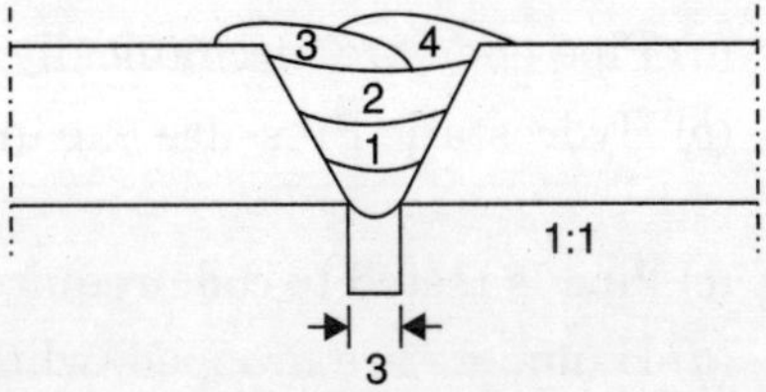			

11.6 PIPELINES (CROSS-COUNTRY)

This section deals with cross-country transmission pipelines which conduct natural gas or liquid products such as crude oil.

Pipes of reasonably long lengths are produced in a tube mill. They are either seamless or electric-resistance welded, or submerged-arc welded. Laying of pipelines involves only circumferential welding in the field.

Seamless pipes are made from solid round billets of proper diameter and length. Surface defects of the billets are initially removed by scarfing. The billets are heated and pierced to make a hole in the solid billet. The so-formed pipe is passed successively through a plug-rolling mill to elongate it and reduce the wall thickness to the desired dimension. The pipe is rounded and smoothed on the inside and outside surfaces by passing through a reeling machine. The pipe is finally sized by passing through sizing rolls, straightened, expanded, hydrostatically tested and beveled at the two ends.

Resistance-welded pipes are made from rolls of steel strip in a tubemaking machine. In this machine, the continuously fed strip is passed through forming rolls to form a straight O-shaped section, which is electric-resistance welded at the seam. The emerging pipe is tested continuously by means of a non-destructive testing device and cut to the desired length. A coiler is used if a long length of pipe is to be supplied in coil from.

The operation of producing large diameter pipes by the submerged-arc process is best understood by referring to the procedure followed by a firm in the U.S.A.

The firm produces mild steel pipes up to 13 m length and diameter between 500 and 900 mm and thickness between 6.3 and 12.7 mm in the following stages :

1. Shearing the edges to exact widths, bevelling the edges and pre-forming the plate by an initial bending of the edges.

2. U-ing press.

3. O-ing in a semi-cylindrical die with another top semi-cylindrical die activated by two massive hydraulic rams of 6,000 tons capacity.

4. Tack welding and tack grinding.

5. Cleaning the pipe in degreasing bath.

6. Tab is weld at each seam end to assure proper lead-in and cut-off of finish welds.

7. The pipe is welded finally by the submerged-arc process, one run on the inside and another run on the outside. For the first pass, water-cooled backing is used.

8. The finished pipe is moved on to the expander, where it is surrounded by locked restraining dies, while water at extreme pressure is pumped in, expanding the pipe against the enclosing dies. The expander does the following functions:

(*a*) Pipe ends are mechanically expanded to size.

(*b*) Hydrostatic Pressure expands the pipe to the exact size of the mechanically locked dies.

(*c*) Pipe is tested to code requirements.

(*d*) Hammers are dropped, while pipe is under maximum code pressure.

(*e*) Inspector examines welds for leaks.

Two 13 m long pipes may be welded to make 26 m lengths, again using submerged-arc welding.

Finally, there is end facing and bevelling.

The forming is at the rate of 20 m/min and output is up to 3,000 tons in eight hours.

A typical boom welder used for the internal welding of pipe by the submerged-arc process is shown in Fig. 11.8. It is fitted with a television monitor. The 375 mm diameter boom enables pipes of 450 mm and large diameters and lengths up to 10 m to be welded internally.

Pipes are also welded by the submerged-arc process, using the so-called spiral welding technique. The main advantage is that with a given width of plate or coil, a wide range of pipe diameters can be fabricated.

In this technique, the edges of plates or coils are trimmed to the required width and bevelled. They are then subjected to a modified three-roll bending arrangement supported by internal or external cage rolls, and the result is a continuous helix. The first welding pass is laid on the internal diameter of the seam and then on the external diameter, 180° away. The conventional single electrode or two electrodes in tandem may be used for the submerged-arc process. To feed the stock continuously into the machine, ends of plates or coils are welded only on the inside by the submerged-arc process prior to forming. After seam welding, the required length of pipe is cut off and the external cross-weld is completed. The maximum outside diameter of seamless pipes is 650 mm. High frequency resistance seam welding is used to produce pipes and tubes of diameters ranging from 12.5 mm to over 1,250 mm and with wall thicknesses of between a fraction of millimetre and 25 mm. Submerged-arc welding is best suited for large diameter pipes, which can be internally and externally. Penstock pipes of 10 m diameter and above have been welded by this process.

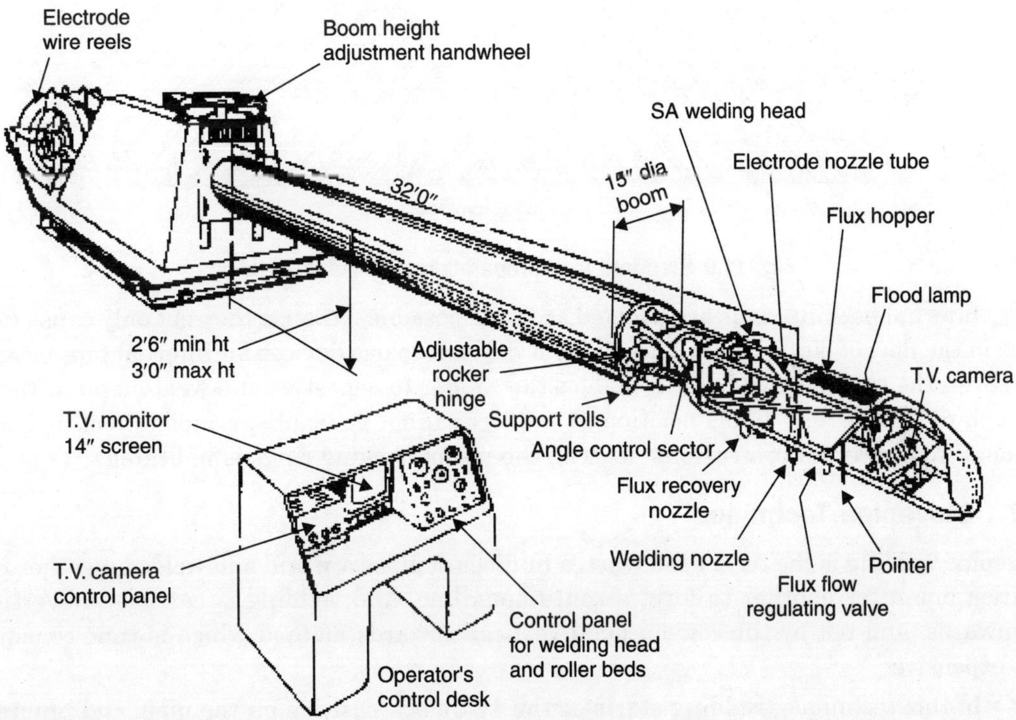

Fig. 11.8 Diagrammatic arrangement of boom and controls for internal pipe welding equipment

Generally, pipes for the transmission of liquid products are smaller in diameter than pipes meant for natural gas. The common diameters used for gas transmission are 600, 750 and 900 mm (24, 30 and 36 inch), though recently these have been increased to 1,400 or 1,500 mm.

Transmission pipelines are usually manufactured to the API specifications for Line Pipe. They specify, among other things, the strength levels of various steels to be used, working stress levels and longitudinal joint efficiency of pipes, and tests for the qualification of procedures and welders.

11.7 PIPELINE WELDING

Most pipeline welding involves girth welding from external side only, because the diameters are too small to permit welding from the inside. The commonly used joint design is shown in Fig. 11.9. It is well suited for the stovepipe technique described below. In special cases, the angle of bevel is increased from 30° to 37.5°.

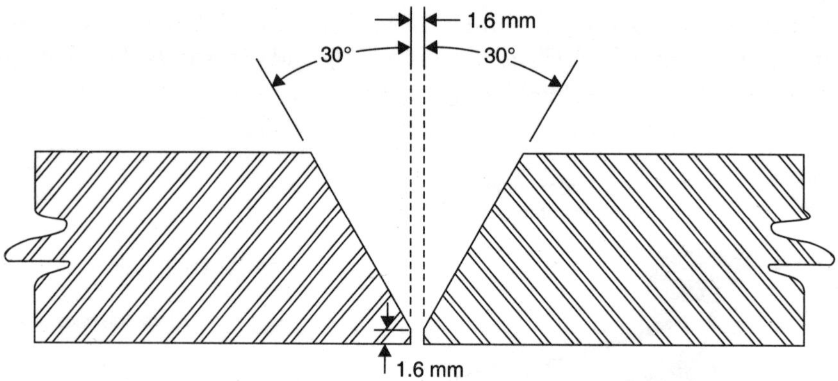

Fig. 11.9 Standard joint preparation for pipeline welding

Internal backing rings are avoided as far as possible, because they not only cause turbulence in the flow of material, but also make it difficult to use devices for internal pipe cleaning. Moreover, the stovepipe technique enables the welder to deposit sound weld-metal at the root through the entire 360° in 5G position. If welders cannot guarantee complete root fusion and freedom from internal protrusions (icicles), the use of backing rings is indicated.

11.7.1 Stovepipe Technique

Stovepipe welding is the term used when a number of pipes are laid and welded together in G5 position one after another to form a continuous line, and welding is carried out vertically downwards, and not by the conventional vertical upwards method which is time consuming and expensive.

In this technique, welding starts at the 12 o'clock position on the pipe, and progresses vertically down until the 6 o'clock position is reached. On completion of one half of the pipe, the opposite side is welded in the same manner, thus producing an endless root run known in the field as a stringer bead.

The second run, known as the hot pass, is then put into the joint. Its name comes from the fact that a high current is used to deposit the run, so as to burn out any defects that may be present from the stringer bead. With the exception of the final run, all subsequent runs after the hot pass are termed filler beads. Their purpose is to bring the weld deposit to just below the level of the pipe surface. The number of filler beads required will depend largely on the pipe-wall thickness and the preparation.

There are times, however, when it is necessary to deposit a filler bead all round the pipe periphery, especially as the weld nears completion. In most cases only the areas between 2 to 4 and 10 to 8 o'clock on the joint (see Fig. 11.10) will require additional weld-metal. These concave areas are rectified by the quick deposition of a weld run called a stripper bead, which brings the concave areas flush with the remaining weld-metal elsewhere in the joint. To finish the pipe weld the final run is made, which is appropriately called the capping bead.

The joint preparation and fit-up is as shown in Fig. 11.9. Welding is done with AWS E6010 and E7010 class electrodes. These are chosen because the small volume of stiff, thin slag coating deposited on the weld bead, together with the forceful arc, facilitates rapid changes of electrode angle during vertical-down welding on fixed pipes.

To compensate for the thin slag coverage, extra protection from the atmosphere is provided by a gaseous shield of carbon monoxide and hydrogen evolved from the cellulosic coating during welding.

For stovepipe welding, the maximum current specified by the producer for the size of electrode is increased by approximately 10%. DC supply with electrode positive (positive polarity), is often recommended. There may be occasions, however, where scale on the pipe causes surface porosity. In such cases, changing the electrode polarity from positive to negative tends to reduce this problem.

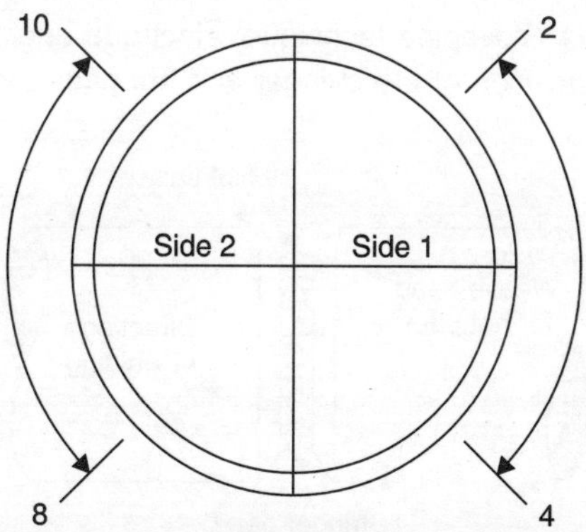

Fig. 11.10 Stovepipe technique; positions for stripper beads

For deposition of the stinger bead (root run), once the arc has been established, the cup of the electrode must be literally pushed into the root of the joint. No weave of the electrode is necessary, only a light drag action as welding proceeds, to ensure that the arc is allowed to burn inside the pipe. An electrode angle of 60° in the direction of travel to the pipe tangent (see Fig. 11.11) must be held throughout.

This practice produces a very small root run, which allows for a controlled penetration bead. If one or more burn-throughs (windows) occur during the laying of the stringer bead, they can be quickly rectified by the remelting process of the second run.

Immediately following the stringer bead and while it is still warm, the hot pass is put down with an electrode angle held at 60° to the pipe tangent. A short arc must be held with a light drag, together with a forward and backward movement of the electrode (see Fig. 11.12), in order to fuse out any undercut and/or wagon tracks, caused by the stringer bead. In addition to remelting the portions containing windows, the higher current used for this run prevents the formation of slag lines at the toes of the stringer bead.

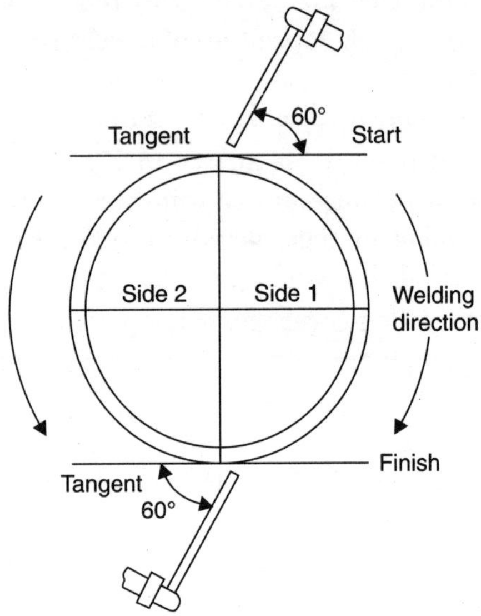

Fig. 11.11 Stovepipe technique; electrode angle during deposition of the stringer and hot pass runs

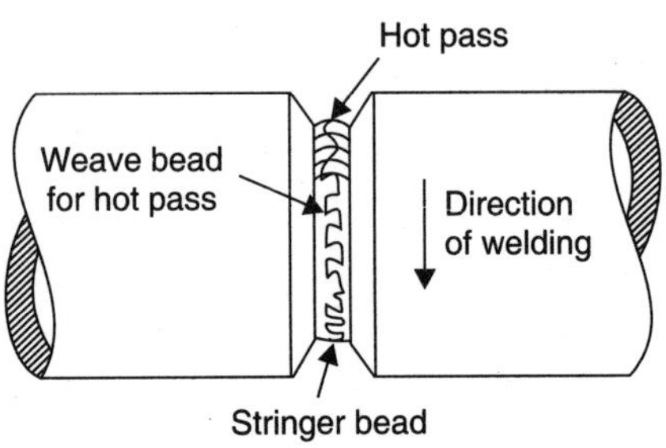

Fig. 11.12 Stovepipe technique; electrode manipulation during deposition of the hot pass

For the filler bead deposition, it is necessary to alter the electrode angle from 60° to 90° to the pipe tangent. However, on reaching the 4 o'clock (8 o'clock on side 2 of the pipe) the electrode angle is increased from 90° and reaches 130° at the 6 o'clock position of the pipe (see Fig. 11.13).

From the 12 o'clock down to 4 o'clock (8 on side 2), a normal arc length with a rapid weave across the weld face is required, pausing memontarily at the toes, from 4 o'clock (8 o'clock) down to the 6 o'clock position, the electrode manipulation is changed from a weave to a lifting or vertical movement of the arc away from the deposit on to the weld pool. By adopting this technique on the filler beads, flat weld faces with the absence of undercut are produced.

For the stripper beads, a medium to long arc is required to spread the weld deposit. A slight weave of the electrode may be found beneficial, depending on the current setting and width and depth of the bead required. The angle of the electrode is held at 90° to the pipe tangent, irrespective of the position on the pipe periphery.

Finally the capping bead completes the joint, using a medium to long arc length, with a rapid side-to-side movement of the electrode tip. The angle is maintained at 90° to the pipe tangent except from 4 to 6 and 8 to 6 o'clock positions when the electrode angle is increased to 130°.

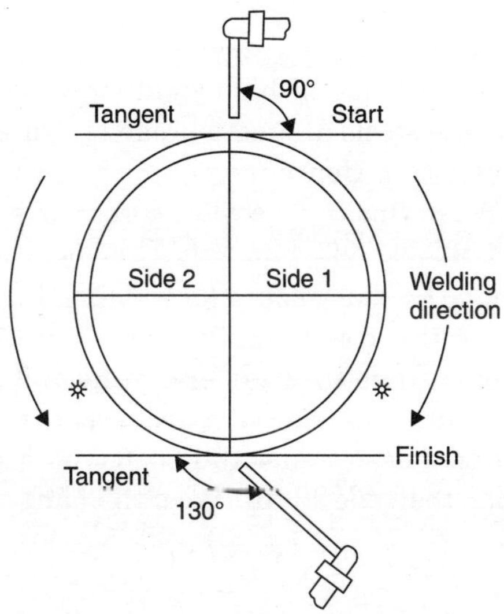

Fig. 11.13 Stovepipe technique; electrode angles for filler and capper beads. From positions* electrode angle changes from 90° to 130°

For these sections, the electrode should be manipulated to produce a lifting and flicking action. To achieve best results, the capping bead should be restricted to the width and depth of ~19*1.6 mm. Weld beads wider than this are somewhat difficult to control.

The electrode size for various passes depends on wall thickness. For depositing the stringer bead, for example, 3.25 mm diameter electrode is used for wall thickness below 6.3 mm, and 4 mm diameter for larger thicknesses. For first and second filler passes, 4 mm diameter electrode is commonly preferred. For third filler, stripper and cover passes, 4 or 5 mm diameter electrodes are used depending on wall thickness.

It is difficult even for a normally well-experienced welder to use stovepipe technique successfully, unless he is given special training with suitable electrodes on actual pipe joints. Experience has shown that only about 20% of the otherwise skilled welders are capable of mastering the stovepipe technique.

The adoption of stovepipe technique in pipeline construction demands a well-planned disposal of the crew, in order to ensure that welding operations take place rapidly along the line. The pipes are first lined up by the line-up crew with the help of an internal line-up clamp. A good joint fit-up is the necessary condition for a flawless, well penetrated stringer bead, and it is the responsibility of the line-up crew to ensure it. Two welders then complete the stringer bead (first pass). The line-up men and these welders then move on to the next joint, while a second group of welders deposit the hot pass (second pass). They then shift to the next joint, while the third group of welders completely fill the joint. The third group, called firing line, includes a larger number of welders, since more welding is involved in completing the joint. The stringer welders and the hot pass welders work in groups of two or four.

Stovepipe technique is not possible with rutile type (E6013 class) electrodes, because the relatively large volume and high fluidity of the slag render vertical downward welding difficult with these electrodes, good joints can be made by welding vertically upwards. But the technique is slow and results in lower productivity.

11.7.2 LH Electrodes

In recent years, increasing use is made of high-yield steels for pipeline, for example, the SL × 60 and SL × 52 steels. These steels are more prone to hydrogen-induced cracking in the HAZ than the conventional mild steel. Hence the pipe ends need to be preheated when E6010 – E7010 electrodes are used. When this is done, the stringer pass and the hot pass have to be made with an increased speed of 230 – 300 mm/min. This increases the strain on the welder.

Special LH electrodes have been developed for welding SL × 52 and SL × 60 steels using the stovepipe technique, without the need for preheating. With these electrodes, the root gap is increased to 2.5 mm to accommodate the heavier coating and the welding speed is kept as low as 150 mm/min. The disadvantage of reduced speed is more than made up by the thickness of the root pass, which is twice that deposited with E6010 type. The deposition efficiency of the LH electrode being 20% higher than the E6010 type, the joint can be completed with fewer layers and in shorter arc time.

11.7.3 MIG/CO_2 Process

The inherent advantageous features of this process could make it preferable to MMA welding, but there are several difficulties. The normal spray transfer technique which is capable of giving high deposition rates would give rise to burn-through and considerable spatter when CO_2 is used for shielding. The dip transfer technique using argon/CO_2 mixture for shielding is better suited for 360° welding, but the shallow penetration of this process can lead to incomplete fusion. Moreover, the upkeep of the equipment at site demands the services of properly trained mechanics and a regular supply of spares.

For the welding of pipes large enough to accomodate a MIG/CO_2 welding head inside, fully automatic equipment has been developed. A typical piece of equipment consists of four welding heads, mounted at 90° spacing, for internal welding and two welding heads for external welding. The two top internal welding heads proceed simultaneously from the top of the pipe downward to make the weld. The two opposite internal heads then counter rotate to complete the joint. The external welding units are light and portable, and they are used in conjunction with a tracking band, which is attached around the pipe at a fixed distance from the weld. The two units operate simultaneously on each side of the joint, proceeding from the top of the pipe downward. It is also possible to use the external units simultaneously with the internal units.

For the internal weld which is made first, a small V-groove is provided. For external welding, a V-groove with 20° included angle is adequate to ensure complete fusion. This means reduced weld-metal required to complete the joint. The welding wire is of 0.8 mm diameter and the shielding gas is 70% argon – 25% CO_2. This argon-rich shielding reduces spatter to the minimum.

The system may also incorporate a pipe-end preparation machine, which is used ahead of the welding operation. The internal welding machine may be combined with a line-up clamp.

Such systems have been used with success for various onshore and offshore construction projects in the U.S.A., Canada and England.

11.7.4 Flux-cored Process

A typically system utilising this incorporates an end preparation machine and makes all the weld passes from the outside. It uses two welding heads, mounted 180° apart, for the root pass and four welding heads, spaced at 90°, for the subsequent passes. The root pass is deposited over a copper back-up attached to a specially designed internal line-up clamp. All welding proceeds from the top to the bottom. The flux-cored welding wire is of 2 mm diameter. No external gas shielding is used, which is a welcome feature for site welding. The joint consists of 58° included angle, 1.6 mm root and 2.5 mm root face.

11.7.5 Underwater Pipelines

Pipelines for underwater service are laid in marshy land, shallow waters or in considerable water depths. MMA process is commonly used for welding. The welders work at stations located on barges. The pipe laying starts from the land or shore and proceeds towards deeper waters. As many as five welding stations may operate on several barges, followed by two radiographic stations and a coating station. Coating is meant for corrosion protection. Large diameter pipes are preferably concrete coated to provide corrosion resistance as well as negative buoyancy. In offshore construction, the completed pipe sections is lowered gradually by means of a semi-buoyant stringer, which holds the pipe until it has neared the sea bed. After laying, the pipe is buried in the sea bottom.

11.7.6 Inspection and Testing

For important pipeline construction, the welding procedure as well as the welders must be qualified. The necessary guidance is obtained from any of the following or equivalent standards:

(*a*) API Standard 1104, Standard for Welding Pipelines and Related Facilities

(*b*) ASME Boiler and Pressure Vessel Code, Section IX

(*c*) ANSI B 31.8, Code for Gas Transmission and Distribution Piping

In the qualification test a sample pipe is welded in accordance with the procedure adopted and coupons are removed by gas cutting; they are then subjected to various tests such as tensile, nick break, root and face bend tests. If these tests meet the code requirements the welder or procedure is taken as qualified.

Inspection is carried out both during and after welding. During welding, the points to be checked are: (*i*) edge cleanliness, edge preparation and joint fit-up; (*ii*) physical condition of the electrodes; (*iii*) functioning of the power source and current setting; (*iv*) soundness and penetration of the stringer bead; (*v*) soundness and quality of hot passes; and (*vi*) interpass cleaning. After welding, the joints are subjected to visual and radiographic inspection. The latter is carried out with X-rays or gamma-rays. Special radiographic equipment has been designed for large diameter pipelines, which enables the X-ray or gamma-ray source to be propelled through the pipeline on a battery driven or engine-driven crawler unit. The unit is provided with a mechanical or radiological device to locate and stop at a welded joint. Film belts are wrapped around the joint circumference to radiograph the entire joint in one exposure. The unit is programmed for speed, exposure time and other radiography parameters before insertion into the pipeline. Such an equipment can travel several kilometres through a pipeline, thus enabling the contractor to proceed continuously with welding without waiting for radiographic inspection to catch up with him.

For small diameter pipe, radiography has to be done from outside. In this case, the source is placed on one side and the film 180° opposite. At least three exposures are necessary to cover the entire joint, and increased exposure time per exposure is required. Hence external radiography is more time-consuming than internal radiography. Other NDT methods are rarely used. Ultrasonics, for example, cannot perform reliably because of the irregularities of the manual-arc welded stringer bead and cover pass.

Sometimes the completed pipeline needs to be pressure-tested prior to being placed in service. The common practice is to test it hydrostatically with water to stress levels equal to the actual yield point of the base metal.

■ QUESTIONS

11.1 What do you mean by the term 'piping'? What is the difference between 'pipeline and piping? What type of guidance is provided in standard codes regarding welding of pipings, selection of materials, design, fabrication, erection, testing and inspection?

11.2 (*a*) With neat sketches briefly explain the joint design, and edge preparation of pipe end for MMA welding.

(*b*) What is a backing ring? With neat sketches explain the joint fitup using consumable insert for Tig Weding of butt joints. Also explain briefly the fitups for fillet welded joints.

11.3 What is the significance of heat treatment in the welding of pipings? Briefly explain the common methods of heat treatment. How the welded pipes are supported during heat treatment to prevent deformation and distortion? Briefly explain how off-shore pipework is carried out.

11.4 Briefly describe the stages in which mild steel pipes are fabricated before welding.

11.5 Briefly describe with neat sketches the procedure commonly followed for the welding of pipe-lines on site, what is stove-pipe technique of welding pipelines?

11.6 What is the importance of low hydrogen electrodes?

CHAPTER 12

Life Prediction of Welded Structures

12.1 INTRODUCTION

1. All welded structures are expected to have an **estimated service life.** The **actual service life** may be more or less than the estimated period.

2. To ensure safe service and avoid unexpected failure, it is customary to **inspect** the welded components/structures at regular intervals.

3. Welded structures suffer from **defects/discontinuities** leading to failure.

4. The defect which most commonly leads to failure is some or the other form of **crack,** which when attains a **critical length** runs at unbelievably high speed leading to catastrophy.

5. Once a crack has been detected, it is imperative to repair it.

6. If repair is not possible steps are taken to assess the residual life of the component/ structure so that steps are taken to replace it quickly before its life expires.

7. If **unexpected failure occurs,** causes are investigated, so that steps are taken to eliminate such causes from **future structures.**

There are two aspects of the problem for structures in-service with cracks having initiated in them *viz.*

1. Residual life Assessment
2. Failure analysis.

12.2 RESIDUAL LIFE ASSESSMENT OF WELDED STRUCTURES

Chemical process plants and power plants are constructed in accordance with some construction codes and tested according to the relevant inspection codes.

Construction and inspection codes for major components of chemical and power plants are given in the following table (Table 12.1).

Table 12.1. Construction and inspection codes for major components of chemical/power plants

S. No.	*Type of equipment*	*Construction Code (design + manufacture)*	*Inspection code warnings notes on environmental induced damage*
1.	Pressure vessels	ASME Boiler and Pressure Vessel code sec. VIII	API standard 510
2.	Piping	ANSI code B 31.3	API standard 570
3.	Storage tanks	API standard 620	API standard 653

These codes do not talk about guidelines to assess the fitness of the equipment or determining its remaining useful life. They provide only the **design rules** and **method of construction and inspection.**

It has been found that a large proportion of process equipments have failed in service due to **manufacturing defects** or **severe working environment.**

12.2.1 Fitness for Service (FFS)

It is the ability of a structure to serve satisfactorily under a given set of process conditions for a reasonable period economically. This means the determination of accepable critical sizes of cracks (or other defects) or extent of material deterioration beyond which equipment cannot be adjudged as suitable for continued service.

Residual Life Assessment (RLA)

It is the time period during which the equipment shall retain the fitness-for-service characteristics. Fitness-for-service thus becomes very important for residual life assessment.

Extensive and expensive inspection programs are undertaken, in addition to routine inspections, **to monitor the extent of in-service deteriorations.** These inspections are more rigorous than routine ones and are needed with a view to find out whether a particular material condition was service induced or existed since the structure was built.

Deterioration of the material properties which is important for assessing the safety and reliability, must be assessed before an effective analysis for FFS or RLA is considered.

12.3 INVOLVEMENT OF EXTERNAL AGENCIES IN FFS AND RLA

Govt. bodies and jurisdictional agencies get involved in FFS and RLA if the welded structure concerned is critical and its failure may cause hazard to life and heath of the people living around. The trend is towards their increased interest in performance-inspection frequency, acceptance standards, repair procedures, and record keeping. In some countries it is mandatory to establish FFS and RLA after a stipulated service period.

FFS **criterion** leading to RLA should satisfy the following conditions:

1. Should be sound, practical and based on latest know-how.

2. It should be acceptable to owners and operators both.
3. Acceptable to relevant jurisdictional and certification authorities.
4. Should be based on proven inspection techniques.
5. Should be based on material properties that account for in-service degradation specific to the situation concerned.
6. It should be adaptable to short and long term needs.

12.3.1 Development of Expertise on FFS and RLA

1. **Historically** industry itself gives top priority to safe operation of process equipment by setting concensus guidelines and implementing various inspection requirements based on existing knowledge and experience available in that period.
2. Over a period of time with increased experience and improved knowledge regarding material behaviour and stress analysis a number of FFS analysis and RLA programmes and guidelines have been developed by individual organisations and by professional and standardisation bodies.
3. While individual programmes and guidelines are being updated periodically, it may take some time before a common set of guidelines based on concensus of all the agencies involved is developed.

12.3.2 Justification for FFS and RLA Studies

Any fabricated metallic component has imperfections/discontinuities as recognised by code of construction which lay down the allowable limits of such imperfections. But there is no consensus procedures in industry that categorically spells out the methodology for accurately judging the **Fitness-for-purpose** for any vessel or piping components with defects beyond the code limits. The **important elements of fitness for service** approach are as follows.

1. Understanding the origin of the imperfection.
2. Knowing its present status.
3. Knowing the size, orientation, location and other relevant characteristics of the imperfections.
4. Establishing the stress acting at the location of relevance.
5. Characterisation of the material.

FFS and RLA in Presence of Service Induced Defects

Incase the defect is service induced, past records will not provide sufficient justification and safety margins to be employed. For such complex situations a higher level of analysis and data base is needed.

12.4 NATURE OF DAMAGE IN SERVICE

There are various types of damages in service and each type needs to be dealt with separately. In refineries, for example, the following types of deteriorations may be encountered:

- General corrosion
- Pitting attack
- Hydrogen damage (Hydrogen attack—Blistering, sulphide stress—Corrosion cracking (SSCC)—Hydrogen induced cracking (HIC) embrittlement.
- Stress corrosion cracking (SCC)
- Metallurgical degradation
  - — Temper imbrittlement
  - — Secondary precipitation
  - — Carburisation
  - — Graphitisation
  - — Spheroidisation
- Fatigue/corrosion fatigue
- Creep/creep fatigue
- Oxidation

While the nature of the above mentioned damages are different, these can be grouped on the basis of the mechanism by which these affect the health of the equipment. Table 12.2 shows the defect categories and assessment of equipment fitness.

Table 12.2. Defect type and assessment of Equipment Fitness

	Nature of Defect	*Effect on Reliability*	*FFS and RLA Approach*
I.	General corrosion • Pitting (closely spaced) • Hydrogen attack • Oxidation • Blistering • Spheroidisation	Decrease in load carrying capacity " " " "	Increase in inservice stress
II.	Pitting scattered	Leakage	Nozzle opening stresses
III.	Blistering (sulphide stress corrosion cracking) HIC/SOHIC SSC, Fatigue/corrosion Hydrogen attack (linking of fissures to form cracks)	Linear defect, liable to cause rupture or leakage " " " "	Fracture mechanics
IV.	Creep/Creep Fatigue	Rupture	1. Creep damage accumulation model. 2. Fatigue crack growth
V.	Hydrogen Embrittlement	Decrease in ductility	Toughness characterization and/or fracture mechanics

12.5 INSPECTION TECHNIQUES APPLIED FOR FFS/RLA STUDIES

Based on the past experience on detailed examination of cracks and other damages observed in storage tanks and pressure vessels, the following points in regard to inspection techniques must be considered.

1. Use improved techniques to detect sub-surface flaws, dimensions, locations, depth and number of cracks.
2. Improved technique should be able to :
   (*a*) inspect the entire vessel inside and outside.
   (*b*) inspect it while in operation.
   (*c*) monitor and measure flaw on-line.
   (*d*) have sizing accuracy adequate to identify the margins to critical flow size.

Analysis of Available Data on Plant History

1. Analysis of data includes: review of original design, past operating conditions, inspection and maintenance records. This helps in locating and ranking and analysing the critical areas. For this purpose operators, plant inspection and maintenance staff are interviewed to assess plant and process upsets, fires, modifications, repair. This could affect the residual life assessment.
2. Once the material deterioration mechanism is recognised and state and extent of flaws through appropriate inspection methods have been established, the next step is to establish the critical condition of material degradation beyond which it would be unsafe to operate the structure. Next step is to determine the rate of growth of flaw/deterioration so that the time period required for reaching the critical limits of flaw size or material condition could be estimated.
3. A number of approaches to determine the critical sizes of flaws have been developed and are available in ASME Sec. XI approach, BS-PD 6493 approach, and CEGB R–6 Methods. Present metal condition can be established by destructive tests, where feasible, or in situ non-destructive metallurgical tests.
4. With these inputs the extent of life spent and the remaining life can be worked out on case to case basis.

12.6 WELD FAILURE

Failure is a term in which a member is subjected to plastic deformation, leading to failure, causing heavy losses to life and property. These losses are of two types: (*i*) Direct losses, (*ii*) Indirect losses, as shown in Fig. 12.1 below.

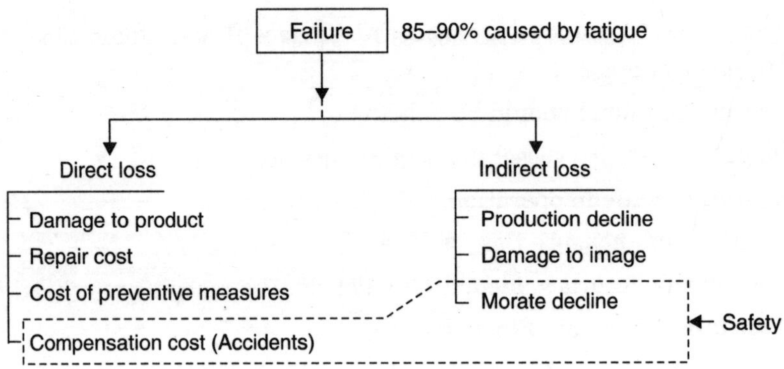

Fig. 12.1. Weld failures types

Chapter 13

Welding of Plastics

13.1 INTRODUCTION

Most commonly used plastics are either thermoplastics or thermosetting plastics. Thermoplastics could be compared to wax. They are capable of remelting and changing shapes. Thermosetting plastics could be compared to an egg. When boiled, an egg becomes solid and sets, it can not be brought back to liquid condition and cannot be reshaped. Thermoplastics are weldable thermosetting plastics are not weldable but can be joined by adhesive bonding processes.

A number of widely used plastics can be welded as they are thermoplastics. The most common of these are polyvinyl chloride (PVC), polyethylene, polypropylene, acrylonitrile budadiene styrene (ABS) and acrylics. Such plastics can be welded by melting the surfaces to be joined and allowing them to solidify while in contact. Plastics containing volatile components may form gas bubbles which cause the formation of defects in the welds made.

Friction welding machines can be used to produce excellent welds in circular cross-section components. The most common method of welding plastics uses hot gas as a source of heat and uses torches similar to an oxy-fuel torch.

Welding torches for plastics are designed to let a compressed gas flow through electrically heated coils which raise the gas temperature to between 175° and 315°C. This hot gas passes through an orifice forming a narrow gas, stream which can be directed to the surfaces to be joined. See Figs. 13.1 and 13.2.

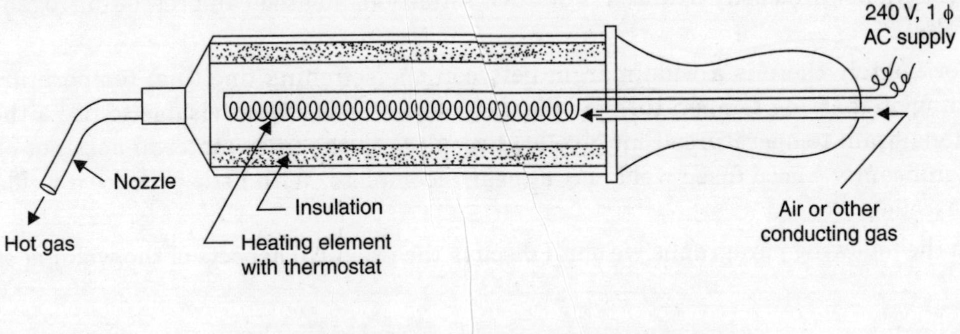

Fig. 13.1 Electrically heated plastic welding torch

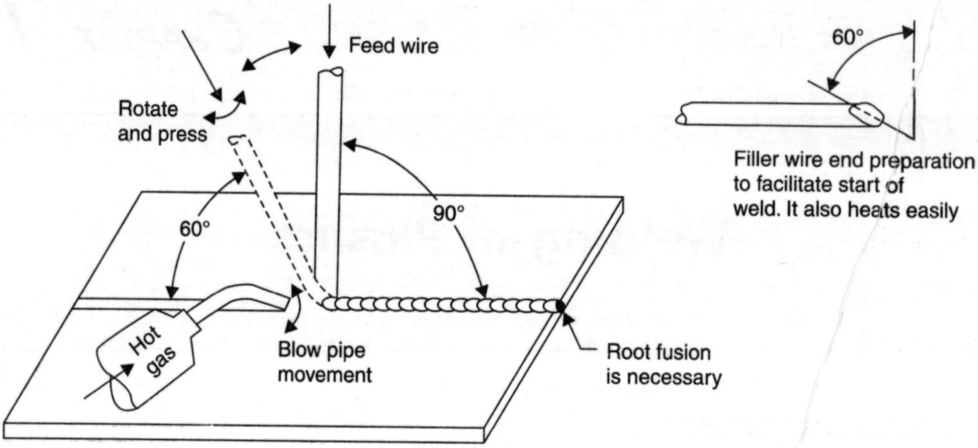

Fig. 13.2 Manual hot-gas torch welding

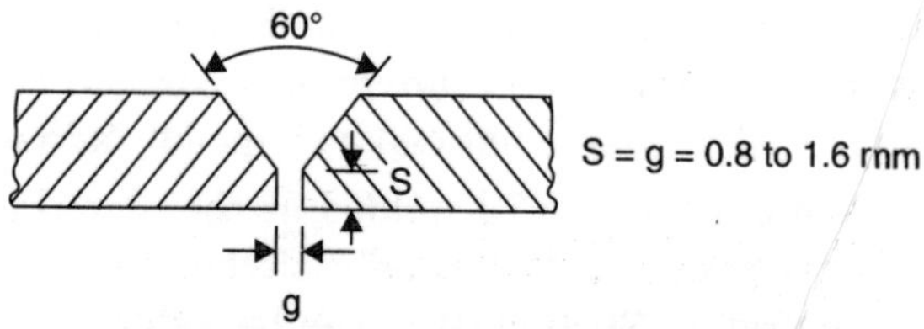

Joint preparation for welding

Table 13.1. Manual welding force on filler rod (intermittent)

Filler rod dia. mm	*Approx. load (kg)*
2.4	1 kg.
3.2	1.8 kg.
4.8	3.0 kg.

Power requirements rarely exceed 500 W for the heating element. Gas/Air flow is of the order of 280 l/min which can be supplied by 1/4 horsepower compressor motor. Some plastics (*e.g.* polyethylene) are easily oxidised. For such situations heated compressed nitrogen gives best results.

Fortunately there is a wide margin between the softening (melting) temperature and the burning or charring temperature for thermoplastics. It is still advisable to use a thermostat and maintain temperatures that give best results. As the filler material does not change shape significantly a good fused weld may appear incomplete. With little practice a welder can deposit excellent beads.

In the following paragraphs we shall discuss the practical aspects of the welding of PVC plastics.

13.2 HOT AIR WELDING OF PVC PLASTICS

Plastics are finding surprisingly new and diversified applications replacing metals and ceramics. From ordinary toys and utensils to the complicated precision heart valves, the plastics have proved not only to make life more comfortable but also to extend it.

Plastics have a combination of desirable properties. They have high strength to weight ratio, corrosion resistance against most of the corrosive media, low cost and ability to take good finish.

Plastic structures can be fabricated by welding. Thermo-plastics are the only weldable plastics as they maintain their molecular structure even after repeated heating. Among the common thermo-plastics are: acrylics, fluorocarbons, shellac, asphalt, nylon, polyethylenes, polyvinyles and protein substances. Among the above the rigid polyvinyle chloride has sufficient resistance against corrosion, strong acids alkalies and organic solvents. It is, therefore, the most common thermoplastic in use these days. The term 'welding of plastics' is still rarely known amongst the engineers because of the fact that the use of plastics is still not very common in many industries and the plastics which are used can normally be joined by organic solvents like carbon tetrachloride and adhesives like areldite. PVC, however, is almost insoluble in most of the organic solvents. Though, it is slightly soluble in carbon tetrachloride but, the action is very slow. There are certain other limitations too, in the way of joining plastics by the methods other than welding. With the help of welding adequate strength at the joint is achieved in minimum time. Within a few minutes after welding, any welded joint can be handled with reasonable care, facilitating rapid and economic fabrication of plastic structures.

13.3 WELDING ACTION

Unlike metals, the welding action in plastics takes place due to the adhesive bonding at high temperatures, between the parent material and the filler rod. There is no mixing or puddling action as is common in the metallic weld pools. The melt of plastic is quite viscous and has poor flow properties, while good flow properties are essential for obtaining homogeneous welds. The surfaces of the parent material and the filler rod are heated and brought near to the melting temperature and by the application of pressure the filler rod gets adhered to the adjoining (weld bead) surfaces to be joined. Thus a homogeneous weld bead is not obtained but the filler rod gets adhered to the material in its neighbourhood and thus, gives a defect free non porous joint.

13.4 EQUIPMENT

The tool used for hot gas welding resembles in appearance with the ordinary welding torch (Fig. 13.1). Direct flame chars the material (PVC) and, therefore, hot gas is used for welding purposes. The torch consists of a main body which contains a heating element. At one end of the body there is an inlet hose connector for the gas and a handle for gripping the torch while the other end has a nozzle through which the hot gas is available for use. The welding gas (usually air) enters the torch at some pressure and gets heated while passing over heating element and comes out of the exit nozzle at a desired temperature. The gas temperature is controlled by providing in the heating element circuit, a thermostat valve which controls the 'on' and 'off' period of the current fed to the element, thus regulating the temperature of the gas to a desired value depending upon the parent plate thickness. The torch may also be heated by using a fuel gas. A sectioned view of the torch used is shown in Fig. 13.1.

WELDING OF PVC PLASTIC USING HOT AIR TECHNIQUE

For the welding of PVC sheets, hot-air technique is commonly used. Air is easily available and gives good results with PVC. Air flow needed for the process can be obtained by using a small air compressor, with automatic tripping device to obtain constant pressure. Supply air pressure can be measured by a mercury manometer shown in Fig. 13.3.

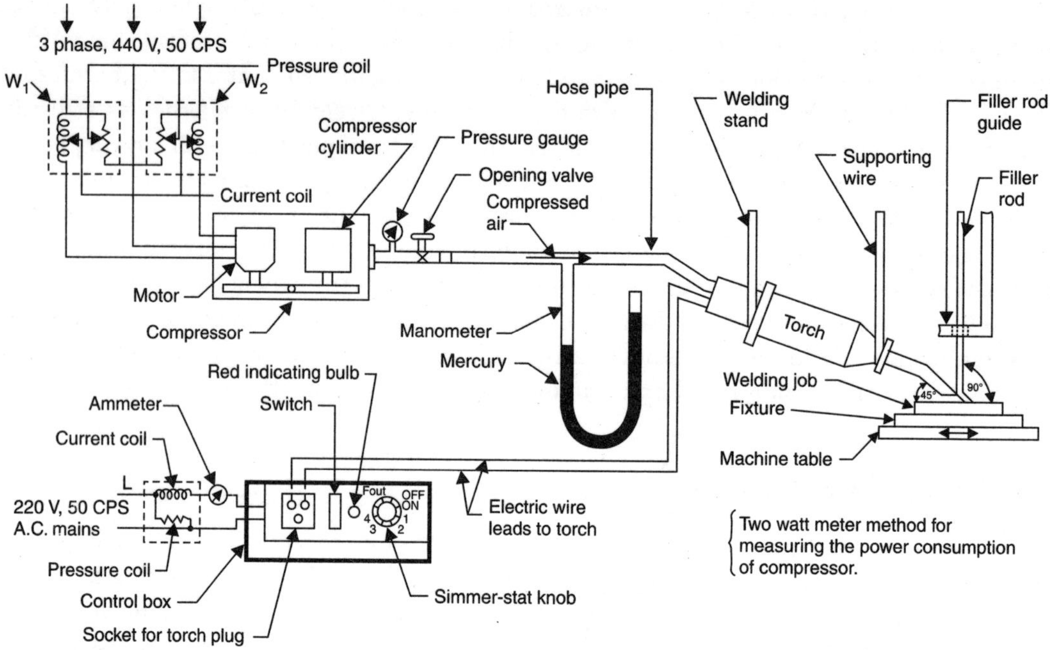

Fig. 13.3 Block diagram of welding set-up

Rigid PVC sheets in common use are of 3 mm thickness and can be welded by using 3 mm filler rods. Air temperature was controlled by using a simmer-stat that controls the amount of current in the heating coil (Fig. 13.3). Edge preparation for different plate thicknesses is given in Table 13.1.

Welding traverse speed. It depends upon air, temperature, nozzle distance from plate and filler rod. It is manipulated by the experienced welder to obtain quality welds. Rod is fed to the plate at an angle of 90°. A fixture can be made if required to guide the filler rod at 90° and keep the torch nozzle at an angle of 45° with the joint line (Fig. 13.3). Milling machine table could be used to obtain uniform traverse speed. A large number of traverse speeds are possible with this arrangement.

Gap Distance. There is a slight variation of temperature with change of gap distance. This could be noticed from Fig. 13.4. Thus a slight variation of gap distance between the torch nozzle and plate due to hand welding will not appreciably affect the weld quality.

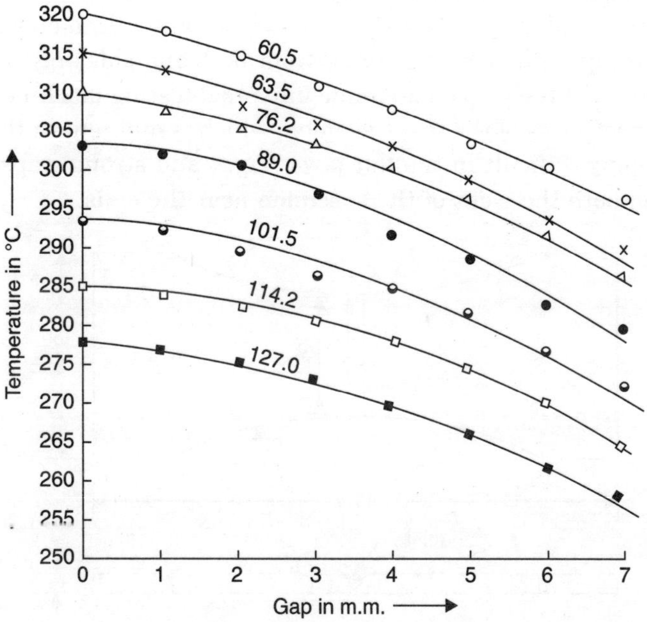

Fig. 13.4 Gap distance between torch and the job versus temperature of hot air

Welded joints. Two types of welded joints in general use are:

(*a*) Butt joints.

(*b*) Double strap fillet joints (see Fig. 13.6).

To obtain a butt joint, the plates to be joined are bevelled (60° V groove angle), cleaned, assembled over a backing plate and clamped to the machine table-vice. The compressor is started, torch is switched on, air pressure is regulated to about 100 mm of mercury. When a constant temperature of the system is achieved, a tack weld is made at the starting end by simultaneously heating the base plate and filler rod. As the mating surfaces fuse, it will be possible to slightly rotate the filler wire in-place, slight pressure is applied to the filler rod to affect proper adhesion. The table is then moved away from the torch. The pressure on the rod is maintained with slight rotary motion on the filler wire as shown in Fig. 13.2. This manoeuvre is a matter of practice on the part of the welder. After completion of one pass, the table is stopped and filler rod is cut. The process is repeat for subsequent passes as needed to fill the joint groove completely. Satisfactory welds have been obtained at a traverse speed of 50 mm/min.

Similar procedure is adopted for obtaining double strap fillet joints except that the assembly of the piece to be welded is tilted through an angle of 45° to facilitate the heating of fillet properly. The fillet in this position served as a 90° V-groove angle and heat is equally distributed to the plates to be joined.

13.5 TESTING OF JOINTS

Dumbell type test specimen has been proposed in the literature[4] for finding out the strength of plastic sheets with no mention about the testing of the joint strengths in welds. Dumbell type specimen as shown in Fig. 13.5, has been used for testing the strengths of the parent plate as well as that of the butt welded specimen by some investigators. End effects can be avoided by removing and discarding a strip 35 mm wide from both the sides of the welded test piece. These test specimen can be tested on a 20 tonne universal testing machine using flat grips and 2 tonne scale. The smoothness of the test specimen, which is inherent in the rigid P.V.C. sheet, may render the gripping difficult in the flat jaws. Tight and strong grips can be obtained by making cerrations on both the sides of the specimen near the ends.

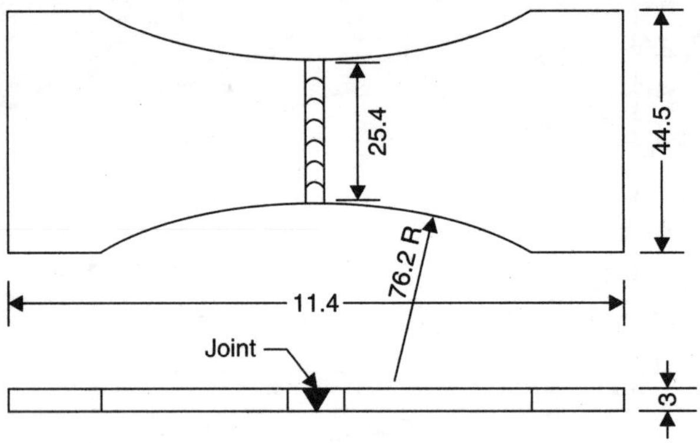

Fig. 13.5 Test specimen for butt joint

Straight test pieces are used for testing the strengths of double strap fillet joints, as shown in Fig. 13.6. The testing procedure is the same as in the case of butt welded joints.

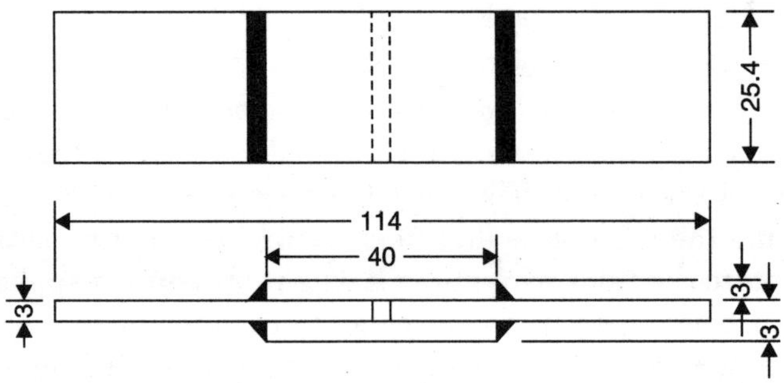

Fig. 13.6 Test specimen for double strap fillet joint (all dimension in mm)

Chapter 14

Welding Under the Influence of External Magnetic Field

Super imposition of magnetic field has been reported in the literature to affect the characteristics of the welding arc and the properties of the welds produced.

Magnetic field can be applied to the welding arc in three different modes. If the direction of the magnetic field is parallel to the direction of electrode travel, it is considered to be a parallel field and if the field is perpendicular to the direction of electrode travel and electrode axis, it is referred to as a transverse field. Finally, if the field is parallel to the axis of the electrode it is termed as longitudinal field or axial field.

Factors which affect the arc behaviour during the application of a magnetic field can be summarized as follows:

1. Distance between the electrodes
2. Type of shielding gas used
3. The magnetic field intensity
4. The electrode material
5. The electrode geometry
6. Arc current

To calculate the influence of the above factors in conjunction with the different types of magnetic fields on the arc the following two basic approaches have been suggested in the literature:

1. Ampere's rule (flexible conductor)
2. Force on electrons

The second approach is more accurate as it takes into account the variation in shielding gases and electrode materials, but the physical constants (*e.g.* mean free path of the electron, the temperature of ions etc.) needed to substitute in the mathematical equations obtained are not available. The first approach is, therefore, used quite often to study the behaviour of a welding arc under externally applied magnetic field. In the following paragraphs, the effect of the superimposition of the above three types of magnetic fields on the behaviour of the welding arc and the characteristics of the welds obtained, will be discussed.

14.1 PARALLEL MAGNETIC FIELD

According to the Flemming's left hand rule, the arc, under the influence of parallel field will be deflected towards right or left across the weld bead length depending upon the direction of the parallel field (forward or backward). Keeping this in mind the findings of the earlier investigators may be analysed. Deminskii and Dyatlov have reported work on aluminium-magnesium alloys using the GMA process and alternating parallel magnetic field. They found the arc oscillated across the weld axis. Bachelis & Mechev found that on increasing the magnetic field-strength, penetration into the parent metal decreased and weld-width increased. Serdyuk confirmed the above findings and found further that with parallel field fine droplets transferred with improved heat distribution perpendicular to weld seam.

14.2 TRANSVERSE MAGNETIC FIELD

According to the Flemming's left hand rule the arc under the influence of this type of field will be deflected forward or backward depending upon the direction of the magnetic lines of force and the polarity of the welding system. Work of the earlier investigators may be analysed keeping this in mind.

Kovalev showed that the transverse magnetic field can be used for automatically regulating the depth of penetration. Hicken and Jackson found beneficial effects of constant transverse magnetic field when the arc was deflected forward with respect to the electrode travel speed. It was possible to increase welding speed four times and still obtain welds free from undercuts. Weld width was found to reduce with increase in magnetic field during stainless steel welding. For aluminium, however, weld width increased with increase in magnetic field (0 – 50 gauss). Mandelberg successfully increased the welding speed of submerged arc welding process. Kornienko found they for hard facing, required depth of penetration on higher currents and deposition rates could be obtained using transverse magnetic field. Sheinkin found the application of transverse magnetic field to increase the productivity of the submerged arc welding process used for making butt joints between prepared edges.

14.3 LONGITUDINAL MAGNETIC FIELD

A magnetic force acts on the arc, in this system of magnetic field, only when the angle between the direction of the electron stream and magnetic lines of force is not zero. As the arc has a conical shape and the current carrying electrons also move along the surface of the arc, their motion can be resolved in two components, one along the axis of the arc and the other perpendicular to it. The component along the arc does not contribute to the magnetic movement. The component perpendicular to the arc exerts a force on the arc causing the arc (molten particles of the metal in the arc) to rotate clockwise or anticlockwise depending upon the direction of magnetic field and polarity used.

The first work on the influence of the external longitudinal magnetic field was reported to have come from Erdman-Jesnitzer and associates, who worked on coated electrodes and for MIG welding of steel. They found that this field influenced the droplet formation and metal transfer. Gvozdetskii and Mechev carried out basic studies on the behaviour of MIG arc in external longitudinal magnetic field. Longitudinal magnetic field has been found by Gupta to increase weld-width, decrease depth of penetration and increase reinforcement height. The bead has been found to deflect in one side in MIG welding while no such effect was found in submerged arc welding. Alternating longitudinal magnetic field has also been found to increase weld width, decrease depth of penetration and increase reinforcement height with increase in the intensity of longitudinal magnetic field.

Regarding the mechanical properties of welds, Erdmann-Jesnitzer *et al.* in 1959, reported no increase or decrease in HAZ hardness due to the application of magnetic field. Gupta has also reported results which agreed with Erdmann-Jesnitzer. On the basis of Hall and Petch relation it has been postulated that tensile strength of the welds made with high current welding arcs under longitudinal magnetic field superimposition should be higher because of grain refinement.

The first report regarding the effect of external longitudinal magnetic field came from Erdmann-Jesnitzer and associates who studied the effect of such field on metal transfer and welding parameters such as arc-current arc-voltage, rate of metal deposition and arc temperature etc. during welding with coated and uncoated electrodes as well as for gas shielded arc welding. In 1967 they gave a method of modifying, through the action of magnetic field, the phenomena associated with the operation of the electric arc.

The effect of longitudinal magnetic fields on the shape of the transferred metal droplets in gas-shielded-arc welding has also been reported recently.

Erdmann-Jesnitzer and associates have also the credit of introducing, for the first time in the history of welding, the concept of pulse magnetic field similar to the pulse current arc welding. The effect of magnetic field on droplet formation and metal transfer, special possibility of arc control and basic principles of Lorentz force have been considered by them. To study the droplet transfer phenomena during welding Erdmann-Jesnitzer and associates used various methods and Rehfeldt in 1966 developed a wonderful device the "Analyser Hannover" for this purpose.

14.4 IMPROVEMENT OF WELD CHARACTERISTICS BY THE APPLICATION OF MAGNETIC FIELD

By the application of external transverse magnetic field, the arc may be deflected either forward towards the direction of welding or backward. Forward deflection can be used to advantage for welding thin sections. With forward deflection of the arc the weld width increases and penetration is decreased, weld metal spreads because of arc deflection.

This effect can also be used to advantage in the welding of plates at higher welding currents and higher welding speeds. Normally higher welding speeds and higher currents cause undercuts to develop on the weld deposits. Because of arc deflection forward direction weld metal spreads and fills up the undercuts formed. Jackson C.E. has used this effect in the welding of aluminium and welding speeds upto 2 times the normal welding speeds could be reached with no undercuts. The strength of the welds was not only unaffected but was a little on the improvement side.

Forward deflection of the arc has also been used to advantage by the author in the hard facing by arc welding. Forward deflection caused sallow penetration, the dilution of the weld deposit with the base plate was reduced and a weld deposit rich in alloy content and improved overall properties was obtained.

Arc deflection by the proximity of multiple arcs can also be used to advantage. A two-or three-wire submerged arc utilises the magnetic fields of neighbouring arcs to obtain higher travel speeds without undercuts.

Backward deflection causes heavy undercutting and extensive reinforcement. This has little use in practical welding.

Alternating (transverse fields, however, cause the arc to oscillate back and forth across the weld axis with a frequency equal to that of the applied field. This effect is used to advantage in the gas tungsten arc welding GTAW process using hot wire. Higher welding speeds with good penetration and absence of undercuts were the advantages associated with this type of field.

The weld deposit microstructure showed fine grains. Weld strength was also improved.

Axial magnetic field rotates the arc. This field has been used by the author in improving the weld deposit characteristics of underwater welds. Constant external axial field causes arc rotation. The metal drops do not fall straight but they also rotate in a circular path before depositing on the plate. Rotation of the drop in circular path causes centrifugal forces to act on it. The drops fall on the plate in a large area causing weld width to increase. Higher welding speed and higher currents could be used with the absence of undercuts. The mechanical properties of the welds are not changed. Welding production rate can thus be doubled without affecting the weld deposit properties.

With axial field and consequent rotation of arc the penetration is reduced under similar welding conditions. This can also be used for welding thin plates and for hard facing of metals.

Alternating axial magnetic field has been found (by the author) to be of good value. Alternating axial field causes the arc to oscillate in a circular path. The arc twists rightward and leftward. This effect causes stirring of weld pool which causes the formation of finer grains and consequent improvement of mechanical properties. The author has found improvement in mechanical properties upto 30% of that obtained without field, in underwater welding.

14.5. MAGNETIC IMPELLED ARC WELDING

Thin-walled steel tubes, hollow sections, flange and other assemblies may be joined by an arc process which closely resembles flash welding in the type of apparatus employed. The workpieces are held in clamps, one of which can be moved on the axis of the work. Between the clamps and the joint line two solenoids are placed around the work, one on each side of the joint. These solenoids are energized by a direct current in a manner to produce the same pole on each side of the joint and, to allow them to be placed over the work and removed after welding, they must be split. With the workpieces initially in contact a.d.c. welding source with a range of 28–48 V is connected across the gap. On withdrawing the workpieces from contact an arc is struck across the gap which is then opened to 1–2 mm. The magnetic field created by the solenoids is radial with respect to the axis of the work and this causes the arc to motor around the outer edges of the workpieces (see figure below) which in a few seconds become molten. The gap is then closed rapidly by the moving platen to squeeze out the molten metal and consolidate the weld. A normal machined end is all that is required at the joint and no special treatment of the surfaces of the workpieces is necessary. Welds can be made without any shielding but, if desired, to improve the appearance and quality of the upset metal a shield of argon, nitrogen or other reducing gas may be provided.

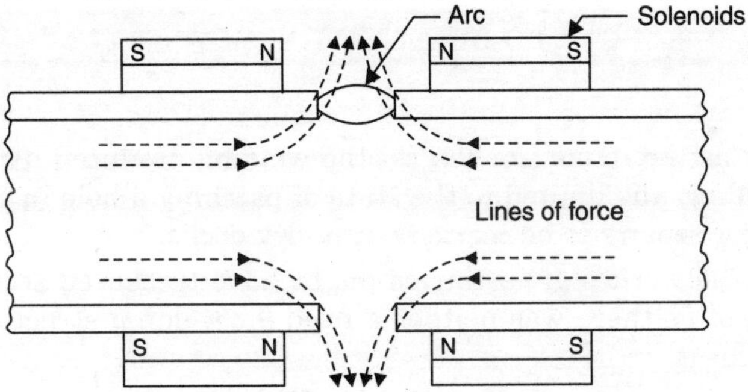

Fig. 14.1 Magnetic impelled arc welding. Diagram does not show platen clamps or arc supply circuit

The similarities with flash welding are obvious but there are important differences. With flash welding the source of heat is form both resistance heating of molten bridges and short-lives arcs when the bridges are broken. Molten metal is expelled from the joint in the process and there are comparatively long periods of inactivity when no current is passing and there is therefore no heating. With the magnetically impelled arc, however, heating is continuous, little metal is expelled and the process is therefore more efficient and the heating cycle considerably more rapid. As the arc tends to adhere to the periphery of the joint this limits the process to welding relatively thin hollow sections of up to 5 mm wall thickness and makes it generally unsuitable for solid sections. Upset forces tend to be less than for flash welding but, because of the rapid heating and smaller heat-affected zone, the rate of upset must be higher. The flash of expelled metal is smaller, smoother and more uniform than with flash welding.

CHAPTER 15

Fundamentals of Underwater Welding Art and Science

Underwater welding, as the name implies, is "the welding produced inside water". A decade back underwater welding was limited to the state of patching a hole in a sunken ship, just to get her afloat for major repairs to be carried out in dry docks.

One or two of the world's great navies might have treasured secrets about sub-ocean welding but for most of us there was neither a need for welding structures under water nor was there a solution for it.

The recent intensification of efforts in the field of exploring the seas for the natural resources beneath its beds has aroused the interest of welding engineers to develop tools and techniques for obtaining reliable welds under water.

The present techniques for underwater welding are far from complete and have limited applications in salvaging operations. Because of the high cost of dry habitat welding the primary thrust in research and development has been with open water (wet) welding.

Underwater welds suffer from defects like undercuts hard and brittle HAZ, microcracks due to hydrogen embrittlement, solidification cracking, stress corrosion cracking, etc.

15.1 COMPARISON OF UNDERWATER AND NORMAL AIR WELDING

Underwater arc welding differs from air welding in the following features:

1. Electrodes are painted for waterproofing.

2. Electrode core wire is usually the same as in air welding but in the case of the welding of high strength steels inside water using wet welding technique, a core wire of stainless steel or special steel is preferred.

3. The flux coating in common use is that of rutile type. Iron-oxide covering, which is not very common in air welding, has been found to be more advantageous (Khan, 1979).

4. In air welding a gap is maintained between the electrode and the parent plate. This gap cannot be maintained in water as soon as the electrode is lifted for maintaining a gap the arc extinguishes. For maintaining an arc in water, it is necessary to keep the electrode in contact with the plate. A slight pressure is also maintained. Cooling action of water on flux coating and waterproof paint results in the formation of a barrel at the end of the electrode. Arc burns inside this barrel space (see Fig. 15.1).

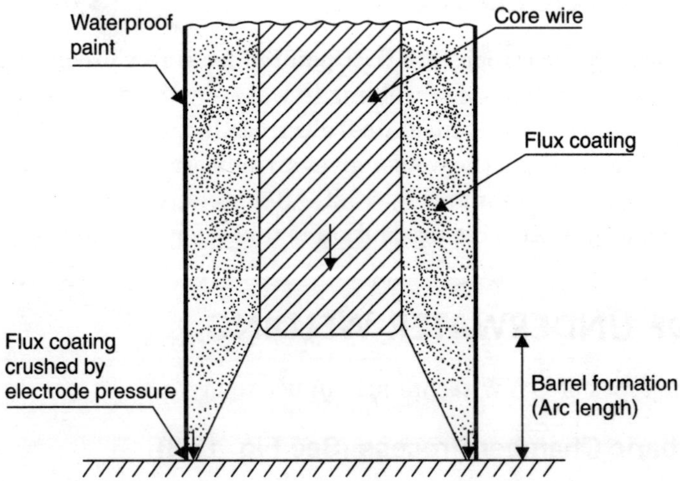

Fig. 15.1 Barrel formation during Wet-welding

5. Underwater arc is surrounded by a bubble of steam and gases. The pressure on the arc equals the atmospheric pressure plus the pressure of the water column above the arc as shown in Fig. 15.2. The pressure around the arc, thus, increases with depth. This affects arc behaviour and equilibrium of chemical reactions which affects weld chemistry. Carbon, silicon and manganese content of the weld metal increases with depth with corresponding change in properties.

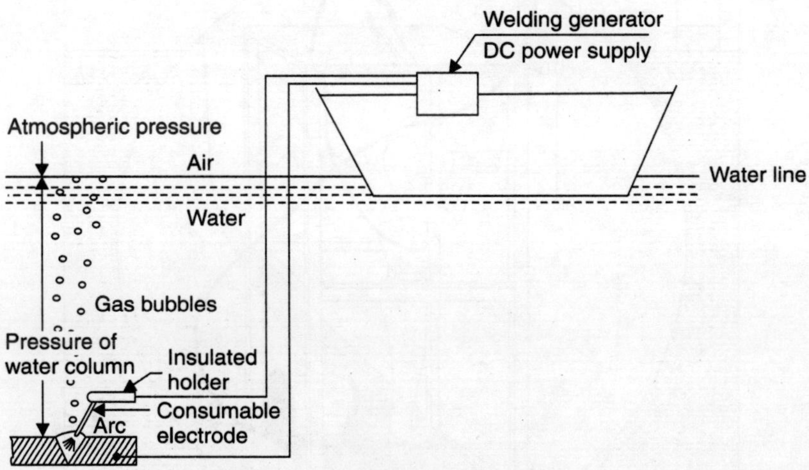

Fig. 15.2 Underwater wet-welding

6. Cooling rates in air welding could be controlled by change in arc-energy input. There is far less scope for doing this as the voltage and current during underwater welding have a close range.

7. Hydrogen and oxygen levels are normal in air welding while weld-metal and heat affected zone hydrogen and oxygen levels are well in excess of those in air-welding. This is due to increased amounts of hydrogen and oxygen in arc bubble.

8. Electrode holder is insulated.

15.2 WELDING PROCEDURE

While welding in water the electrodes are first painted for water proofing, kept in waterproof containers and are taken to the place of welding in water by the diver-welder. During welding the electrode is held in a special (fully insulated) electrode holder. When the electrode is brought to the plate in the welding position, the welder gives an indication to the operator of the generator called "tender" to put the generator on Fig. 15.2. After weld bead is completed another signal is given to put the generator off. This precaution is taken for the safety of the welder.

15.3 TYPES OF UNDERWATER WELDING

There are four basic types of UWW techniques in use today.

15.3.1 Dry Hyperbaric Chamber Process (See Fig. 15.3)

(*i*) Weldment and welder completely enclosed.

(*ii*) Weld properties similar to air welds.

(*iii*) Equipment–bulky, costly, and complex.

(*iv*) Fit-up time is more.

(*v*) Two or more support ships and a crane are needed.

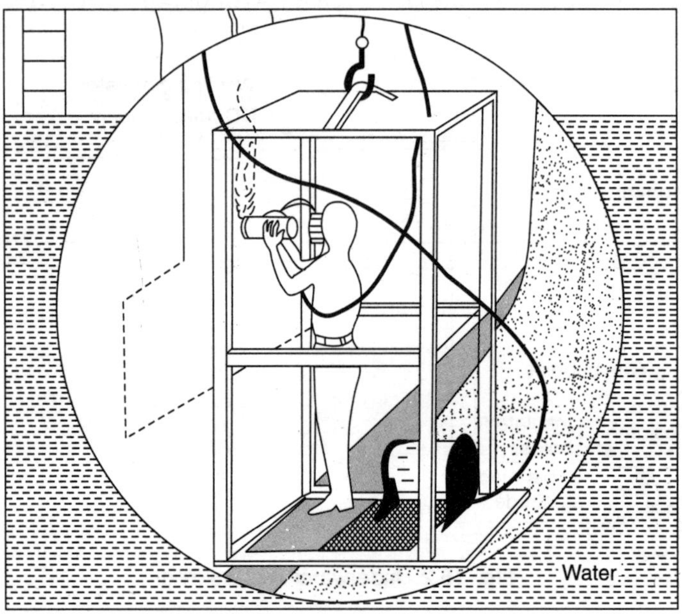

Operational Views "Habitat Welding"

(*a*) Ship repairs

Fig. 15.3 Use of Hyperbaric chambers (Habitat welding)

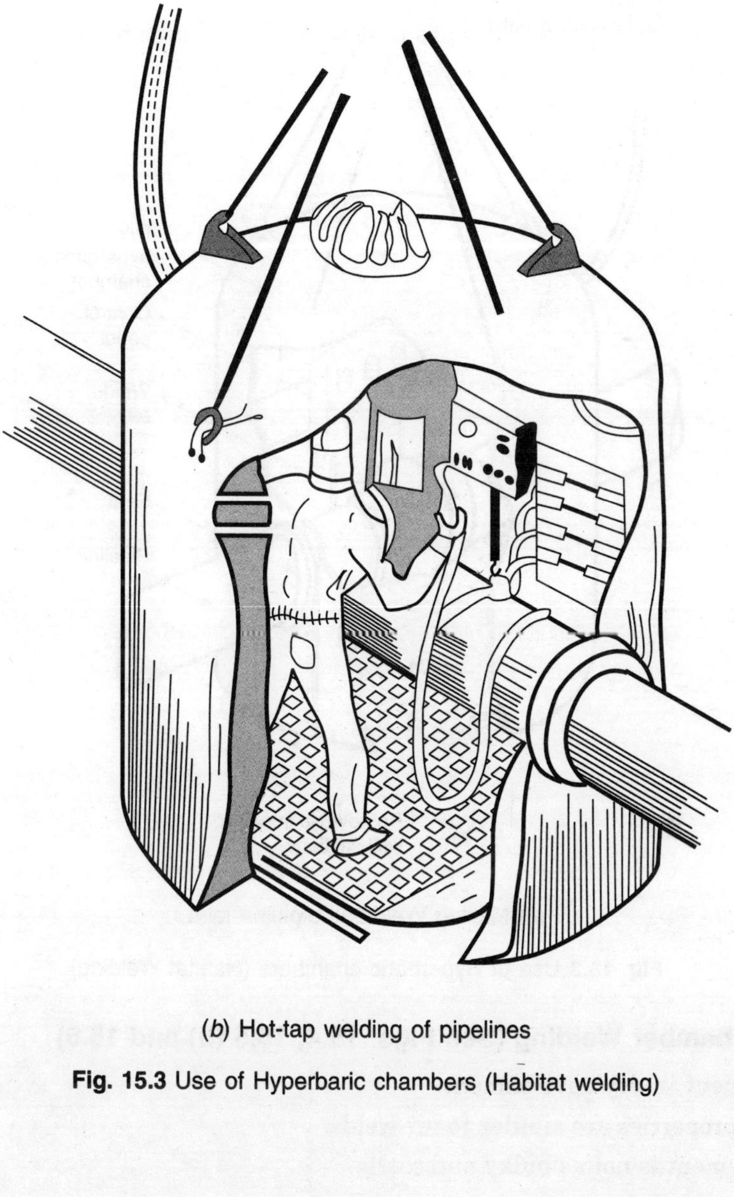

(*b*) Hot-tap welding of pipelines

Fig. 15.3 Use of Hyperbaric chambers (Habitat welding)

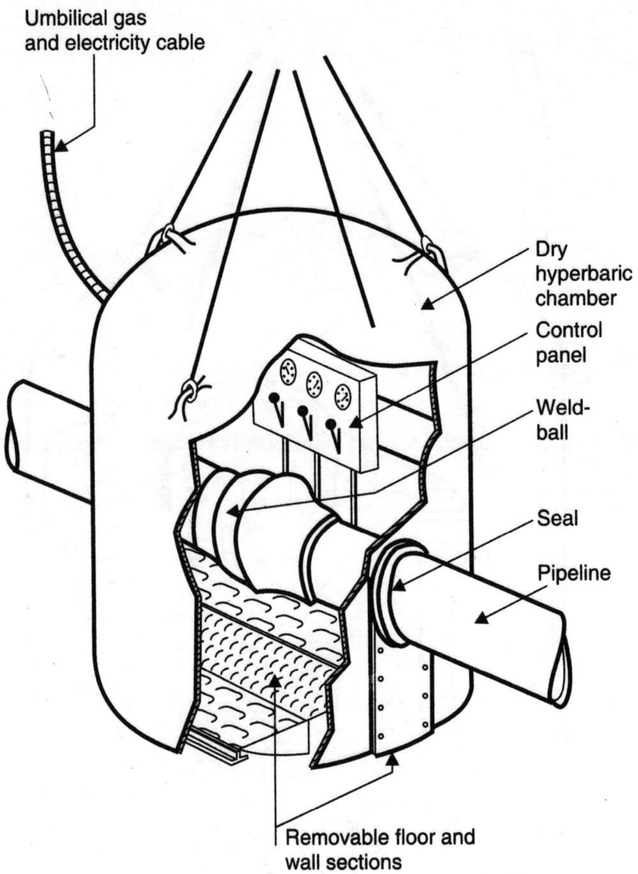

(*c*) Making Weld-ball pipeline joint

Fig. 15.3 Use of Hyperbaric chambers (Habitat Welding)

15.3.2 Local Chamber Welding (See Figs. 15.4, 15.5 (b) and 15.6)

(*i*) Weldment in dry environment.

(*ii*) Weld properties are similar to air welds.

(*iii*) Equipment is not as bulky and costly.

(*iv*) Fit-up time is less.

(*v*) Usually requires a small crane.

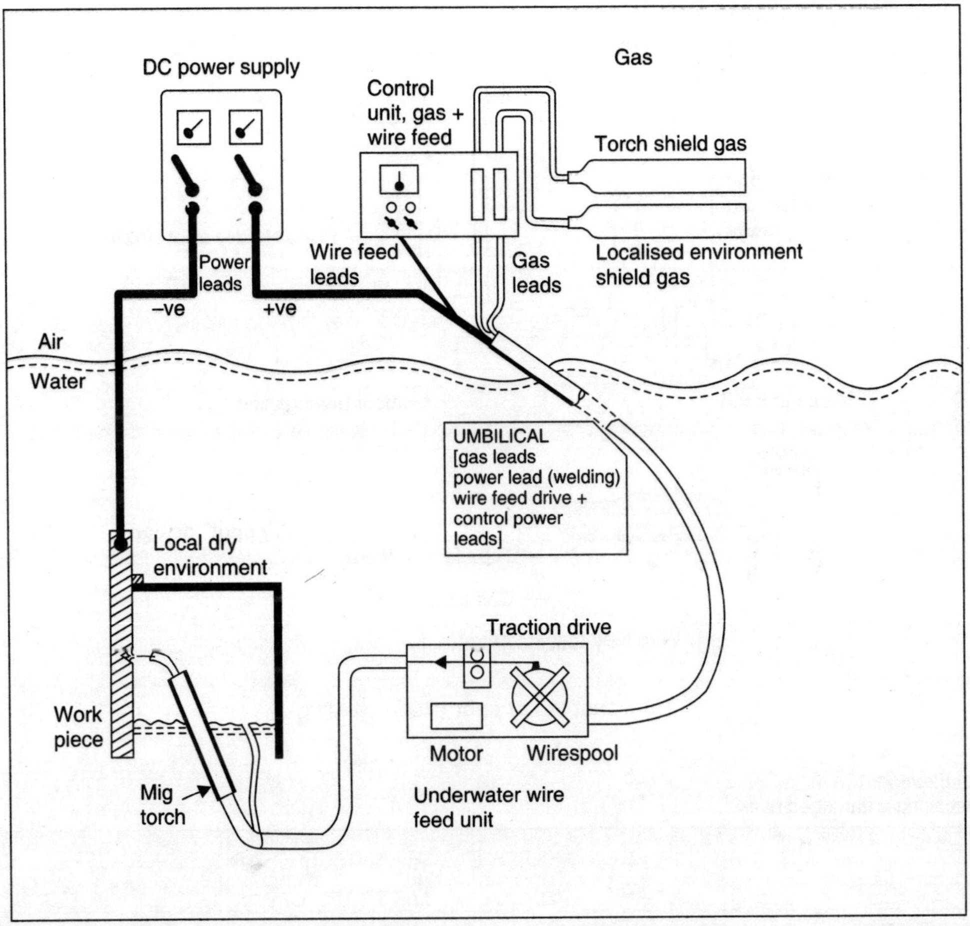

Fig. 15.4 Schematic diagram of continuous wire MIG welding underwater using local dry environment

15.3.3 Portable Dry Spot (see Fig. 15.5)

(*i*) Weldment is enclosed in dry environment (transparent plexiglass box) and welder is submerged in water.

(*ii*) Weld properties are similar to air welds.

(*iii*) Equipment: No heavy equipment is needed.

(*iv*) Gas and wire feeding is difficult as MIG is mostly used.

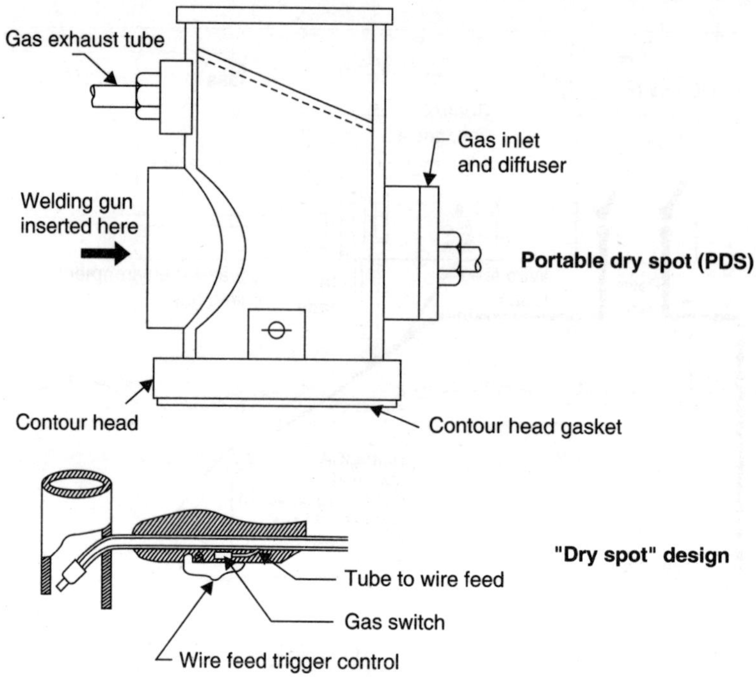

(*a*) Portable dry spot (PDS) welding

(b) Example 1
Repairing a damaged riser

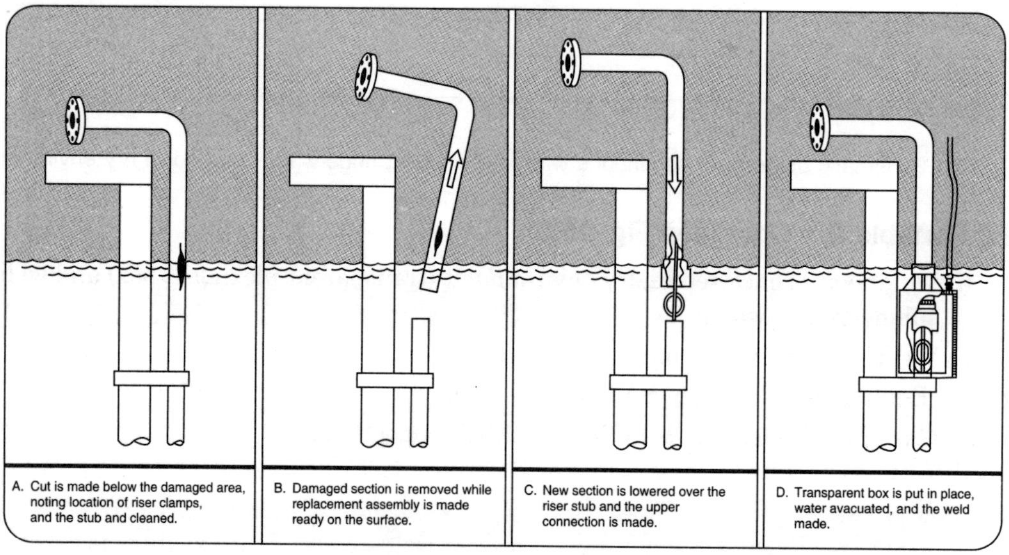

(*b*) Stages in the repair of damaged riser using Local Dry Environment "Hydrobox"

Fig. 15.5 Underwater dry welding

Hydrobox in use for a Vertical Riser Repair

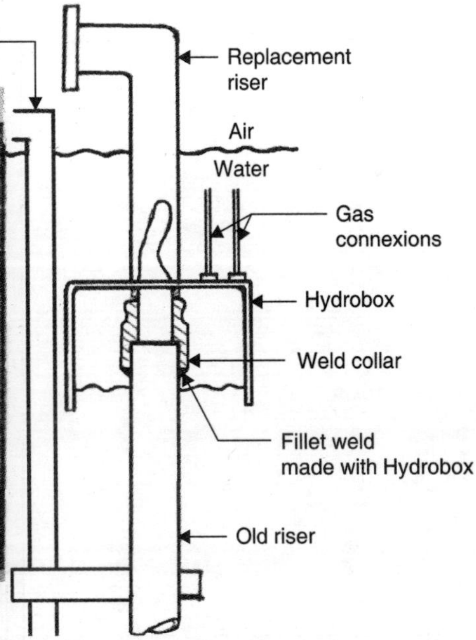

Fig. 15.5 (c) The Hydrobox Showing Schematic Arrangement for making a Riser Repair (details) (Kirkley, Lythal, 1974)

Fig. 15.5 Underwater dry welding

15.3.4 Wet Welding

(*i*) Weldment and welder both exposed to water.

(*ii*) Weld properties are inferior to air welds.

(*iii*) Standard air welding equipments can be used.

(*iv*) No fit-up time or negligible fit-up time.

(*v*) Process is convenient.

Advantages of Wet-Welding

1. Welders can reach positions inaccessible by other methods.
2. Process is fast.
3. Cost of welding is very low.
4. More freedom of repair design and fit-up.
5. Standard welding equipment could be used.

Disadvantages of Wet-Welding

Due to direct contact of the arc and the molten weld-pool with water, there is a 'Quenching' effect that increases tensile strength but reduces ductility. The porosity and hardness also increase.

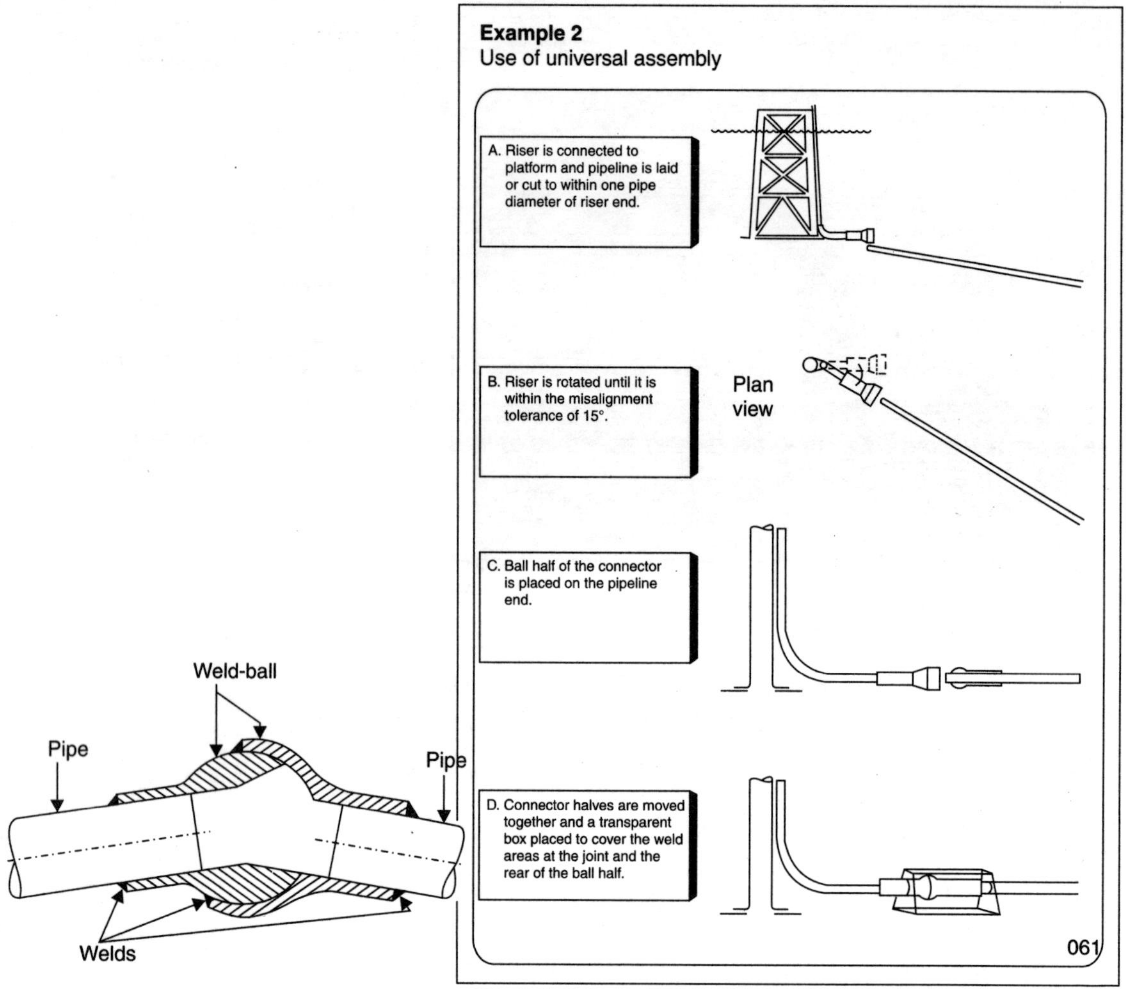

Fig. 15.6 Use of universal assembly being welded in a dry chamber (transparent perspex) (Kirkley, Lythal, 1974)

15.4 UNDERWATER MMA 'WET'-WELDING PROCESS DEVELOPMENT

Deposition of stringer beads (see Fig. 15.7) has, generally, been recommended in the literature. Necessary strength can be achieved by superimposing additional beads. The advantages of stringer-bead technique include:

1. Easy control over travel speed.

2. Uniform bead surface.
3. Good arc stability.
4. Reduced risk of slag inclusions.
5. Reduced chances of undercutting.
6. Consistent and satisfactory penetration.
7. Ease of welding in low visibility conditions.

The following precautions are taken to produce good welds:

1. The joints should be well fitted.
2. Should be free from rust, oil, paint etc.
3. No abrupt changes in weld contours.
4. The ends of the short welds or tacks should be thoroughly cleaned and hammered to give a smooth surface.
5. The bead or layer deposited should be cleaned of slag, spater or globules before superimposing additional runs. Van der Willingen (1946) described the use of a special wrapped heavy coated iron powder electrode which gave high deposition rate and excellent touch welding characteristics. Fig. 15.7 shows the types of beads made in underwater welding. Table 15.1 shows the effect of the type of underwater welding conditions mentioned above on weldability of steels commonly used.

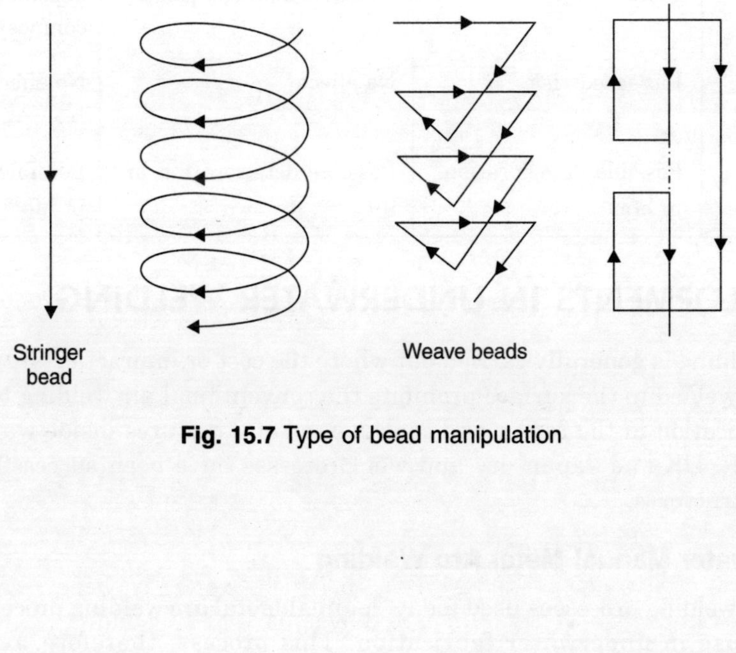

Fig. 15.7 Type of bead manipulation

Table 15.1 Summary of likely effect of underwater welding conditions on potential weldability

Aspect of weldability	*Wet welding*	*Local chamber welding*	*Habitat welding*
1. Hydrogen cracking	Very high increased risk of cracking	Probably some increased risk particularly at great depths	Probably some increased risk particularly at great depths
2. Solidification cracking	Some increased risk with depth	Some increased risk with depth	Some increased risk with depth
3. Lamellar tearing	Possible increased risk particularly at depth	Possible increased risk particularly at depth	Possible increased risks particularly at depth
4. HAZ toughness	Probable deterioration	Little effect anticipated except possible slight deterioration immediately after welding	No effect anticipated except possible slight deterioration immediately after welding
5. Weld metal toughness	Deterioration	Possible effect at depth dependent on composition	Possible effect at depth dependent on composition
6. Stress corrosion	Increased risk	No effect	No effect
7. Fatigue	Possible deterioration in life	Possible deterioration in life	Possible deterioration in life

15.5 DEVELOPMENTS IN UNDERWATER WELDING

Underwater welding is generally carried out where the cost or impracticability of bringing the structure to be welded to the surface prohibits the conventional air welding to be carried out. It finds its application in the repair and construction of structures inside water. In countries like USA, USSR, UK and Japan dry and wet processes have been successfully used in the fabrication of structures.

15.5.1 Underwater Manual Metal Arc Welding

Among the wet welding processes used today, manual metal arc welding process is still finding its maximum use in underwater fabrication. This process, therefore, requires especial consideration. The major parameter, for study in this process is the type of electrode. Waterproof coating has already been discussed earlier.

A critical review of literature indicates that almost all the varieties of electrodes have been used with varying degrees of success. From their results and our own experience on underwater welding some basic conclusions have been drawn and reported in this text. The discussion would logically start with the underwater welding arc.

15.5.2 Underwater Arc

Underwater welding arc is exposed to two basic mechanisms of compression and constriction. Underwater arc is surrounded by a bubble. Hydrogen content (about 93%) of the arc bubble atmosphere together with water surrounding it compresses the arc and at the same time it has a severe cooling effect on arc column compared to normal air welding. This causes arc-constriction. This compression and constriction of arc column result in a higher current density in underwater arc. Further, in straight polarity welding, the limited geometrical dimensions of the electrode end prevent the free expansion of the cathode spot with increase in welding current. The arc is thus constricted. This apparently explains the fact that the volt-ampere characteristic curves of an underwater arc are concave or rising. Due to these compressive forces the increase in the cross-sectional area of the arc lags behind the given increase in the welding current, thereby raising the current density or field intensity (this distinguishes underwater welding with air welding). Thus to maintain same arc conditions the current should be increased by 10% per atmosphere (10 meters of water) of additional pressure. These higher current densities produce higher arc temperatures. Temperature of arc column at different currents and depths is given in Table 15.1.

15.5.3 Arc Shape

Madatov found that the basic shape of the arc column was cylindrical for metal-arc welding and truncated cone with its base on the work for thin wire CO_2 welding. Metal transfer characteristics for the two types of welding processes are given in Table 15 .2.

15.5.4 Arc Atmosphere

A peculiar feature of underwater welding is an arc bubble which is maintained around the arc. The size of the bubble fluctuates between a small bubble barely covering the arc column and a large bubble of 10-15 mm diameter, that eventually breaks away from the weld puddle and floats to the surface, leaving behind a nucleus bubble with a diameter of 6–9 mm. This phenomenon of bubble growth and its break away occurs at an approximate rate of 15 times per second at 150 mm of water depth. Gases generated per second for E–6013, E–6027 and E–7024 are 40 cc, 50 cc and 60 cc respectively. The gas-bubble consists of 62–82 percent hydrogen, 11-24 percent carbon monoxide, 4-6 percent carbon dioxide, and the remaining 3 percent is nitrogen and metallic and mineral salt vapours.

15.5.5 Arc Stability

During underwater welding the arc-voltage and current values fluctuate. A stability factor for comparing arc performance was defined by Madatov as maximum current divided by minimum current. The arc is considered to be stable for values of this factor near one. For values much higher than one the arc is considered unstable. One cause of these fluctuations is the variation in voltage due to changes in arc length during metal transfer. Another cause of fluctuation is collapsing of thick flux covering occurring every 0.3 second or less during the arc welding. Different electrodes produce different levels of stability. Silva has found E-7024 more

Table 15.2 Temperature of Arc Column at Different Currents and Depths

Welding condition			*Temperature of arc column °K*	
Depth m	*Current Amps.*	*Effective dia. of arc column cm.*	*Thin wire electrodes*	*Stick electrodes*
10	100	0.202	8400*	9300
10	200	0.205	9200*	10200
10	300	0.210	9750	10700
10	400	0.260	10150	11100
10	500	0.317	10650	11500
20	300	—	10000	11000
40	300	—	10300	11300
60	300	—	10400	11500
80	300	—	10600	11700
100	300	—	10800	11800

*Calculations based on assumption that arc column is a cylinder of arc length 2 mm. Stick electrode air-arc temperature is 6000 °K.

Table 15.3 Rates of Metallurgical Reactions in various methods of underwater welding

Characteristics of Metal Transfer	*Thin wire without CO_2*		*Thin wire with CO_2*	*EPS-52 covered electrode*
	Salt Water	*Fresh Water*	*Fresh water*	
Drop Transfer* per second	12	16	23	44
Life time of drop,** second	0.1700	0.1305	0.0575	0.0254
Average weight of one drop, gm.	0.1670	0.0804	0.1100	0.1100
Volume of one drop in mm^3	21.4	10.3	14.1	14.1
Coefficient of reactivity of the process, Cn	21.8	16.77	7.37	3.26
Arc Voltage, Volts	39	40	39	39 (S.P.)
Arc Current, Amps.	240	250	240	240

*Drop-Transfer throughout.

**Lifetime of drop has largest apparent effect.

stable than E-6027, while E-6013 was found comparatively unstable because of its coating being thinner than the other two. Arc has been found to be more stable in salt water than in fresh water. This is due to the ease of ionization of sea water. But there is more current leakage in sea water (upto about 65-110 amp. at an open circuit voltage of 83–99 volts).

15.5.6 Metal Transfer

Normally, the metal transfers in droplets (globules). Occasionally a large drop forms and short circuits the arc. Drop transfer frequency as reported by Brown is 80 to 100 drops per second for the coated electrodes used by him. Madatov reports the frequency to be 44 drops per second for the type of electrode he used. Thus the drop-transfer frequency depends upon the type of electrode in addition to other factors. Underwater arc is constricted and produces a high arc core temperature of 9000°K to 1100°K at 10 m depth) as compared to 5000°K to 6000°K for air welding (Table 14.1). This increased temperature causes fast melting ra' e for plate as well as electrode. The weld puddle which would otherwise have been uncontrollable solidifies rapidly due to the quenching effect of water.

With the above background of underwater arc and metal transfer mode in mind let us now analyse the work carried out by the various underwater welding investigators on different types of electrodes.

15.5.7 Electrodes Used

Electrodes used by various investigators along with their findings have been listed in chronological order in Table 14.3. Each type of electrode will now be discussed in detail.

Cellulosic. These electrodes give a harsh digging arc resulting in a high penetration. In underwater welding the currents used are high to maintain the arc. This has been found to aggrevate the situation and produce more undercuts and convex bead. The results are not good even with reverse polarity. E-6010 has been found to spatter violently, gives irregular beads, and produces clouds of black smoke while E-6011 (which contains potassium silicate also in its coating) gives almost no spatter, produces continuous bead. It means that the presence of a substance which ionizes easily improves the electrode performance.

Rutile. Rutile electrodes have been found to be superior to cellulosic and second to acidic but Silva and Hazlett have found plain rutile electrodes inferior to iron powder type. Light coated rutile electrodes E-6013 have been recommended by the U.S. Navy in their manual on underwater welding and cutting in 1953.

Oxidizing. Oxidizing electrodes give satisfactory welds but the welds are inferior in strength and ductility as compared to acid and rutile electrodes. A comparison of various electrodes electrodes is given in Table 15.3.

Table 15.4. Strength characteristics of various coated electrodes used underwater

Sl. No	*Investigator*	*Type of Electrodes*	*Water proofing coating*	*Yield strength N/mm^2*	*Ultimate tensile strength N/mm^2*	*% age reduction in area*	*Impact strength Joules*
1.	Berthet and Kermabon	(*i*) Acid	Vinyl lacquer	460.6	490	8.5	40-28
		(*ii*) Rutile	-do-	416.5	436.1	5.5	33.6–27.4
		(*ii*) Oxide	-do-	372.4	436.1	17.5	34.4–33.6
2.	Hibshman and Jensen	(*i*) Oxide coated	–	279.3	387.1	–	–
		(*ii*) Organic Coating	–	343.0	377.0	–	–
3.	Silva & Hazlett	(*i*) Rutile E-6013	–	470.4	558.6	14.3	23°C :13.6 0°C:9.6 – 15°C:8.0
		(*ii*) Heavy coated rutile E-7023	–	470.4	588.0	16.0	23°C:19.2, 0°C:12.8, – 18°C:9.3
		(*iii*) Iron oxide E-6027	–	509.6	646.8	13.1	23°C:24.45 0°C:10.64, – 18°C:8.32
4.	Grubbs	Multipass stick rutile E-6013	–	509.6	588 – 656.6	6.10	70°F, 32.8 30°F, 29.92 0°F, 23.2 – 30°F, 13.6
5.	Madatov	Iron Powder	–	372.4–490	–	–	–
6.	Meloney	Rutile E-6013	–	–	539 – 705.6	14 – 19.3	68°F, 37.28 32°F, 20.48 – 60°F, 9.92

Iron Powder. In 1946, Van Der Willingen developed an electrode with a substantial amount of iron powder in its coating and a high coating material to core wire ratio. These electrodes were found easy to use in low visibility conditions, had excellent drag-welding characteristics and higher deposition rates.

Madatov in 1962 found these electrodes to give stable arc and fine droplet transfer with occasional short circuits. Silva and Hazlett found them to be superior to rutile. Masubuchi in 1974 found heavy coated rutile E-7024 and Iron-oxide E-6027 to give higher heat inputs than basic and rutile. For E-6013 better coating has to be designed to eliminate chiping of the outside of the coating during welding. Arc elongation effect is more serious in E-7024 and E-6027 and therefore the discrepancy between the machine current setting and the actual measured value is 15-25 amp. for E-6013 and E-7014 electrodes and 50-150 amps for E-7024 and E-6027 electrodes. This arc elongation effect is to be avoided.

Acid. Acid electrodes are those electrodes which have higher ratio of (silica + titenia) to Iron-oxide-Manganese-oxide. Acid electrodes have been found to give good results by Berthet. Nobody else reported on acidic electrodes. More work is required to study these and basic electrodes in detail before arriving at a final conclusion.

Basic. The covering has been found to be very brittle. The weld deposit has often been found to contain surface porosity.

From the above discussion it can be concluded that none of the existing electrodes for air welding can be directly used for underwater welding and special electrodes have to be developed to avoid the difficulties encountered in the use of the existing air welding electrodes. In the following paragraphs we shall discuss the characteristic requirements for underwater welding electrodes.

15.6 CHARACTERISTICS DESIRED IN ELECTRODES FOR MMA WET-WELDING

Flux covering for underwater welding electrodes should have some special characteristics in addition to the usual characteristics required in air welding. Because of arc constriction effect, the current density of underwater arc column is more and therefore deeper penetration is obtained in underwater welding. The arc should therefore have soft behaviour. Purely cellulosic electrodes are unsuitable for underwater welding as their arc is harsh and has digging tendency. Arc should have high stability to counter the extinguishing effect of water. Because of poor visibility conditions the coating should give easily removable slag to assist in multipass welding. Coating should be made non-conducting and non-hygroscopic by applying suitable insulating and water-proof paint on flux covering.

Soft arc behaviour

Rutile and iron powder coatings give soft-arc. These rods have therefore been used quite successfully in underwater welding. Iron-oxide coated electrodes give better strength and ductility than plain rutile ones in flat and horizontal position. For multipass, all position welding these rods fail because the solidified flux on the bead surface is difficult to remove for subsequent pass to be made. Rutile electrodes are therefore preferred. Hibshman and Jensen have however found welds stronger in tension than base plate when they used cellulosic electrodes. Their results have not been confirmed by other investigators. Rutile electrodes are therefore preferred by most of the underwater welders these days.

High arc stability

Because of the extinguishing effect of cold water surrounding the arc, the problem of arc stability in water deserves special attention. Compounds having low ionization potential (*e.g.* salts of potassium and cessium etc.) or compounds that promote electron emission tend to stabilize arc in shielded metal arc welding. By manipulating electrode coating composition an arc with better stability can be obtained. With a very stable arc, weaving of the weld bead may also be possible. This will permit larger heat inputs to the weld per unit length, larger bead size (mm^2) and lesser hardening. This will further improve the strength properties.

Non-conducting and non-hygroscopic coating

Ordinary coatings, which are invariably porous, absorb water when immersed in water. The moist coating gives porous welds and permits current leakage (through electrolysis). To protect the electrode from these two effects waterproofing non-conducting paints are used. Hrenoff in 1934 used shellac, Peillon process recommended paraffin wax, underwater cutting and welding manual of US Navy recommends Shellac, Ucilon or Celluloid dissolved in acetone for this purpose. Waugh and Eberlein 1954 recommended shellac as good coating. Avilov in 1955 used Kuzbass Varnish and bitumin dissolved in petroleum spirit, Karmabon and Berthet in 1962 settled for Vinyl lacquer on the basis of their experience. Because of varied opinion on this issue, this aspect has also been thoroughly studied by Khan in 1979.

15.6.1 Special Electrodes

Iron powder additions are sometimes made to the flux covering to increase the electrode deposition rate

Hrenoff et al. in 1934 used special flux covering coating (chalk and water glass: first layer; iron oxide and water glass: subsequent layers. They used Shellac as a water proofing coating. They found that the electrode was successful in fresh water but sea water required water proofing. Tensile strength was lower than that found by Hibshman and Jensen when they used cellulosic and oxide electrodes. This may be due to poor visibility in his experimental set-up.

Van der Willingen in 1946 used "self made Iron Powder heavy coated electrodes." He found these electrodes to be easy to use in low visibility conditions, have high deposition rate, and excellent drag or contact welding characteristics.

15.7 POLARITY

Electrode negative polarity produces less undercuts and spatter, better, bead shape, more regular welds and minimum corrosion damage to the electrode holder. Polarity made little difference to weld appearance or visibility. Barrel length was however more with electrode positive. Electrode positive or negative polarity and alternating current could all be used for underwater welding.

15.8 SALINITY OF SEA WATER

Electrical conductivity of water was found to increase with salinity. For bottom sea water it was approximately 0.03 mohs per cubic centimeter. It was easy to initiate and maintain the arc in saline water, penetration was increased. In MIG welding, salinity was reported to increase droplet size, reduce the number of drops per unit time and power consumption. Madatov in his work of 1962 concludes that salinity improves bead shape of underwater welds.

15.9 WELD BEAD SHAPE CHARACTERISTICS

Madatov in 1969 studied the weld shapes obtained in underwater welds using 5 mm EPS 52 (iron powder) electrodes and represented these in terms of weld penetration shape factor or simply "shape-factor" defined as the ratio between the weld width and depth of penetration. He reported that as welding current increased, weld reinforcement remained constant but the width of the weld increased and the penetration decreased with the result that the "shape factor" increased from 3.50 at about 200 amp to 5.00 at about 300 amps. Decrease in penetration was explained by stating that the travel speed increased on the mechanised feeding arrangement used. He also found, using a GMA process with 1.2 mm wire at 34 to 43 volts that the penetration shape factor varied between 2.5 to 5.00.

Increase in salinity or hydrostatic pressure reduced the shape factor. As the angle of torch nozzle changed from a leading to a trailing angle, the bead became narrower and taller with decreased penetration. A larger lead angle was supposed to increase post heating to the weld puddle and increase the metal flow back into the sides of the weld crater.

Silva in 1971 also investigated under water shielded metal are welding and reported shape factor of 4.2 to 5.4. He claimed that sufficient energy was required to bring the heat-affected-zone to approximately the size as in air. He found that the penetration did not decrease under water as claimed by other investigators. Rutile electrode E(6013) gave a semicircular penetration profile whereas with iron powder electrodes (E-7024 and E-6027), penetration was deep in the centre and tapered off rapidly towards the edges of the bead. Increased in penetration might be due to long barrel in iron powder electrodes.

Masumoto et al. in 1971 using a 4 mm coated iron powder electrode obtained underwater welds at 150 to 180 amps. The penetration shape factors were found to be between 5 to 7. Gas metal arc welds at 120 to 210 amps gave shape factor between 3 to 5.5.

Billy in 1971 investigated GMA welding and found that at a voltage of 36 to 42 volts, the shape factors varied from 2.1 to 2.9 reflecting quite good penetration that was obtained.

Hasui et al. in 1972 developed a plasma arc welding process that gave excellent welds. For welds without shielding liquid, the ratio was 1.7 to 4.2 and with shielding the ratio was between 1.8 to 2.3. The plasma welding appears to give better weld shape than either shielded metal arc or gas metal arc welding processes.

15.10 MICROSTRUCTURE OF UNDERWATER WELDS

Non-equilibrium microstructures were obtained in underwater welding due to the fast cooling rates which resulted in the formation of martensite and bainite in the heat affective zone (HAZ) adjacent to the fusion line. The HAZ of under water-welds was not wide as that of similar air welds. The width of coarse grains zone of air welds was much smaller than the width of the corresponding zone of under-water welds. This was because of higher arc and metal temperatures. It was found that the microstructure was dependent upon the waterproof coating used, type of electrode and the number of passes used.

Micro-examination of the welds was conducted in 1971 by Silva which reveal ferrite-pearlite structures in the weld metal and a narrow band of bainite/martensite adjacent to the fusion boundary in the HAZ. With rutile electrode, the martensite band was wider (0.2 – 0.6 mm) than with ironpower type (0.1 mm).

Grubbs and Seth in 1972 reported the presence of a martensitic band adjacent to the fusion boundary with austenitic deposits. According to them alloying elements like chromium and nickel diffused into the base material to give compositions which readily transformed to martensite on cooling.

Masumoto et al. in 1971 reported similar results with 4 mm iron powder electrode at 180 amps. Maximum hardness of 300 Hv (1 kg) in a band less than 1 mm and a partially hardened weld bead and a heat affected-zone of 4 mm, GMA welds at 120 amps and 26 volts showed a peak hardness values of 400 HV (1 kg) and a heat-affected zone width of 6 mm. Hasui, et al. in 1972 reported that for single pass welds the micro hardness approach 400 VHN (200 gm) in a narrow region of 0.5 mm adjacent to opposite side of the plate reduced the original peak hardness to 300 VHN (200 gm). Total heat-affected zone extended for a total of 4.5 mm from the fusion line.

Stalker et al. 1974, indicated that there was a wide range of measured hardness values within one sample and from one weld to another which was partially because of a mixed (hard and soft) microstructure which is typical of mild steel heat affected zone. Despite these differences there was no trend for the heat affected zone at toe of the weld (closer to water) to be harder than the under bead position of the weld. They also reported that there was no apparent relationship between the incidence of cracking and the level of hardness in the heat affected zone.

Brown et al. in 1974 expressed the opinion that the best comparative measure for predicting cooling rate would come from measuring the heat input per unit weld bead size. E 6013 rutile electrodes appeared to result in the lowest heat input while E 7014 rutile iron-powder electrodes were slightly hotter than E-6013, E-7024 rutile iron powder heavy coated and E-6027 super heavy coated iron powder, both gave much higher heat inputs than E-7014 and were approximately equal to each other in heat input. Localized martensitic transformations appeared in almost all underwater welds immediately adjacent to the fusion line, but extended upto 0.5 mm or less into the heat affected zone. Maximum hardness of 400 HK (100 gm) in SP and 500-600 HK (100 gm) in RP was obtained with E-6013 electrode 4 mm diameter with an energy input of 10-13 kJ/in and 9-11 kJ/in respectively.

15.11 NEW DEVELOPMENTS

M. Hamasaki (Government Industrial Research Institute, Japan) and M. Watanabe (The Welding Institute of Japan and Osaka University) have described the development in UWW in Japan. Among the methods being used are gravity welding and firecracker welding (also known in Europe as the Elin-Hafergut Method). In the latter process either one or two electrodes are set horizontally in weld joint and covered with grooved copper blocks before ignition. Good results were claimed for the firecracker method. Still more interesting, perhaps, is the water curtain type of CO_2 (Mig) welding method which has been developed at the Government Industrial Research Institute. This method uses a dual nozzle which provides a shielding gas flow from an inner nozzle and a concentric flow of water from an outer nozzle. Both flux-cored and solid wire electrodes have been used, and the maximum speed achieved with each was 1.2 and 1.3 m/min respectively. Butt and fillet welds were made experimentally. Work has also been done elsewhere on the use of shielding gas introduced at a slightly greater pressure than that of the ambient water at depth.

Investigations have been conducted at the Japanese Institute of Metals with a technique called "water plus gas shielding" for plasma-arc welding. Basically, the principle is the same as water-curtain Mig, but in this instance water flows from 12 holes in the bottom of the nozzle. Arc voltages could be as low as 20 V but increased with a greater depth of water. The process is claimed to give good results down to depths of 300 m. Its disadvantages are slow running speed, 60-80 mm/min, and the fact that it cannot be applied to rimming steel.

A. W. Stalker, 1974, dealt with tests carried out to assess the underwater running characteristics and crack susceptibility of various electrode types. The most promising were found to be a ferritic electrode with an oxidizing iron flux covering and a high nickel austenitic type. Even with these electrodes, however, it was necessary to apply continuous heating during the welding cycle to avoid hydrogen cracking in butt welds on carbon-manganese structural steels. In a second series of tests designed to give a preliminary assessment of arc behavior in a hyperbaric environment Tig, Mig and Manual metal-arc welds were made at pressures upto 32 bar.

Takemasu et al. in 1982 conducted laboratory tests on fire cracker welding simulating pressures down to 100 m. The beads deposited with commercial electrodes had both good appearance and sound mechanical properties.

Shinada K, et al., 1982 reported the use of remote controlled fully automated MAG welding process for underwater welding 12 mm thick pipes at 10 m water depth using 1.2 mm diameter solid electrode wire and 75% Ar 25% CO_2 gas mixture in 3 passes. Power source was d.c. electrode positive. The quality of underwater welds was equivalent to that obtainable on land.

Stevenson A.W. in 1983 discussed the techniques for off-shore repairs and strengthening procedures including underwater welding, highlighting the ways in which an underwater contractor can help.

Allum C.J., 1982 discussed the role played by TIG welding in underwater applications. The process has been reported to give good results upto 500 m depth. The nature of TIG arc significantly changes with increasing depth (pressure around the arc). Above 30 bars arc appearance becomes highly distorted due to refractive index variations between the arc and the observer (distance of about 70 cm). Manual arc manipulation becomes difficult. It is a matter of speculation on whether TIG is suitable for mediterranean waters (2,500 m deep).

Allum C.J. 1983 discussed the scope of the process of dry hyperbaric underwater welding. Automated welding appears to be a possible solution in deep waters because of low stability and poor visibility and manoeverability limiting the use of manual process. It has also been pointed out that the arc could be stabilized by using magnetic field.

Delaune, P. T. Jr., in 1987 reported the use of AWS D 3.6 specifications for conveniently specifying and obtaining underwater welds of predictable performance level. These specifications enable a designer to choose the weld type for a given situation and formulate a fracture control plan.

15.12 SUMMARY

The following summary projects the important aspects of underwater welding from the point of view of a welding engineer:

1 . Underwater welding is carried out where the cost or impracticability of bringing the structures to be welded to the surface prohibit the use of conventional air welding.

2. Shielded metal arc wet-welding is most convenient and economical process among the processes used.

3. Underwater welding electrodes should have softer arc behaviour to eliminate undercuts.

4. The coating should be such that it shields (shrouds) the underwater arc to eliminate current leakage and rapid quenching of the weld pool. This can be achieved by selecting a suitable water-proofing coating.

5. The coating should burn or fry out easily so that the feed rate is uniform and there is no jerky movement of electrode.

6. The coating should contain ingredients which give highly stable arc so that weaving of the weld bead is possible.

7. Water-proofing coating should be non-conducting and non-hygroscopic. This will avoid current leakage from electrode to electrically conducting sea water and the electrodes will not absorb moisture during welding.

8. Iron powder electrodes have been found useful but due to the arc elongation effect they do not give good results. With plain rutile coating this effect is not dominated, but the strength of welds is inferior to the values obtained with iron powder electrodes. A coating in between the two would prove useful.

9. Rutile or iron-oxide flux covering water proofed by cellulosic lacquer gave best arc stability, and good mechanical properties of the wet-welds.

10. A bubble of steam and gases is formed around the arc during wet-welding. This bubble protects the arc and weld pool from water.

11. Salinity of water improves arc stability and penetration.

12. Underwater arc core temperatures are around 11000°K (at 10 m depth), while air-arc temperature is around 6000°K the droplet transfer frequency is 44 for iron-powder and 80–100 for rutile–electrode during underwater welding.

13. Weld microstructure contains ferrite-pearlite structures in the weld metal and a narrow band of bainite-martensite adjacent to the fusion boundary in the heat affected zone.

15.13 POSSIBLE FUTURE DEVELOPMENTS

Work on underwater arc welding is still under development stages. U.S.A., U.S.S.R., U.K. and Japan are still working upon the ways to improve the quality of "wet" welds in water. Underwater arc wet welding is the cheapest and most convenient of all the welding processes available to-date. Future work may be carried out in the following areas:

1. There is no electrode as yet which can be said to be the final answer for underwater "wet" welding. Basic work is still needed to develop a special "underwater arc welding electrode".

2. Underwater welds are produced at fast cooling rates. Distortion of the plates is low. The process could be used to weld at places where it is desired to have low distortion.

3. It is expected that hardfacing of metals if carried out underwater will deposit very hard beads. The process could be tried.

4. Hydrogen is a serious problem in underwater welds. Work is necessary to develop electrodes and welding precesses that could give low hydrogen weld deposits.

5. Salinity of sea water affects the weld characteristics. A systematic research work could be conducted to explore the effect of different levels of salinity on weld characteristics.

6. Work can also be done to study the effect of depth of water on underwater welds.

7. Special tools and techniques can be developed to shield the underwater arc from the effect of surrounding water.

8. Some technique can be developed for preheating the plates before welding or post heating-treating the welds for improving the metallurgical characteristics of welds produced in water.

References

Amstead, B.H., P.F. Ostwald, and M.L. Begeman (1979), Manufacturing Processes, John Wiley, N.Y.

Anon (1917), A Better Way to Weld Underwater, Ocean Industry, Feb.

Anon (1953), Underwater Cutting and Welding, U.S. Navy Technical Manual USN Supervisor of Diving, Naval Ship System Command, p. 692–9, Navy Dept., Washington D.C.

Anon (1973), Underwater Welding and Cutting, Welding Design and Fabrication Vol. 46, VNo. 10, pp. 64–66.

Allum C.J. (1982), G's Underwater Role: Present and Future, Welding and Metal Fabrication, April, 1982, pp.124–132.

Allum C.J. (1983), Overcoming Problems in Hyperbaric Welding, Welding and Metal Fabrication, Sept. 1983, pp. 315–320.

Amson J.C., and Salter, G.R., (1966), An Analysis of the Gas Shielded Consumable Metal Arc System, Physics of the Welding Arc, A symposium, 11 W, London, pp. 133–147.

Baker, H.W., (1956), Modern Workshop Technology, B.I. Publications.

Billey, A.F., (1971), The Effect of Gas Metal Arc Parameters in Seawater on Welding Current, Arc Voltage, Bead Geometry and Soundness, NCEL Preliminary.

Brown, A.J. (1974), Fundamental Research on Underwater Welding. MIT Report No. MITSG 74–29.

Cooksey C.J. and D.R. Milner (1966), Metal Transfer in Gas Shielded Consumable Metal Arc Welding, Physics of Welding Arc Symposium, IW, Cambridge U.K.

De Garmo, E. Paul, Materials and Processes in Manufacturing, Mc Millan N.Y.

Delaune, P.T., Jr., (1987), Offshore Structural Repair using Specification for Underwater Welding AWS D 3.6, Welding Journal, Vol. 66, No. 2, pp. 32–43.

Dennary F., (1966), Electric Field Distribution in Welding Arc, Physics of Welding Arc, Sym. IW, Cam. U.K.

Erohin, A.A., Rykalin, N.N., (1966), Heat Balance of Electrode Droplet Melting Process in Arc Welding, Physics of Welding Arc, Sym., 1W, Camb., U.K.

Foster, V.R. (1958), Guide to Underwater Welding, Welder, Vol. 27, No. 136, Oct. Dec. pp. 88–92.

Galerne, A., (1967), Underwater Welding and Painting, Sym on Underwater Welding Cutting and Hand Tools, held at Battelle Memorial Institute pp. 43–53.

Ghosh, A. and A.K. Mallik, (1985), Manufacturing Science, East-West Press.

Giachino, J.W., William Weeks and Elmer Brune, (1967) Welding Skills and Practices American Technical Society.

Gourd, L.M., (1980) Principles of Welding Technology, Edward Arnold, London, Grubbs, C.E. and O.W. Seth (1972) Multipass All-position Wet-Welding—A new underwater Tool Paper OTC 1620, Offshore Tech. Conf. Houston Texas

Haim, G., and H.P. Zade (1947), Welding of Plastics, Crosby Lockwood and Soan Ltd.

Hamasaki, M. et al., (1973), On the studies of underwater Fire Cracker Welding, J. of Japan Welding Society, Vol. 42, Report 1/2 No. 7/9, 1973, pp. 666–73 and 897–906.

Hart, P.H.M., (1976), The Potential Welding Offshore Engg. Materials, Underwater Welding for Offshore Installations, Inst. Sem. London.

Hasui, A., et al. (1972) Development of Underwater Plasma Welding Process, 2nd, Int. Ocean Development Japan, Vol. 2, pp. 1005–1036.

Hibshman, N.S., C.D. Jensen, and W.E. Harvey, (1933), Electric Arc Welding Underwater, Welding Journal, Vol. 12, No. 10, pp. 4–9.

Herenoff, K, and M. Livshitz, (1934), Electric Arc Welding Underwater, Welding Journal Vol. 13, No. 4, pp. 15–18.

Ishizaki, K, ; A. Oishi, R. Kumagai (1966), A Method of Evaluating Metal Transfer Characteristics of Welding Electrodes, Physics of Welding Arc Symp IW London.

Ishizaki, K., (1966) on the Formation of the Weldbead, Physics of Welding Arc Symp. IW, London.

Kalpakjean, S. (2005) Manufacturing Engineering and Technology, Pearson Kermabon, R. and P. Berthet, (1962), Underwater Arc Welding, Saudage et Techniques Connexes, Vol. 16, No. 5/6, pp. 179–192.

Kirkley, D.W., (1975), Welding Underwater—Problems and Solutions, Welding Construction No. 123, No. 6, pp. 99–102.

King, L.A., and J.A. Howes, (1962), Material Transfer in Welding Arc, Physics of Welding Arc, Symp. I.W. London.

Kinney, G.F., (1957), Engineering Properties and Applications of Plastics, John Wiley and Chapman and Hall Ltd.

Kirkley, D.W., (1975), Welding Underwater—Problems and Solutions, Welding Construction Vol. 23, No. 6., pp 99–102.

Khan. M.I., and S. Lal (1976), State of Art in Underwater Welding, J. of the Inst. of Engr's. Vol. 56 No. Pt ME 6 pp. 288–291.

Khan M.I., (1976), Requirements of Electrodes for Underwater Welding, Welding and Fabrication, Vol. 1, *n*7.

Khan, M.I., (1977), Study of Underwater Welding Parameters, Ph.D., Thesis, University (IIT) Roorkee.

Khan, M.I., S., Lal and P.C. Gupta, (1978), Underwater Wet Welding with Coated Electrodes, Afro-Asian Conf. on Welding and Metals Technology, Delhi, India, Paper 26.

Khan, M.I. S.B. Garg (1969), Performance Rating of Metal Arc Welders, The Inst. of Engrs., Roorkee Sub-centre Annual, No. Dec. 1969, pp. 56–58.

Khan, M.I. and R.A. Agarwal (1970), New Plasma Arc Applications, Machine Building Industry. May-June 1970, pp 1–3.

Khan, M.I. and Vijay Kumar (1971), Development of Welding Procedures for Rigid P.V.C. Plastic by Hot. Air Technique, Journal of Inst. of Engrs. (India) Vol. 51., Feb., pp. 46–49.

Khan, M.I. and C.L. Raina, (1972), Double Fillet Weld Test for Assessing Hot Cracking Tendency of M.S. Welds, 5th MTDR Conf. Proc. University of Roorkee, India April 1972.

Khan, M.I. (1978), Investigation of the Effect of Welding Parameters and Waterproofing on Characteristics of Underwater Welds J. of Institution of Engineers (India). Vol 58, Pt MES, March 1978, pp. 215–220.

Khan, M.I., (1978), Underwater Wet-Welding with Coated Electrodes, Afro-Asian Conf. on Welding and Metals Technology, Delhi India, Paper 26, Feb. 78.

Khan, M.I. (1978), Investigation of the Effect of Welding Conditions on the Metallurgical Transformations and Mechanical Properties of Underwater Welds, Afro-Asian Conf. on Welding and Metals Tech Delhi, India Paper 27, Feb. 78.

Khan, M.I. (1979), A Study of Hardfacing Under Magnetic Field, Proc. ISME Conf. New Delhi, Paper E.P. 4.1 Dec., pp. 174.–176.

Khan, M.I. Khalifa. M. Othman, (1990), Increasing Productivity of Submerged Arc Welded Spiral Pipes by the Application of Axial Magnetic Field, Proc. Cairo Conf. on Prod. Engg. and Design for Development, Cairo, Egypt, Dec. 27–29, 1990, pp. 73–79.

Khan, M.I., M.R. Abau-Zeid and P.C. Gupta (1991), A Systematic Study of the Effect of Electrod Flux Coating Composition on the Properties of Underwater Shielded Metal Arc Welds, First Inst. Conf. on Mfg. Technology. The Hong Kong Inst. of Engrs. Hong Kong, Dec. 27–29.

Khan, M.I., (1992), Study of Effect of Superimposition of Transverse Magnetic Field on Arc Characteristics, Bead Geometry, Microstructures and Mechanical Properties of Underwater SMA Wet-Welds, 5th Inst. Conf. on Prod. Engg. Design and Control, University of Alexandria, Egypt, Dec. 1992.

Khan, M.I., (1993), Magnetic Control of Welding Arc to Improve Sub-Ocean Structures Fabrication, Proc. Mediterranean Petrolium Conf. Int. Energy Foundation, Tripoli, Jan. 18–21, 1993, p. 243–252.

Khan, M.I., (1993), Investigation of the Effect of Externally Applied Magnetic Field on the Properties of Air and Underwater Shielded Metal Arc Wet-Welds. International Conf. on Advances in Materials and Processing Technology (AMPT 93) Dublin 24–27 August, pp. 1279–1287.

Lancaster, J.F. (editor) (1984), The Physics of Welding, Pergamon Press, Oxford, U.K.

Lancaster, J.F., (1980), The Metallurgy of Welding, Brazing, and Soldering, Allen & Unwin London.

Lancaster, J.F. (1962), Influence of Heat Flow on Metal Transfer in MIG Welding Aluminium, Physics of Welding Arc Sym. 1962, I.W.

Leven, M.L. D.W. Kirkley, (1972), Welding Underwater, Metal Const. and Brit Welding Journal, Vol 40, No. 5, pp. 167–170.

Little, R.L., (1976), Welding and Welding Technology, Tata Mc Graw Hill, New Delhi.

Madatov, N.M. (1962), Special Features of Underwater Touch Welding, Aut, Weld. Vol. 15, No. 9, pp. 63–66.

Madatov, N.M., (1965), The Properties of Bubble of Steam and Gas around the Arc in Underwater Welding, Aut Weld. Vol. 18, No. 12, pp. 25–29.

Madatov, N.M., I.K. Pokhodnya, B.A. Kostenko, (1965), High Speed X-Ray Cine Photography of the Underwater Welding Arc Weld Prod. VII, No. 9, p. 72–73.

Madatov, N.M. (1966), Energy Characteristics of the Underwater Welding Arc, Weld Prod. Vol. 12, No. 3, pp. 11–14.

Madatov N.M., (1969), Shape Relationships for Underwater Welding, Weld. Prod Vol. 15, No. 3, pp. 11–13.

Masubuchi, K. (1980), Analysis of Welded Structures : Res Stresses, Dist. and their Consequences, Oxford, NY

Mandal, N.R. (2004), Welding and Distortion Control, Narosa.

Mandal, N.R. and Adak M. (2001), Fusion Zone and HAZ prediction through 3D simulation of Welding Thermal Cycle, J. of Mech. Bahaviour of Materials, Vol. 12. No. 6 pp. 401–414.

Masumoto, I. et al., (1971), Study of Underwater. Welding, J. of Japan Welding Soc. No. 40(7).

Needham, J.C., (1966), Control of Transfer in Aluminium consumable Electrode Welding, Physics of Welding Arc, Symp. IW, Cambridge London.

Needham, J.C., (1978), (Tech. Dir.) Advances in Welding Processes, 4th Int. Conf. Herrogate, IW Cambridge, London.

Neumon J.A., and F.J. Baackoff (1959), Welding of Plastics, Reinhold Chapman & Hall.

Nikolaev, G., and N. Olshansky, (1977), Advanced Welding Processes, Mir Pub., Moscow.

Potapevskii, A.G., N.M. Madatov, (1967), Melting and Transfer of Metal during Underwater Welding with Fine Electrode Wire, Aut. Welding Vol. 2, No. 12, pp. 110–113.

Rykalin, N.N., A. Uglov, A. Kakora, (1978), Laser Machining and Welding Mir Pub., Moscow. Rykalin, N.N., I.D. Kulagin, A.V., Nikolaev, (1966), Vaporized Electrode Material and Energy Balance in Welding Arcs. Physics of Welding Arc, Symp. IW London.

Serdjuk, G.B., (1966), Magnetic Forces in Arc Welding Metal Transfer, Physics of Welding Arc, Sym. IW, Cambridge, London.

Shinada, K., and M. Tamura, (1982), Welding and Cutting (FGR) Welding and Metal Fabrication, July-Aug., p. 295–297.

Silva, E.A. and T.H. Hazial (1971), Shielded Metal Arc Welding Underwater with Iron Powder Electrodes, Weld. J., Vol. 5, No. 6, 1971.

Smith, A.A., J. Pintard (1966), Characteristics of Short Circuiting Transfer, Physics of Welding Arc Sym. I.W., Cambridge.

Stalker, A.W., P.H. W. Hart and G.R. Salter, (1974), A Preliminary Study of Manual Metal Arc Welding, Report : 3412/58/74 (RR/SMT/R. 7406).

Stevenson, A.W., (1983), Offshore Options Reviewed, Welding and Metal Fabrication, June, pp. 244–252.

Stockman, K., (1968), Underwater Cutting and Welding, Norgos, Vol. 32 No. 3, pp. 44–47.

Sunnen, J.F. (1966), Electrical Parameters during Metal Transfer, Physics of Welding Arc, Symp. I.W., Cambridge.

Tweeddale, J.G., Welding Fabrication, Vols. 1–3, Elsvier Pub. Co.

Takemasu, M.N. Yasaki, S. Fukui, (1982), Underwater Welding Application Gives New Life to Firecraker Welding, Welding and Metal Fabrication, July/Aug. pp. 287–292.

Udin. H., E.R. Funk and J. Wulf (1954), Welding for Engineers, J. Wiley and Sons N.Y.

Van der Willingen, P.C. (1946) Contact Arc Welding, Weld. 1. Vol. 25 No.5, pp. 313s-320s.

Waugh, R.A., and O.P. Eberlien, (1954), Underwater Metallic Arc Welding, Weld. J. Res. Supp., Vol. 33 No. 10 pp. 531s–534s.

Welding Handbook, Sec 1–5, American Welding Society, 1976.

Index

A

B

C

D

E

F

G

S

T

U

V

W

X